CAMBRIDGE LIBRARY COLLECTION

Books of enduring scholarly value

Life Sciences

Until the nineteenth century, the various subjects now known as the life sciences were regarded either as arcane studies which had little impact on ordinary daily life, or as a genteel hobby for the leisured classes. The increasing academic rigour and systematisation brought to the study of botany, zoology and other disciplines, and their adoption in university curricula, are reflected in the books reissued in this series.

Paxton's Flower Garden

Best remembered today for his technically innovative design for the Crystal Palace of 1851, Joseph Paxton (1803–65) was head gardener to the Duke of Devonshire at Chatsworth by the age of twenty-three, and remained involved in gardening throughout his life. Tapping in to the burgeoning interest in gardening amongst the Victorians, in 1841 he founded the periodical *The Gardener's Chronicle* with the botanist John Lindley (1799–1865), with whom he had worked on a Government report on Kew Gardens. *Paxton's Flower Garden* appeared between 1850 and 1853, following a series of plant-collecting expeditions. Only three of the planned ten volumes were published, but with hand-coloured plates (which can be viewed online alongside this reissue) and over 500 woodcuts, the work is lavish. Volume 1 includes colour plates of orchids, Lindley's speciality, along with a pitcher plant and Moutan peony, both still unusual and exotic at the time of publication.

Paxton's Flower Garden

Volume 1

Joseph Paxton
John Lindley

CAMBRIDGE UNIVERSITY PRESS

Cambridge, New York, Melbourne, Madrid, Cape Town,
Singapore, São Paolo, Delhi, Tokyo, Mexico City

Published in the United States of America by Cambridge University Press, New York

www.cambridge.org
Information on this title: www.cambridge.org/9781108037259

This edition first published 1850–1
This digitally printed version 2011

ISBN 978-1-108-03725-9 Paperback

The original edition of this book contains a number of colour plates, which have been reproduced in black and white. Colour versions of these images can be found online at www.cambridge.org/9781108037259

PAXTON'S

FLOWER GARDEN.

BY

JOHN LINDLEY AND JOSEPH PAXTON.

VOL. I.

LONDON:
BRADBURY AND EVANS, 11, BOUVERIE STREET.
1850-1.

PREFACE.

IT was stated at the commencement of this work that its design was to supply, in monthly parts, as full an account of all the New and remarkable Plants introduced into cultivation as is necessary to the Horticulturist, and as the price and extent of a periodical will permit; the history of such plants being sought in the Botanical Works published on the Continent, to which English cultivators have little access, as well as in those of our own country, and in the Gardens or Herbaria from which they are derived.

It was expected that by this means the English reader would be able by degrees, by mere reference to the indexes of matter which will accompany each part, to ascertain the real Horticultural value of the numberless so-called novelties with which the lists of dealers are crowded. The abundance of double names, which botanists call synonymes, but which in common parlance are termed *alìases,* would also, it was hoped, be gradually referred to their true denomination, and the purchaser thus be spared the mortification of finding that after procuring half a dozen different names he is still in possession of but one Species, and that perhaps one with which he was previously familiar.

To effect this purpose it was proposed to separate each Number into two distinct Parts. In the FIRST PART would be found Three Coloured Plates of Plants, which from their beauty, or remarkable tints, especially demand this expensive style of illustration. Here it was not proposed to introduce any species which can be as well represented without colour; by which means a large part of the cost of Botanical periodicals would be saved for the purpose of being applied to the embellishment of the Second Part. The title of the SECOND PART, "Gleanings and Original Memoranda," fully explained its purpose. It was announced as consisting of Notices, long or short, according to the importance of the subject, of as many plants published in contemporary publications, or observed by the authors, as could be enumerated in eight or ten pages. Unimportant species were to be merely mentioned; those of higher interest to be described at greater length; and of the most remarkable there was to be introduced Woodcuts, in which an attempt would be made to combine accurate representations with some pictorial effect. In the selection of species for full illustration, it was intended to divide the plates as nearly as possible between Stove, Greenhouse, and Hardy Plants; so that each department of the Flower Garden might be equally cared for.

It was also stated that since the work was intended for English readers, the English language would be adopted, as far as possible, in all familiar names and descriptions. English names of the plants represented in the coloured plates were to be given in preference to technical Latin ones, in the hope that by degrees the ear may be relieved from the necessity of dwelling upon sounds, which, even to the learned, are often harsh and unpleasant; for, there seemed to be no valid reason why the system of talking Greek and Latin without understanding it might not be banished from familiar Natural History. At the same time, for the convenience of Foreign Naturalists, and of those who prefer technical to familiar words, the names employed in strict science were to be given, and the distinctive characters of the species to be added in Latin.

The authors now, at the conclusion of this first volume, venture to hope that all their intentions have been carried into effect. Thirty-six plants of great beauty have been represented in colours; 120 of inferior interest, have formed the subject of woodcuts; and 229 others have been made the subject of comment, or of sufficient notice for all general purposes. On the whole 385 species have been collected into the volume, of which 156 have been more or less illustrated. This they trust will be accepted as a satisfactory guarantee that succeeding volumes will be produced in the same spirit, the fidelity and excellence of the plates increasing with the advancing skill and experience of the artists.

PLATE 1.

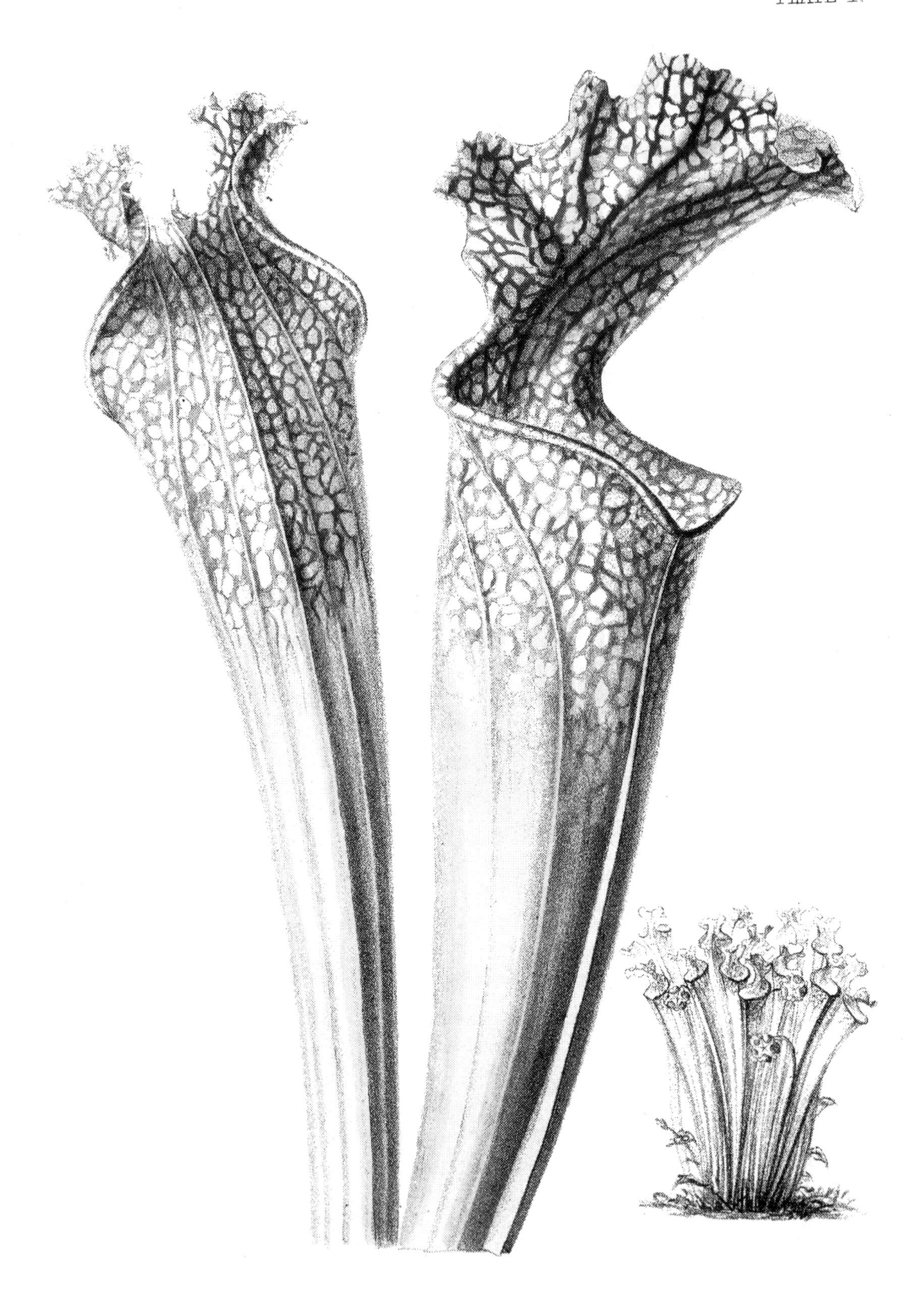

I. Constans Pinx & lith

Printed by Maclure, Macdonald & Macgregor, London.

[Plate 1.]

DRUMMOND'S SIDE-SADDLE FLOWER.

(SARRACENIA DRUMMONDII.)

A Stove Marsh Plant from Florida, *belonging to the Natural Order of* Sarraceniads.

Specific Character.

DRUMMOND'S SIDE-SADDLE FLOWER.—Pitchers long, straight, dilated upwards, angular, tapering much to the base ; furnished with a sharp projecting rib in front, with an undulating inflexed roundish blade, which is covered with long hairs in the inside. Flowers purple.

SARRACENIA *DRUMMONDII;* ascidiis strictis angulatis extùs glabris basi angustatis apice dilatatis, anticè in costam angustam rectam productis, laminâ subrotundâ undulatâ crispâ apiculatâ intùs hirsutâ, flore fusco-purpureo.

Sarracenia Drummondii. *Croom's Observations on the genus Sarracenia,* No. 3, *with a plate, in the Annals of the Lyceum of Natural History of New York,* vol. 4.

Visitors to Chatsworth, in the summer and autumn of last year, were scarcely more surprised at the glorious aspect of the Victoria Lily, than at the exquisite beauty of the plant now represented, many large specimens of which decorated a neighbouring stove among rare Orchids of the richest hues and the most interesting forms.

It was, we believe, originally introduced by the late Mr. Drummond, who met with it in Florida, near the town of Appalachicola. It has since been found abundantly, by Dr. Chapman, on the western borders of the river of the same name, below Ocheesee. It, therefore, inhabits the swamps of a region, which, during summer, experiences a tropical heat, as is in some measure indicated by the presence of Orchidaceous Epiphytes, such as Epidendrum Magnoliæ and tampense.

The pitchers of this plant are from eighteen inches to two and a half feet long, perfectly erect and straight, with very much the form of a postman's horn. Their colour is of the most vivid green, except at the upper expanded end, where they are brilliantly variegated with white, red, and green. The rim of the orifice of the pitchers is slightly folded back, from the front towards the back, where it expands into a broad roundish arched cover, much undulated and crisped. In the inside this cover is clothed with long hairs, which partially disappear towards the entrance of the pitcher, at which point there is a considerable exudation of sweet viscid matter, apparently secreted by the hairs which exist there. The flower is of a dingy purple colour, roundish, about two and a half inches in diameter, with five blunt acuminate sepals, five obovate inflexed petals, and a pale green dilated five-angled membranous stigma, which is nearly as long as the flower itself; each angle is divided into two short lobes, beneath which, in a fold, lies the real stigmatic surface. These flowers have little beauty, and are by no means the object of the gardener's care.

The so-called pitchers are in reality the leaves of this plant, in a very singular condition; the pitcher itself being the leafstalk, and the cover its blade. By what mode of development this kind of structure is produced has never yet been conclusively shown. It has been thought that the pitcher is formed by the folding together, in its earliest infancy, of the two sides of a flat leafstalk, the line of which union is indicated by a firm elevated rib, which proceeds from the base to the opening of the pitcher, as if to stiffen and sustain it; but this is not certain, and it is more probable that the pitcher is the result of a hollowing process, coeval with the first growth of the pitcher itself, and analogous to that which produces the hip of the rose, or the cup at the bottom of the calyx of Eschscholtzia, or the cups that appear accidentally upon cabbage leaves.

If the exact nature of the pitcher is thus undecided, we are still further from a knowledge of the use for which so singular an apparatus is destined. To the common idea, that nature intended it to hold water, arise these objections: that water is not found in the pitcher except after rains or heavy dews, and that plants which grow naturally in bogs can hardly require any unusual apparatus for supplying them with water. Others think that the pitcher is a contrivance for detaining insects in captivity till they perish and decay, the putrefaction of these creatures conducing to the nutrition of the plant. But there is no apparent reason why the Side-saddle flower should require this sort of special nutriment more than its neighbours in the same bogs, which have no pitchers. This, however, is certain, that if the pitchers were intended for fly-traps, they could hardly have been more ingeniously contrived. It is the honey of the mouth of the pitcher that tempts the insects to their destruction; and, accordingly, they are found in abundance at the bottom. In the plant now before us we count, in the month of February, about a dozen, two of which are wasps; and Mr. Croom says, that he found in one of his a large butterfly, (*Papilio Turnus*). Reversed hairs keep them there without hope of escape. As the sides of the pitchers consist of very lax cellular tissue, containing large cavities in every direction, and as starch grains in abundance escape from the sides when wounded, it is a question whether this starch, converted into sugar by the vital force of the pitcher, may not serve to sweeten the water in which the imprisoned insects meet a miserable end?

The manner in which the North American Side-saddle flowers are grown at Chatsworth is explained in the following memorandum, which is applicable to the more common species as well as to that which is the immediate object of the present article:—

The stove is decidedly the most suitable place for these species making and maturing their growth, at which time they require much warmth and moisture. A temperature of from 80° to 100°, with plenty of water at the roots, and syringing three times a day, from March till September, we have found to suit them the best. During their season of rest, a greenhouse would probably answer the ends of cultivation better than the stove; at all events, the plants should be kept in a dry cool atmosphere, from 40° to 60°, not higher. The best time for potting is January, and the best material for that purpose is silver sand and Sphagnum, well mixed with a portion of peat and potsherds, broken quite small. It is important to have plenty of drainage, and no fear need be entertained of excess in this particular. It has been customary at Chatsworth to place the pots in saucers which have been kept full of water during the whole of the summer season. We do not, however, attach any importance to this practice. The plants will thrive equally well without saucers. Pitchers are usually formed in October, and continue perfect for three months. The number of pitchers on an individual plant of S. Drummondii varies from fourteen to twenty-three. We have measured individual pitchers of this species, and find the maximum length two feet three inches, and the maximum girth at the *top*, six inches. Flowers usually open in March and April. By removing the flower-buds as they appear, the succeeding pitchers become much finer.

PLATE 2.

L. Constans Pinx & Lith.

Printed by Maclure, Macdonald & Macgregor, Lith. London.

[PLATE 2.]

THE GLITTERING GLAND-BEARING TRUMPET-FLOWER.

(ADENOCALYMMA NITIDUM.)

A Stove Plant, from BRAZIL, *belonging to the Natural Order of* BIGNONIADS.

Specific Character.

THE GLITTERING GLAND-BEARING TRUMPET-FLOWER.—A smooth climber. Leaves on rather long stalks, mostly in threes, or in pairs with an intermediate simple tendril, the leaflets on short stalks, (the middle one longest,) elliptical-oblong. Racemes axillary, or nearly terminal, shorter than the leaves, velvety. Bracts, oblong or linear, as long as the calyx, glandular below the point. Calyx with five short teeth, velvety, irregularly glandular, sometimes slit on one side. Corolla rather velvety.

ADENOCALYMMA *NITIDUM;* scandens, glabrum, foliis longiusculè petiolatis plerisque trifoliolatis, aut bifoliolatis cirrho simplici intermedio, foliolis breviùs petiolatis, (impari longiùs) oblongis glabris, racemis multifloris foliis brevioribus axillaribus terminalibusque, bracteis oblongis linearibusque apice glandulosis calyci æqualibus, calyce 5-dentato nunc fisso irregularitèr glanduloso, corollâ subvelutinâ.

Adenocalymna nitidum: *Martius in De Candolle's Prodromus,* vol. 9, p. 200.

WE received a specimen of this very pretty climber from Messrs. Knight and Perry of the King's Road, in the beginning of February. They obtained it, about five years since, from Mr. Makoy of Liége, under the name of Fridericia Gulielma, which belongs to a totally different plant, belonging, however, to the same natural order.

It is found wild in the Empire of Brazil, in various places, of which Mons. Alphonse De Candolle gives the following enumeration. "Thickets and dry places, near Rio Janeiro; on the Corcovado mountain, near the Mandioc farm; in the province of Bahia, near Maracas." We also possess it from a more inland station, but without any precise locality. It is said to vary much in appearance, unless, indeed, more than one distinct species is comprehended under the same name. Professor von Martius has a plant called Adenocalymma sepiarium, which is said to be one of the supposed varieties.

That which is in cultivation is a thin-leaved, smooth, climbing plant, with a yellowish tint. The leaflets grow in pairs, with a simple tendril between them; or else in threes without an intermediate tendril; they are shining on each side, from three to five inches long, and of an oblong figure with a sharp tapering point; when in threes, the central one has a much longer stalk than the others. The flowers grow in clusters, which in the plant before us are not more than an inch and a half long, supporting seven blossoms, but in the wild specimens they sometimes occur as many as thirty on a raceme full five inches long; only a part of them, however, are open at a time. The bracts are velvety, narrow, and placed close to the calyx; they have usually a small shining gland or two below their point. In like manner the calyx, which is also velvety, has several glands of the same kind dispersed irregularly below its five short teeth; it is also often slit down one side. The corolla, which is fully two inches long, is of a thick leathery texture, deep yellow, contracted at the base into a narrow tube as long as the calyx, and enlarged upwards into a somewhat curved trumpet, divided at the edge into 5 nearly equal blunt spreading lobes. The stamens are didynamous, arising from a throat covered with thick short hairs; the fifth stamen is a very short hooked body.

The remarkable glands which appear on the bracts and calyx constitute one of the most striking peculiarities of this genus, and have given rise to its scientific name (*αδην* a gland, and *καλυμμα* a covering) which we have translated at the head of this article. Mons. De Candolle writes the word Adenocalymna, which is evidently wrong. What the use or nature of such glands may be, is unknown. They have a definite form, although an indefinite position; they are quite destitute of the short hairs which clothe the neighbouring parts, and they evidently secrete some fluid, as is shown by their moist surface. They are therefore glands in the proper sense of the word, as limited by Professor Schleiden.

The Glandular Trumpet-flowers are confined to tropical America, where they scramble over trees and decorate the scenery with their bunches of yellow, pink, or orange-coloured flowers. Professor De. Candolle admits nineteen species; among which are some of the most beautiful of Brazilian climbers, often opening thirteen or fourteen large trumpet-shaped blossoms before one begins to fade. To gardeners they would be invaluable, and should be diligently sought for in the provinces of Para, Bahia, Piauhy, and even of Rio itself, whence the species now figured appears to have been brought to Europe. Another very handsome species, the Adenocalymma longiracemosum, was introduced by M. de Jonghe of Brussels, and is probably to be found in gardens.

The best way of growing this has not been ascertained. Messrs. Knight and Co. state that, having appeared "a shy flowerer," it has not received the attention it was entitled to, so that they are unable to offer any advice for its culture founded on practice, but they surmise that the treatment most congenial to it, would be to afford it dry stove temperature, and to place it out in a large tub. It roots freely in a mixture of half light loam, quarter peat, quarter leaf mould. They doubt whether it will be a good plant for pot culture, seeing that they have so grown it ever since they possessed it, and have only induced it to produce the flowers communicated on the present occasion.

For ourselves we would suggest that the unwillingness of the plant to flower, will be overcome by a high temperature applied to the soil—perhaps 84°; and a rest of three or four months. There is no natural indisposition in these climbers to produce their flowers, but they are unable to do so in our stoves from want of that stimulus which nature so abundantly supplies in their native woods. Upon this point the remarks on Aristolochia picta, of which a wood-cut will be found at the commencement of our "Gleanings, &c.," may be advantageously consulted. It should also be remembered that in the places where such plants exist little manure accumulates, except that formed by the ever decaying foliage and fallen wood which strews the earth of the tropical forest; what manure does exist is chiefly supplied by birds.

L. Constans Pinx & Lith

Printed by Maclure Macdonald & Macgregor, London.

[Plate 3.]

WALKER'S CATTLEYA.

(CATTLEYA WALKERIANA.)

A Stove Epiphyte, from Brazil, *belonging to the Natural Order of* Orchids.

Specific Character.

WALKER'S CATTLEYA.—Stems oval, stalked, each having one leaf. Leaves oblong, thick, concave. Flower-stalks 1-2-flowered, with a small spathe-like bract. Petals oval, wavy, membranous, twice as wide as the Sepals. Lip smooth, naked, with short lateral roundish lobes, and the middle lobe rounded and two-lobed. Column broad, thick, rounded off at the upper end.

CATTLEYA *WALKERIANA ;* caulibus ovalibus stipitatis monophyllis, foliis oblongis coriaceis concavis, pedunculis 1-2-floris, bracteâ parvâ spathaceâ, petalis ovalibus undulatis membranaceis sepalis duplo latioribus, labelli plani calvi lobis lateralibus brevibus rotundatis intermedio cuneato bilobo rotundato, columnâ latâ crassâ apice rotundatâ.

Cattleya Walkeriana, *Gardner, in the London Journal of Botany,* vol. 2, p. 662 : *aliàs* C. bulbosa, *Bot. Register,* 1847, t. 42.

For the opportunity of figuring this beautiful flower in really fine condition we have to express our obligation to C. B. Warner, Esq., in whose collection, at Hoddesdon, it has lately blossomed. In the *Botanical Register* a small specimen was published some years since, from Mr. Rucker's garden, under the name of *Cattleya bulbosa,* its identity with what the late Mr. Gardner had previously called Walker's Cattleya not having been suspected. Mr. Rucker's plant had, however, a much more richly coloured lip than this, and must have been a distinct variety.

According to Gardner it inhabits the country beyond the diamond district of Brazil, where it was found by Mr. Edward Walker, his assistant, on the stem of a tree overhanging a small stream which falls into the Rio San Francisco.

The stems are club-shaped and furrowed, each having one leathery, concave, blunt leaf, which is by no means wider at the base than apex; when young or ill-grown they are short and oblong, in which state they gave rise to the name *C. bulbosa,* now cancelled. The flowers grow singly, or in pairs, from within a short, narrow, reddish spathe, and are full five inches in diameter, fragrant, and bright, but not deep, rose colour. The sepals are oblong, acute, and membranous. The petals are broad, oblong, acute, slightly wavy, but not lobed. The lip, which is a richer rose than the other parts, is small, roundish at the end, and emarginate, with two narrow, erect, lateral lobes, which fold over the lower part only of the column. The column itself is very broad, fleshy, rounded, with no lobes or notches such as are found in *C. pumila.*

Perhaps the nearest relation of this plant is with *C. superba*, from which, however, its dwarf habit and incomplete lip readily distinguish it.

All known species of this beautiful genus are so highly deserving cultivation that an enumeration of those which are at present grown seems desirable, especially since the list published some years since in the *Botanical Register*, now requires many important additions. The arrangement there proposed seems, however, to answer all the purposes of the cultivator as well as of the botanist, and is therefore followed in the following catalogue:—

CATTLEYA.

SECTION I.—*Lip rolled round the Column.*

SECTION II.—*Lip flat, not rolled round the Column, and without lateral lobes.*

SECTION I.

* Sepals of the same texture as the Petals, the lateral ones being nearly straight.

1. C. superba, *Lindl. Sertum Orchid.*, t. 22; aliàs *C. Schomburgkii*, Lodd. Cat., aliàs *Cymbidium violaceum*, Humboldt and Kunth.—Demerara.—Flowers deep rose-coloured, fragrant, with a deep crimson lip.
2. C. elegans, *Morren, Annales de Gand*, t. 185.—St. Catharine's, in Brazil.—Flowers large, rose-coloured, with a deep purple-violet lip. Very like *C. superba*, except in colour, but the leaves are represented as being much narrower, and the lip is said not to have either wrinkled veins or callosities. Unknown to us except from Professor Morren's figure made from a Belgian specimen in the possession of M. Alexander Verschaffelt.
3. C. Skinneri, *Bateman, Orch. Mex. et Guatemal.*, t. 13.—Guatemala.—Flowers deep rich rose colour, with a crimson lip.
4. C. Walkeriana, *Gardner, in Lond. Journ. Bot.*, vol. ii. p. 662; aliàs *C. bulbosa*, *Lindl. in Bot. Register*, 1847, t. 42.—Brazil.—Sweet-scented, dwarf, with large rose-coloured flowers.
5. C. pumila, *Hooker, in Bot. Mag.*, t. 3656; *Bot. Reg.*, 1844, t. 5: aliàs *C. marginata*, aliàs *C. Pinellii* of Gardens.—Brazil.—A dwarf species with a lobed column, deep rose-coloured flowers, and a rich crimson crisp lip, often edged with white. In *C. Pinellii*, the flowers are much paler.
6. C. maxima, *Lindl. Gen. et Sp. Orch.*, No. 4; *Bot. Reg.*, 1846, t. 1.—Guayaquil and Colombia.—Flowers bright rose, with convex petals, and a lip richly variegated with dark crimson veins traced upon a pallid ground.
7. C. labiata, *Lindl. Collect. Bot.*, t. 33; *Bot. Reg.*, t. 1859; *Bot. Mag.*, t. 3988: aliàs *C. Mossiæ*, *Bot. Mag.*, t. 3669; *Bot. Reg.* 1840, t. 58.—Tropical America.—The two forms to which the above names have been applied, differ in little except colour. In *C. labiata*, the lip is stained with one deep uniform tint of crimson; in *C. Mossiæ*, it is richly variegated with crimson veins upon a yellowish ground. The first is from swamps in Brazil, the latter is from the Caraccas, where it grows at an elevation of three thousand feet above the sea, sporting into many charming modifications of colour.

   There is a *C. quadricolor* in the possession of Mr. Rucker, with which we are not sufficiently acquainted to say how it differs from the last.
8. C. Lemoniana, *Lindl. in Bot. Reg.*, 1846, t. 35.—Brazil.—Flowers pale pink, whole coloured.
9. C. lobata.—Brazil.—Flowers deep rich rose, whole coloured. Of this, which is in the possession of Mr. Loddiges, we shall take an early opportunity of giving some account.
10. C. crispa, *Lindl. in Bot. Reg.*, t. 1172; *Bot. Mag.*, t. 3910.—Brazil.—Flowers white, crisp, with a rich crimson stain in the middle of the lip.
11. C. citrina, *Lindl. Gen. et Sp. Orch.*, No. 8; *Bot. Mag.*, t. 3742: aliàs *C. Karwinskii*, Martius Choix, p. 15, t. 10.—Mexico.—Flowers bright yellow.

** Sepals somewhat herbaceous, or more coriaceous than the Petals, the lateral Sepals manifestly falcate.

12. C. Loddigesii, *Lindl. Collect. Bot.*, t. 37; aliàs *C. intermedia*, Graham, in Bot. Mag., t. 2851; aliàs *C. vestalis*, Hoffmansegg. Bot. Zeitung, 1. 831; aliàs *C. Papeiansiana*, Morren, Ann. Gand, p. 57; aliàs *C. candida* of gardens.—Brazil, in marshes.—The original, *C. Loddigesii*, has pale purple flowers; in *C. intermedia* or *candida*, they are nearly white.
13. C. Harrisoniana, *Bateman, in Bot. Reg.*, sub t. 1919.—Brazil.—Flowers lilac, the lip with a deep blotch.
14. C. maritima, *Lindl. in Bot. Reg.*, sub t. 1919.—Brazil.—Unknown in gardens; probably not distinct from *C. Loddigesii.*
15. C. Arembergii, *Scheidweiler, in Garten-Zeitung*, 1843, p. 109.—Brazil.—Unknown to English botanists. Flowers large, lilac, sweet-scented.
16. C. Forbesii, *Lindl. Bot. Reg.*, t. 953.—Brazil.—Flowers greenish yellow.
17. C. guttata, *Lindl. Bot. Reg.*, t. 1406; aliàs *C. elatior*,

Lindl. Orch., No. 9 ; aliàs *C. sphenophora*, Morren, in Ann. Gand, t. 175.—Brazil.—Flowers greenish yellow, beautifully spotted with crimson.

18. C. granulosa, *Lindl. in Bot. Reg.*, 1842, t. 1 ; and 1845, t. 59.—Brazil, Paräiba.—Flowers, large olive-coloured, with a long white and yellow or crimson lip. Not from Guatemala, as at first reported ; an error corrected by Mr. Hanbury.

SECTION II.

19. C. Aclandiæ, *Lindl. in Bot. Reg.*, 1840, t. 48.—Brazil. —A magnificent little plant with large chocolate flowers variegated with yellow, and a rich rose-coloured lip.

20. C. bicolor, *Lindl. in Bot. Reg.*, sub t. 1919.—Brazil.—Flowers tawny, with a bright purple labellum. Sometimes has eight or ten flowers in a raceme.

The Cattleya (?) domingensis *of the Genera and Species of Orchidaceous Plants* is a Lælia, and perhaps the same as *L. Lindenii*, a charming plant from Cuba, which we saw lately in the fine collection of Orchids formed by M. Pescatore, at his beautiful seat at Celle St. Cloud, near Paris.

The manner in which the specimen now represented was cultivated is thus described by Mr. Warner's gardener, B. S. Williams, who is one of our best growers of Orchids :—

"This fine species of Cattleya blooms twice a year, (February and June,) on the young growth; its blossoms last five or six weeks in perfection, which is a much longer time than any of the other Cattleyas ; they seldom flower longer than three or four weeks at a time; it is also very sweet-scented and will perfume a whole house. It succeeds best on a block of wood surrounded by a little Sphagnum, and it should have a good supply of heat and moisture in the growing season, but after it has made its growth it should be kept rather dry and may be placed in a much cooler house, say about 60°; it should only have just sufficient water to keep the bulbs from shriveling too much. The plant should be fastened to the block with copper wire and suspended from the roof in a place where there is plenty of light, but not too much sun.

"No doubt exists that Cattleyas rank among our finest Orchids. Their flowers are large and beautiful. In their native countries adhering as they do to the projecting arms of living trees or the prostrate trunks of dead ones, they flourish and are dormant alternately with the seasons; at times they are subject to the saturating effects of long continued rains, and again they are dried up by months of warm weather. Almost all Orchid growers cultivate their Cattleyas in the coolest Orchid house, but I grow them in the hottest house I have, along with the East Indian Aërides, Saccolabes, and Dendrobes. I find that they succeed much better in the hottest house, in which they make fine strong bulbs and good foliage, and always flower strongly and vigorously. It is considered that some species are difficult to bloom, such as Superba and Pumila, two of the finest of Cattleyas; but I experience no difficulty in flowering all the kinds here every year, and some of them twice a-year. Loddigesii flowers twice a-year—in July, and again in September, producing thirty and forty flowers at a time; Crispa, a beautiful species, brings forth about sixty blooms at a time; and Mossiæ, another fine thing, fourteen flowers. Labiata, one of the finest of Cattleyas, is a very free bloomer, and so is Skinneri. Loddigesii, Intermedia, Guttata, and Candida, are also all good sorts and free bloomers.

"In cultivating Cattleyas, the method I follow is to give them a good supply of heat while they are growing; but not too much water at the roots; about twice a week when they are in vigorous growth will be quite enough; for Cattleyas are not very thirsty plants, and by giving them too much water the bulbs are apt to rot. After they have made their growths they should be well rested, by keeping them rather dry. During their dormant season only just sufficient water should be given them to keep their bulbs from shriveling. I give them a good season of rest, which makes them grow more strongly and flower more freely, their blooming season being from November to the latter end of February; and during this time I keep them in a temperature of about 60° or 62° by night, and

65° by day. After the resting season is over I raise the temperature from 65° to 70° by night, and from 70° to 75° by day, and during sun-heat the temperature may be allowed to rise still more; 85° to 90° will do no injury, but air should be given to prevent the heat rising too high, and also to dry the house once a day; but do not permit cold air to circulate among the plants. The air on entering, should be warmed by being caused to pass over the hot-water pipes.

"I grow all the varieties of Cattleya in pots except Walkeriana, which, as I have stated, I grow on a block; all the kinds may be grown on blocks with moss, but I find they succeed best in pots, in fibrous peat and broken potsherds mixed together. The peat should be broken into pieces about the size of a hen's egg. The most material point to be attended to in potting is that the pots should be well drained; this may be effected by placing a small pot in the bottom of the other and filling the latter half full of potsherds, and then placing a little moss over them to prevent the superincumbent peat from getting down and stopping the drainage. If this is not attended to, the water will stagnate, the soil sodden, and the plants will become sickly, a condition from which they seldom recover. Pot about two or three inches above the rim of the pot, and use a few small pegs to keep the peat firmly round the plant. When you re-pot remove all the old soil from the roots, if it can be done without injuring them, and water the plants sparingly afterwards.

"Cattleyas are propagated by division; always choose a young bulb having a fresh bud at its base from the outside of the plant.

"They should be kept perfectly clear of insects by sponging them with clean water; they are very subject to the white scale."

GLEANINGS AND ORIGINAL MEMORANDA.

1. Aristolochia picta. *Karsten.* From the Caraccas. A curious and rather handsome stove twiner, belonging to the Natural order of Birthworts. In the nursery of Mr. Van Houtte of Ghent. (Fig. 1.)

A smooth twining plant, with deeply cordate acute leaves, and purple tessellated flowers, whose limb is 3 inches long, and terminated by a short tail. In the centre, leading to the throat, is a rich spot of a golden colour. "This Birthwort requires all the heat and light which the sun can give it; in its own tropical plains it is exposed to extreme atmospheric vicissitudes, for in the day the earth in which it grows is heated to 167° Fahr.; while at night, under a cloudless sky, radiation and evaporation lower the temperature of the surrounding air to 59°. But these variations are little felt by the roots, which are plunged in a soil covered with dead leaves, &c., which check both solar heat and nocturnal cold. And thus its roots are exposed to a warmth which is not only more uniform, but much less diminished than it would be in our colder regions, by the action of continued rain, which, in fact, in tropical countries communicates to the soil a part of the heat with which the air is surcharged. Thus, at Puerto Caballo, on a wet day (December 4), at half-past 6 in the evening, I found the temperature of rain-water to be 78° 25, while that of the air was only 74° 80. Observe, I had previously remarked it to be 77° 25 R for rain-water, and only 76° 80 for the air; and what is more, on the next day (Dec. 5), after a whole night of rain, at half-past 8 in the evening, I found the rain-water still at 75° 37, while the air marked only 75° 25."—*Van Houtte's Flora*, v. t. 521.

2. Oncidium hastatum (*aliàs* Odontoglossum phyllochilum. *Morren*

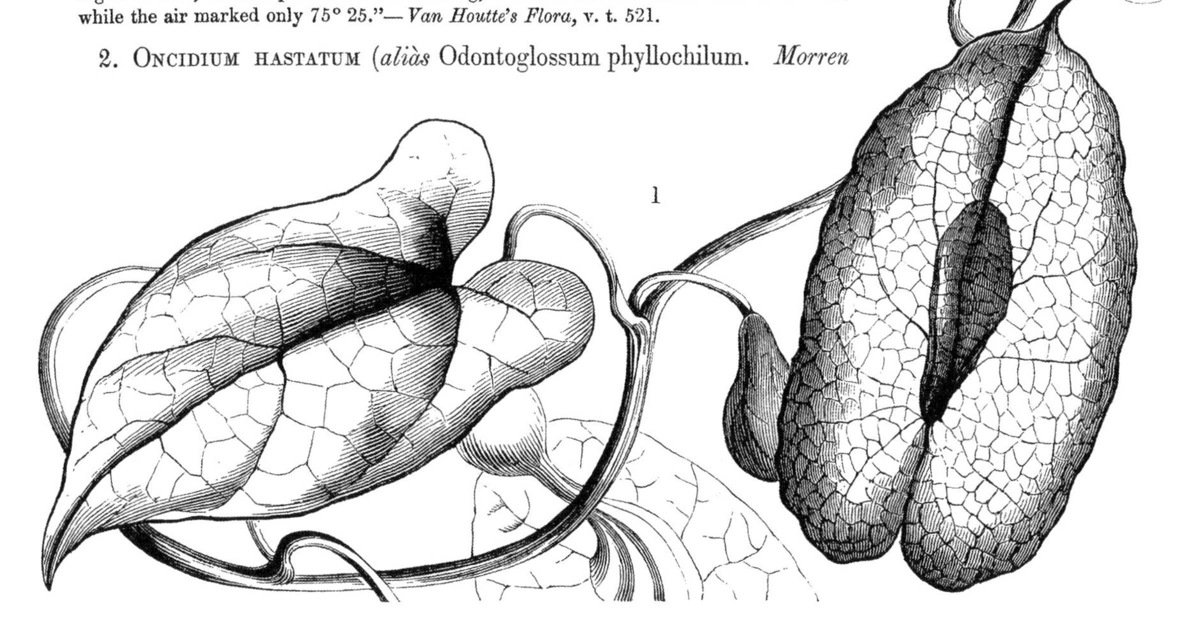

1

in Ann. Gand., t. 271). An orchid from N. Grenada, with large handsome variegated flowers, and a white lip sometimes tinged with green. An old inhabitant of English gardens.

It does not appear that this was published before Professor Morren gave it the name here quoted; but it has long been known in the gardens of this country under the name of Oncidium hastatum. It is a true Oncidium, its column being short and protuberant at the base, and forming an obtuse angle with the lip. In point of value it is about equal to the *Oncidium* (Cyrtochilum) *maculatum*.

3. ECHINOCACTUS RHODOPHTHALMUS. *Hooker*. A Mexican Hedgehog Cactus, with an oblong stem, and handsome red flowers appearing in August.

Received from Mr. Staines, who procured it from the neighbourhood of San Luis de Potosi, in Mexico. In its flourishing state it is extremely handsome, the deep red of the base of the petals forming a ring, as it were, round the densely-clustered stamens and bright yellow rays of the stigma, adding much to the beauty of the blossom. Mr. Smith gives the following account of the manner in which such plants are managed by him at Kew:—"At Tab. 4417, we have said that *Cacteæ* are almost indifferent as to the kind of soil they are grown in, provided it is not retentive of moisture. The present very pretty species will thrive in a mixture of light loam and leaf-mould, containing a small quantity of lime-rubbish nodules, the latter being for the purpose of keeping the mould from becoming close and compact, a condition not suitable to the soft and tender roots of the plant. If cultivated in a pot, it must be well drained; the pot being nearly half filled with broken potsherds, and the upper layer so placed as to cover the interstices, in order to prevent the mould from mixing with the drainage. During winter, Mexican *Cacteæ* do not require much artificial heat: several species are, indeed, known to bear with impunity a few degrees of frost. Where they can be cultivated by themselves, we recommend that the plants and atmosphere of the house should be kept in a dry state during winter, artificial heat being given only during a long continuance of damp cold weather or in severe frost; but at no time during winter needs the temperature of the house to exceed 50° at night. In sunny days in spring the house should be kept close, in order that the plants may receive the full benefit of the heat of the sun's rays. As the summer-heat increases air should be admitted, and occasionally the plants should be freely watered, and in hot weather daily syringed over-head."—*Botanical Magazine*, t. 4486.

4. VALORADIA PLUMBAGINOIDES. *Boissier.—Botanical Magazine*, t. 4487.

This is an aliàs of the now common Plumbago Larpentæ, which is thought by Boissier not to belong to Plumbago. We see very little, however, to characterise a genus in the differences pointed out, and agree with Sir W. Hooker, in thinking that if a new genus is really necessary, the plant ought to bear the older name of Ceratostigma.

5. METROSIDEROS TOMENTOSA. *Achille Richard*. A New Zealand Greenhouse shrub of much beauty, flowering in the summer months. Blossoms rich crimson. One of the order of Myrtle Blooms (Myrtaceæ).

"It inhabits," says Mr. Allan Cunningham (by whom it was introduced to the Royal Gardens of Kew), "usually the rocky sea-coast and shores of the Bay of Islands, where it is called by the natives *Pohutu-Kawa*, and is readily distinguished among other plants by the brilliancy and abundance of its flowers, enlivening the shores of the northern island with its blossoms in December. With us in the greenhouse it has attained the height of six feet, and attracted attention by its copious, compact, but spreading ramification, and the abundance and beauty of its evergreen foliage Its blossoming this year (for the first time) was probably encouraged by planting it out, by way of experiment, in the spring, in a sheltered part of the woods of the Pleasure-ground, in a soil of rich vegetable leaf-mould. During the summer, almost every branchlet was terminated by the vivid scarlet blossoms, and it became a conspicuous object at a distance. In its native country it is described as making its first appearance on other trees, as an epiphyte. By its strong and rapid growth it soon envelopes the parent tree, its woody roots descending till they reach the ground, and there spreading to a great extent, while the main roots, by their numbers and interlacings, ultimately become so combined that they form a trunk of a singular appearance and sometimes of an immense size. The original tree dies, and its decaying trunk becomes food for the parasite; the latter in this respect resembling the fig-trees of the tropics or the ivy of this country. It is also said to form a tree without the aid of others. With us it grows luxuriantly if planted in light loam and kept in a cool greenhouse, and forms a handsome evergreen bush. The figure here represented was made from an individual that had become too large for our greenhouse accommodation. As it afforded the opportunity of testing the degree of cold it would bear, a sheltered situation amongst trees was selected, where it was planted in May 1849. During the summer it flowered profusely, presenting a very striking appearance for an out-door shrub, and continued to flourish till the first frosts; but we observe with regret, that this fine shrub will not live in the open air where the thermometer falls a few degrees below the freezing point."—*Botanical Magazine*, t. 4488.

6. OPHELIA CORYMBOSA. *Grisebach*. A half-hardy annual from the Neilgherries, belonging to the order of Gentianworts.

Of little moment. Stem a foot high, branched. Flowers pale purple, with a white eye, in corymbs. Requires peat.—*Botanical Magazine*, t. 4489.

7. FRITILLARIA PALLIDIFLORA. *Schrenk.* From Songaria. Flowers yellow. In the Garden of Dorpat. Natural order, Lilyworts.

Resembling in habit Fr. lutea and latifolia. Hardy in Livonia, under a covering of leaves.—*Van Houtte's Flora*, v. 518, *e.*

8. MERTENSIA SIBIRICA. *G. Don.* Altai Mountains. A hardy perennial. Flowers blue. Obtained from Siberia by Mr. Van Houtte of Ghent. Natural order, Borageworts.

The glaucous leaves and beautiful blue flowers appear with the earliest spring Easily multiplied by division of the roots.—*Van Houtte's Flora*, v. 518, *c.* Apparently a pretty rock-plant.

9. ANÆCTOCHILUS LOBBIANUS. *Planchon.* A terrestrial Orchid figured in *Van Houtte's Flora*, v. t. 519; appears to be in no respect different from *Anœctochilus Roxburghii.*

10. BERBERIS JAPONICA. *Lindley; aliàs* Ilex japonica, *Thunberg; aliàs* Mahonia japonica, De Candolle; *aliàs* Berberis Beallii, *Fortune.* A magnificent evergreen shrub, with broad pinnated leaves, imported from the north of China, by Messrs. Standish and Co. Has not yet flowered in England. At fig. 2 is represented the upper part of the leaves of the natural size.

A live plant has now been received by Messrs. Standish and Noble, of Bagshot,

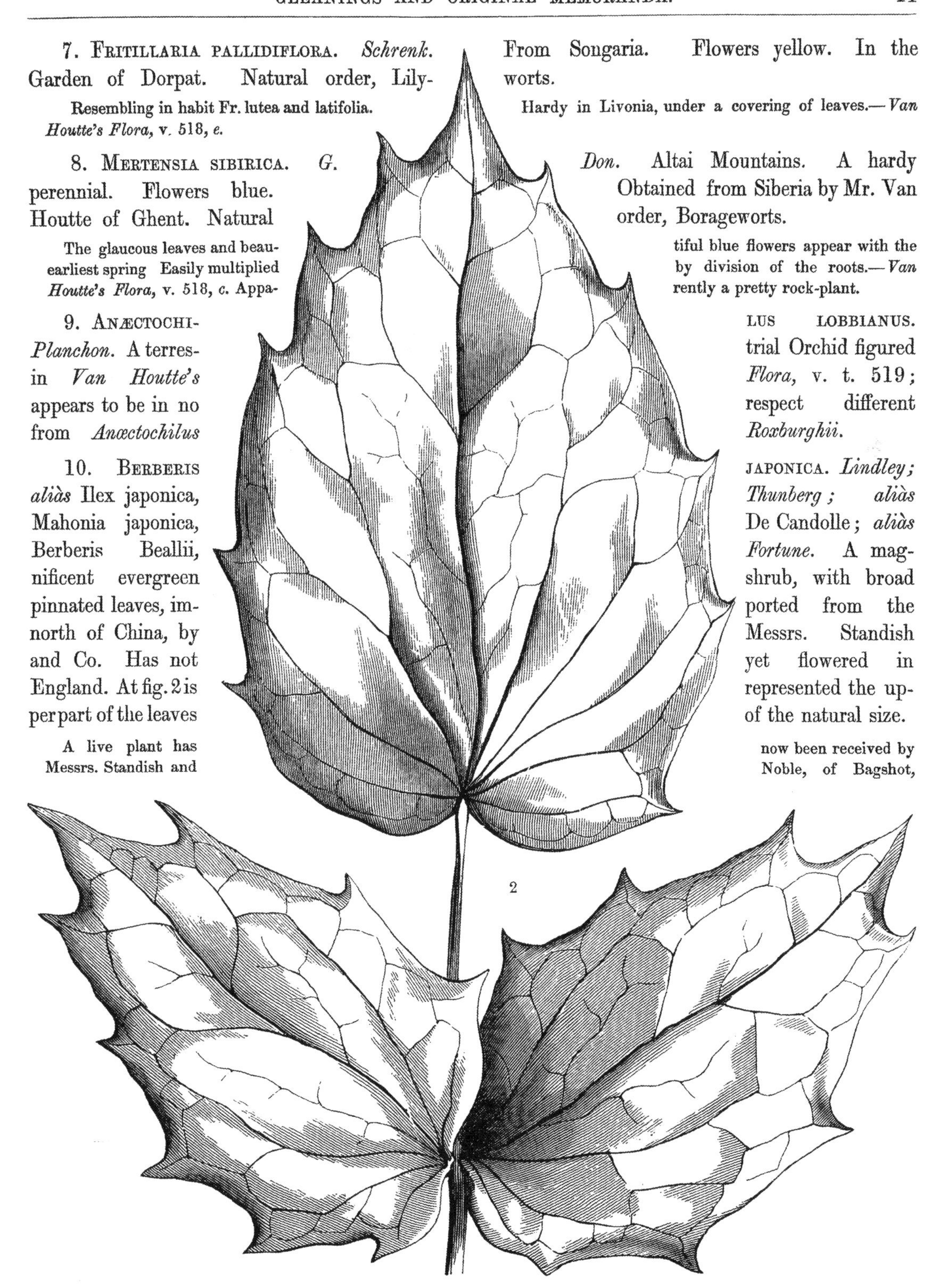
2

from Mr. Fortune, who informs them that it grows from 100 to 150 miles north of Shanghae, and that it is the most gigantic of the Berberries. A leaf, which has been sent me by Mr. Standish, is nearly 15 inches long, and of a stout leathery texture; it originally had four pairs of leaflets, and the usual terminal one; the lower pair has dropped off: the other lateral leaflets are sessile, slightly cordate, about 3½ inches long, with from 3 to 4 strong spiny teeth on each side, and a very stiff triangular point; the terminal leaflet is 5 inches long, and very deeply cordate, with 5 coarse, spiny teeth on each side. This is certainly the finest of the genus, and if hardy it will be the noblest evergreen bush in Europe. There is, however, but one plant of it at present in cultivation, so that its habits are unascertained.—*Journ. Hort. Soc.*, Vol. v., p. 20.

11. GALEOTTIA BEAUMONTII. *Lindley; (aliàs* Stenia Beaumontii. *A. Rich. in Cat. Pescator,* 1849, p. 36.) From Brazil, having been obtained from Bahia by M. Morel, of Paris. An uninteresting stove orchidaceous epiphyte, with the habit of a Maxillaria. Flowers two on a stalk, dull green and brown, with a pale lilac lip. Introduced by Mons. Morel.

G. Beaumontii: pseudobulbis oblongis 2-phyllis, foliis lineari-oblongis aveniis, pedunculo radicali erecto 2-floro floribus haud resupinatis, labelli trilobi laciniâ intermediâ lineari apice deflexâ subulatâ lateralibus truncatis margine anteriore setaceolaceris supra epichilium continuis.

This has flowered in the Garden of Plants at Paris, in the great collection of M. Pescatore, and with M. Morel, in all which places we have seen it. That it is a Galeottia there is no doubt; an obscure genus founded by M. Achille Richard upon a Mexican plant unknown in gardens; and which may possibly be found not distinct from Batemannia. The only difference, indeed, which we see in the present instance is, that this Galeottia has a large ovate gland and short caudicle, while Batemannia has a large ovate gland and no caudicle.

12. TRICHOCENTRUM TENUIFLORUM. *Lindley.* From Bahia. An obscure stove epiphyte, flowering in January. Flowers small, dingy brown, and white. Natural Order, Orchids. Introduced by M. Morel, of Paris.

T. tenuiflorum: foliis . . . , sepalis linearibus acutis, petalis conformibus obtusis, labello obovato emarginato subundulato basi angustato lamellis basim totam occupantibus, columnæ alis semicordatis acutis.

This little plant is of only Botanical interest. It differs from all the known species of the genus in the narrow sepals and petals of its small flowers, and in its almost linear obovate lip with a pair of plates occupying the whole of the base.

The following are the other known species of this genus, none of which deserve the notice of cultivators :—

1. T. fuscum. *Lindl. in Bot. Reg.* 1951.
2. T. maculatum. *Lindl. Orch. Lindenianæ,* No. 127.
3. T. pulchrum. *Pöppig, N. Gen. & Sp.,* pl. ii., t. 115.
4. T. recurvum. *Lindl. in Bot. Reg.,* 1843, *misc.* 17.
5. T. candidum. *Lindl. in Bot. Reg.,* 1843, *misc.* 18.
6. T. iridifolium. *Lindl. in Bot. Reg.,* 1843, *misc.* 178.

13. PHOLIDOTA CLYPEATA. *Lindley.* Imported by Messrs. Low and Co. from Borneo. An unimportant stove epiphyte, belonging to the order of Orchids. Flowers dirty white.

I have only seen the flowers, which resemble those of P. imbricata, but stand in a spike not more than three inches long. The column is very like a three-lobed petal, bordered with brown, and gives the flower the appearance of having two opposite lips. Mr. Kenrick states that the pseudo-bulbs are "about 2 inches long, with a dark-green leaf."—*Journ. Hort. Soc.,* Vol. v., p. 37.

14. BERBERIS WALLICHIANA. *Decandolle; (aliàs* B. macrophylla *of the Gardens; aliàs* B. atrovirens, *G. Don.)* A hardy evergreen bush from the mountains of tropical Asia. Imported by Messrs. Veitch. Has not yet flowered in England.

An evergreen of most beautiful aspect, with brown branches, a very dark green dense foliage, and long, slender, 3-parted spines. The leaves grow in clusters, are about 3 or 4 inches long, with a sharp, prickly point, and numerous fine serratures, ending in a straight point on each side; on the upper side they are a rich bright green, turning to a claret colour in the autumn, and remarkably netted: on the under side they are pale green and shining. With Messrs. Veitch it has stood through three winters without shelter,and is now 4 or 5 feet high. Naturally it is said to grow 10 feet high.—*Journ. Hort. Soc.,* Vol. v., p. 4.

15. BERBERIS LOXENSIS. *Bentham.* A hardy or half-hardy evergreen shrub, imported by Messrs. Veitch and Co., from Peru. Has not yet flowered in England. (Fig. 3.)

It has small palmated spines, and very shining, blunt, obovate, bright green leaves, of nearly the same colour on both sides; they seem to have in all cases a spiny point, and very often several teeth at the sides. The flowers are unusually small, and stand erect in panicled racemes on a long peduncle quite clear of the leaves. Its hardiness is uncertain; but its beautiful foliage makes it worth some protection if necessary.—*Journ. Hort. Soc.,* Vol. v., p. 7.

16. Berberis Darwinii. *Hooker.* From Chiloe and Patagonia. A hardy evergreen bush, of great beauty, imported by Messrs. Veitch. Flowers not yet produced in England. (Fig. 4.)

An evergreen shrub 3 to 5 feet high, of extraordinary beauty, conspicuous for its ferruginous shoots, by which it is at once recognised. The leaves are of the deepest green, shining as if polished, not more than $\frac{3}{4}$ inch long, pale green, with the principal veins conspicuous on the under side, with three large spiny teeth at the end, and about one (or two) more on each side near the middle. Although small, the leaves are placed so near together that the branches themselves are concealed. The flowers are in erect racemes, and of a deep orange yellow. Mr. Veitch informs me that this plant appears to be decidedly hardy: as is probable, considering that it grows naturally near the summer limits of snow upon its native mountains.—*Journ. Hort. Soc.*, Vol. v., p. 6.

17. Berberis tinctoria. *Leschenault.* An Indian sub-evergreen shrub, apparently hardy. Has not yet flowered in England. (Fig. 5.)

The plants in gardens are slender, brown-wooded shrubs, with small slender spines, usually 3-parted. The leaves are thin, not shining, dull green above, glaucous beneath, oblong, blunt, with a spiny point, but scarcely spiny-toothed, except on the seedling plant. The flowers have not hitherto appeared. They are represented by Dr. Wight as standing erect in loose racemes scarcely longer than the leaves, and succeeded by an abundance of dull red fruit.—*Journ. Hort. Soc.*, Vol. v., p. 13. At the lower part of fig. 5 are represented the early leaves of this species, which are cordate and long-stalked, and quite different from the later leaves.

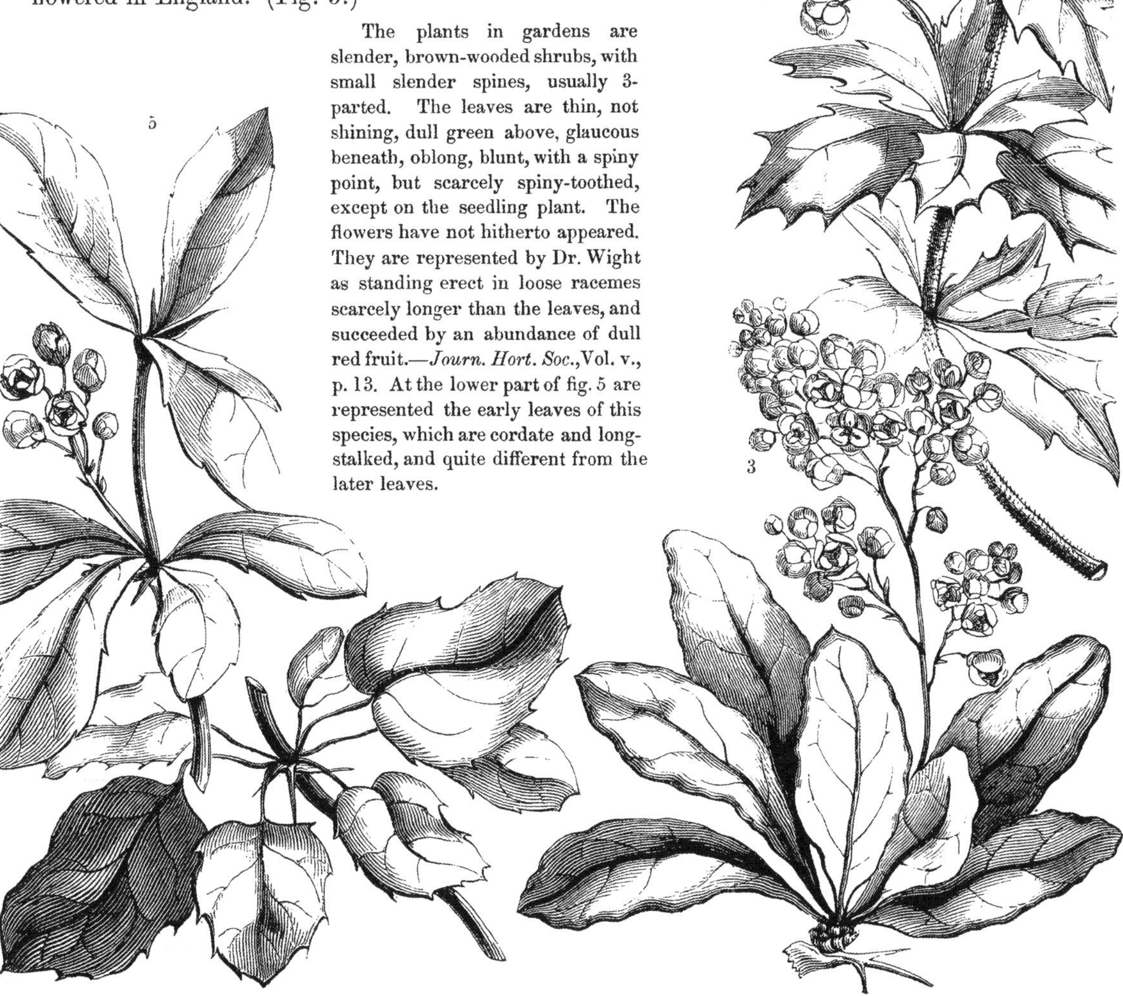

18. Blandfordia flammea. *Lindley.* From New Holland. A beautiful greenhouse perennial, flowering in October. Flowers 2½ inches long, vivid orange scarlet. Introduced by Messrs. Low and Co. Natural order, Lilyworts (Liliaceæ).

This, which is perhaps the finest of the Blandfordias, in a vigorous state is full 4 feet high, and bears 5 or 6 flowers at the end of its graceful stem. The plant which flowered with Messrs. Low, was an unhealthy offset, little more than 6 inches high. The leaves are narrow and stiff; the flowers about 3½ inches long, 1½ inch across the mouth, of the most vivid orange scarlet, with a broad edge of clear yellow. It is even handsomer than *B. intermedia* and *marginata.*—*Journ. Hort. Soc.*, Vol. v., p. 32.

19. Cheirostylis marmorata. *Lindley; (aliàs* Dossinia marmorata, *Morren).* From Borneo? A pretty herbaceous stove plant, belonging to the natural order of Orchids, flowering in September. Flowers white. Introduced by Mr. Hugh Low.

The leaves are of a deep reddish olive-green, with a velvety surface, and are traversed by fine golden veins, which disappear to a great extent when the leaves become old. It is far less beautiful than *Anœctochilus setaceus* or *Monochilus regius.* The flowers are white, with a reddish calyx, in a long, dark, purple, downy raceme. Although destitute of striking beauty, they well repay a minute examination, being covered with pellucid glands, and frosted, as it were, over all the inner surface. Requires damp heat, and a mixture of three parts chopped sphagnum and one-third well-decayed leaf-mould. Increased by the creeping stems.—*Journ. Hort. Soc.*, Vol. v., p. 79.

20. Helianthemum scoparium. *Nuttall.* From California. A small hardy shrubby rock-plant, belonging to the natural order of Rock Roses or Cistaceæ, flowering in September. Flowers yellow. Introduced by the Horticultural Society.

A small prostrate shrub, with wiry branches and linear leaves. The flowers, which are small and bright yellow, grow in twos and threes at the end of the branches on naked pedicels about half an inch long. A hardy little shrub, requiring the same treatment as Cistuses. A very nice species for rock-work, on which it thrives in the full glare of the sun.—*Journ. Hort. Soc.*, Vol. v., p. 79.

21. Calboa globosa. *Lindley; (aliàs* Morenoa globosa, *Llave; aliàs* Quamoclit globosa, *Bentham.)* A Mexican greenhouse twining perennial, of the natural order of Bindweeds. Flowers 2½ inches long, scarlet. Flowers in September.

A rambling perennial, smooth in every part. Leaves thin, on long stalks extremely variable in form; some cordate and acuminate; others sagittate; others completely hastate, with the lobes all narrow, and the lower ones deeply angular. The flowers grow in naked umbels, on a peduncle 9 or 10 inches long; the pedicels are from 1½ to 4 inches long. Each sepal has a long subulate process at the back. The corolla is 2½ inches long, deep rich red, with a curved cylindrical tube, and a campanulate erect limb, divided into 5 erect rounded wavy lobes. This is a strong half woody climber, growing freely in any good rich soil composed of loam and sandy peat. It is easily increased by cuttings of the young shoots, and requires to be kept rather dry in a cool part of the stove during the winter, but should be grown in a cool airy part of the greenhouse during summer, where it will flower from August to October. Although a fine species, it is only fit for growing where there is plenty of room for its tops to spread. It will not flower in a pot, but must be planted in the open ground.—*Journ. Hort. Soc.*, Vol. v., p. 83. *With a figure.*

22. Pentarhaphia cubensis. *Decaisne.* A tender shrub from Cuba, belonging to the order of Gesnerads. Flowers tubular, scarlet, appearing in the summer, handsome.

A shrub with a compact habit, and dark-green, convex, ever-green leaves, obovate, crenated near the point, and netted on the under-side with green veins on a pale ground. The flowers grow singly in the axils of the leaves, on cinnamon-brown stalks an inch long. The corolla is about the same length; tubular, curved and rich scarlet, with a projecting style. The calyx consists of five straight, narrow, sharp lobes, not unlike five brown needles, whence the generic name has arisen; requires a temperature intermediate between the greenhouse and stove; easily increased by cuttings, and grows freely in loam, peat, and leaf-mould.—*Journ. Hort. Soc.*, Vol. v., p. 86. *With a figure.*

23. Pharbitis limbata. *Lindley.* A tender, or half-hardy annual, from Java; imported by Messrs. Rollissons. Flowers very handsome, violet edged with white. Blossoms in the autumn; belongs to the Bindweed order.

This appears to be an annual, and has much the appearance of Pharbitis Nil, from which it principally differs in the great length of its sepals, their excessive hispidity, and the shortness of the flower-stalk. The flowers, equal in size to the old Convolvulus major, but less spreading at the mouth, are of an intense violet, edged with pure white, and have a beautiful appearance.—*Journ. Hort. Soc.*, Vol. v., p. 33.

24. Spiræa decumbens. *Koch.* (*aliàs* S. flexuosa, *Reichenbach,* not *Fischer ; aliàs* S. adiantifolia, of *Belgian Gardens*). A hardy European shrub of the Rosaceous order, with weak twining stems, and clusters of white flowers with a rose-coloured eye. In the Belgian gardens. (Fig. 6.)

This species is a native of the mountains of the Frioul, where it was found by Schiede. It is at present little known, although its graceful habit and abundant sweet white flowers give a claim to the attention of amateurs. It forms a bush about a foot high, and one and a half foot wide, tufted, with numerous shining brown branches. The leaves are obovate or oval, long-stalked, unequally serrated, entire near the base, green above, glaucous beneath. The flowers are in little terminal corymbs. It is perfectly hardy, and is suitable for planting in front of larger shrubs.—*Annales de Gand,* t. 262. To us, it seems to be a very pretty rock-plant.

25. Grammanthes Gentianoides. *De Candolle.* A native of the Cape of Good Hope, and a half-hardy annual. Flowers salmon colour, in hemispherical clusters. Natural order, Gentianworts. (Fig. 7.)

Stems a few inches high, white and brittle. Leaves oblong, blunt, succulent. Flowers numerous, about as large as a sixpence, 5-parted, salmon-coloured, with a pallid stain at the base of the lobes, and a greenish stain somewhat in the form of the letter V. It is rather pretty in a greenhouse, but is not suited for the open air, where it soon rots, even when elevated on rockwork.—Figured in *Van Houtte's Flora,* Oct. 1849, t. 518.

26. Calandrinia umbellata. *De Candolle.* A native of Chili, belonging to the natural order of Purslanes. A very pretty half-hardy annual, with deep rose-coloured flowers growing in clusters opening only under a bright sun. (Fig. 8.)

Stems fleshy, somewhat branched. Leaves very narrow, acute, hairy, those on the stem and next the root alike in form and equally succulent. The flowers when open are about as large as a sixpence, with very round petals ; they grow in many-flowered umbels, and expand in succession during the whole summer. Professor Morren speaks thus of its management in Belgium. Naturally an annual, the seeds are sown in sandy land early in the spring ; this is best done where they have to stand, because such delicate plants do not bear well the operation of pricking out. A soil composed of sand, mixed with decayed vegetable matter, especially rotten leaves, is what suits it best. In order to have large fine flowers, it is as well to give the plants a good watering once or twice during the summer with Guano water. In Belgium the seeds begin to ripen by July. It also makes a very nice pot plant for sitting rooms.—Figured in the *Annales de Gand,* t. 268. We believe this to be one of Messrs. Veitch's many importations, and quite concur with Professor Morren in saying that it is not so much known as it deserves to be, especially in gardens where beauty is in greater esteem than rarity.

27. Trichoglottis pallens. *Lindley.* A stove Orchid from Manilla, bloomed in November at Chatsworth. Flowers green and white, of little interest.

A dwarf erect plant, with oblong distichous leaves, and a lateral flower or two, not quite 2 inches in diameter, pale delicate green, with delicate brownish spots and a white lip. The latter organ was oblong, with a white, shaggy crest on the upper side, and a pair of short, yellowish scimitar-shaped segments standing erect near the base; within these were a pair of forked callosities, one placed before the other in the centre, but no sac or pouch was found between them. The plant is of little beauty, but of considerable botanical interest.—*Journ. Hort. Soc.*, Vol. v., p. 34.

28. MICROSPERMA BARTONIODES. *Walpers* (*aliàs* Eucnide bartonioides, *Zuccarini*). A Loasad from Mexico. Introduced by Mr. Charlwood. A handsome hardy annual, with large bright yellow glittering flowers; the stems are covered with stiffish hairs.

Stems about a foot long, flexuose, succulent, subtranslucent. Leaves ovate-acute, lobed, and serrated. Flower-stalks long, one-flowered, terminal. Petals ovate, or rather obovate, slightly serrated, sulphur-yellow, paler, almost white, beneath. Stamens very long, in five monadelphous fascicles. Its soft, succulent nature, makes it liable to be injured by heavy rain and wind.—*Botanical Magazine*, t. 4491.

29. SPATHODEA SPECIOSA. *Brongniart.* Of uncertain origin—supposed African. A magnificent stove tree, belonging to the Bignoniads, with close panicles of very large pink, trumpet-shaped flowers, stained with crimson. Flowers in the spring.

When this beautiful species blossomed at Ghent, it was about 4 feet high. The panicle appeared at the end of the stem, which was covered with pinnated leaves, seated in threes, each being furnished with oblong-lanceolate, acuminate, serrated, shining leaflets. The corolla is about 2½ inches long, and is protruded from an oblong blunt calyx, which opens on one side to let it pass, at the same time dividing into 2 triangular teeth at the back. Cultivated in a mixture of decayed leaves and rotten dung, mixed with one-third peat and one-third loam; it is represented to be difficult to strike. According to Prof. Morren, it was originally received at Ghent from England.—*Annales de Gand*, t. 260.

30. ODONTOGLOSSUM RUBESCENS. *Lindley.* From Nicaragua, imported by Mr. Skinner. A very handsome Orchid, with fine blush flowers spotted with crimson. Flowers in November.

A charming species, belonging to the beautiful white-lipped section of the genus, and remarkable among them for its flowers being suffused with a tender blush colour. The sepals are very straight and sharp-pointed, richly spotted with crimson. The petals have similar spots near their base; the lip is spotless, crisp, and cordate, but not ciliated.—*Journ. Hort. Soc.*, Vol. v., p. 35.

31. PENTSTEMON CORDIFOLIUS. *Bentham.* A hardy shrub, of the order of Linariads. Flowers rich dull red, in long bunches, rather handsome. From California; flowers in the summer and autumn.

A downy-stemmed half-shrubby plant, with a trailing or spreading habit, so that it is well suited to hang down over stones or rocks. Leaves dark-green, shining, cordate, serrate, slightly downy. Flowers in one-sided, narrow, leafy panicles, which sometimes measure more than a foot in length. The branches of the panicle are hairy, and bear each from three to five flowers when the plants are vigorous. Calyx covered with glandular hairs; corolla not quite an inch and a half long, rich dull red; the tube almost cylindrical; the upper lip straight, nearly flat, slightly two-lobed; the lower three-parted, spreading at right angles to the upper. Hardy, grows in any good rich garden soil, and easily increases by seeds or cuttings. It flowers freely, one year from seeds, and lasts in flower from June to October. It is a very desirable plant.—*Journ. Hort. Soc.*, Vol. v., p. 87. *With a figure.*

32. SPATHOGLOTTIS AUREA. *Lindley.* From Malacca. A pretty terrestrial stove plant, belonging to the natural order of Orchids, flowering in November. Flowers yellow. Introduced by Messrs. Veitch and Son.

Rather handsome, with narrow leaves like those of a Phaius, and a scape 2 feet high, bearing at the very end about half a dozen large golden-yellow flowers, with a few dull sanguine spots on the lip. Mr. T. Lobb found it on Mount Ophir, near the beautiful Nepenthes sanguinea. According to a memorandum by the late Mr. Griffith, it inhabits rocks on Mount Ophir, at places called Goonong, Toondook, and Laydang.—*Journ. Hort. Soc.*, Vol. v., p. 34.

PLATE. 4

L. Constans Pinx & Lith.

Maclure Macdonald & Macgregor Lith. London

[PLATE 4.]

THE TOOTHED CEANOTHE.

(CEANOTHUS DENTATUS.)

A half-hardy Evergreen Shrub, from CALIFORNIA, *belonging to the Natural Order of* RHAMNADS.

Specific Character.

THE TOOTHED CEANOTHE.—A branched evergreen bush, closely coated with ferruginous hairs. Leaves small, oblong, rounded at each end, or almost cordate, coarsely toothed, and revolute at the edge, where they are furnished with distinct slightly stalked glands ; smooth, shining, and deep green on the upper side. Flowers in terminal, stalked, roundish or oblong thyrses or umbels.

CEANOTHUS *DENTATUS.*—Frutex ramosus, tomentosus, sempervirens ; ramis ferrugineis ; foliis parvis penniveniis oblongis utrinque rotundatis v. cordatis grossè dentatis revolutis margine glandulosis : supra lucidis atroviridibus glabris, thyrsis umbellisve oblongis rotundisque pedunculatis, pedicellis calycibusque glaberrimis.

Ceanothus dentatus : *Torrey and Gray, Flora of North America,* vol. 1., p. 268.

DURING Douglas's last journey in California, this plant was first met with, but where is unknown. From specimens communicated to Drs. Torrey and Gray by the Horticultural Society, it was described by those authors. From Californian seeds, procured for the same Society by the Collector Hartweg, it has now been raised in the Society's Garden, whence it has been also extensively distributed among the Fellows. The plant which produced the specimen here represented flowered in February last in Her Majesty's Garden at Frogmore, under the care of Mr. Ingram.

It is a small bush, covered all over with rusty down, except upon the upper side of the leaves. In the cultivated plant the branches are five or six inches long, but in the wild specimens they are not more than a third of that length. The leaves are deep green, shining, wavy, strongly toothed, and rolled back at the edge, quite blunt, and somewhat heart-shaped at the base, on short stalks, furnished with a pair of triangular scale-like stipules. On the edge of the leaves appear many oblong fleshy stalked glands, which in the beginning are pale green, afterwards become yellow, and finally acquire a deep brown colour. To their presence is due a heavy, unpleasant, but slight odour, which is perceptible when the plant is touched; they afford an excellent specific character, but have been overlooked by Messrs. Torrey and Gray. The flowers are bright blue, bordering on violet, and are produced in stalked heads, which are sometimes racemes, sometimes thyrses, and even almost umbels. The authors of the Flora of N. America called them white, assuming such to be the case from the appearance of the dried specimens.

Like all the Californian plants, this naturally endures a hot dry summer, by which its wood is kept short-jointed, and is thoroughly ripened, so as to be enabled to support the severe winters to which it is exposed. It then, also, is loaded with clusters of flowers, twice as long as those here represented, and must become far more beautiful than it now is. Hitherto it has been kept in greenhouses or damp pits, where it has been exposed to none of its natural conditions. Mr. Ingram's specimens are from a spring forcing house.

Not having yet acquired its natural condition, there is a difficulty in judging of its capability of bearing an English winter. All that we at present know about it is that it lives uninjured under a glass frame facing the North, without any aid from artificial heat. We also know that other small-leaved Ceanothes have sustained no injury in exposed places, even though unsheltered, provided the sun has not shone upon them. In the meanwhile it will be desirable to treat this as a frame plant, or to force it with Lilacs, and such things; for which purpose the blue of its flowers, a colour so rare, and so greatly wanted in gardens, renders it peculiarly valuable.

The two uncoloured figures on either side of our plate represent magnified views of the upper and under side of a leaf.

L. Constans Pinx & Lith

Printed by Maclure, Macdonald & Macgregor Lith London

[PLATE 5.]

THE CHANGEABLE ADAMIA.

(ADAMIA VERSICOLOR.)

A Greenhouse Shrub, from CHINA, *belonging to the Natural Order of* HYDRANGEADS.

Specific Character.

THE CHANGEABLE ADAMIA.—Leaves oblong-lanceolate, sharply toothed, entire at the base and narrowed into the stalk, covered with down on the ribs. Panicle pyramidal, downy. Branchlets in cymes. Flowers in seven parts, with about twenty stamens.

ADAMIA *VERSICOLOR;* foliis oblongo-lanceolatis acutè serratis basi integris in petiolum angustatis subtùs in costas pubescentibus, paniculâ pyramidali pubescente, armulis cymosis, floribus heptameris icosandris.

Adamia versicolor: *Fortune in Journal of the Horticultural Society,* vol. 1, p. 298.

A PLANT which had been brought from China by Mr. Fortune and which flowered in the garden of the Horticultural Society in September, 1846, furnished the materials for the accompanying figure. It had been found by him in Hong Kong, growing in ravines about half-way up the granitic mountains of that Island.

It forms a bush with the habit of an Hydrangea, to which genus it is naturally related. The stems and branches are downy; the leaves grow in opposite pairs, are oblong-lanceolate, serrated, sharp-pointed, somewhat convex, with a red midrib, which as well as the other ribs is slightly downy. The flowers appear in pyramidal downy panicles. In the bud state the corolla is pure white; more advanced it assumes a violet and ultramarine tint; at a later period it becomes a clear delicate blue, and upon opening it forms a handsome violet star of six or seven points, inclosing about twenty deep violet stamens, in the centre of which are found five bright blue styles. Thus there is found in the same panicle, at the same time, an infinite variety of tints of clear blue and violet, as well as pure white, the effect of which is extremely pleasing. It is understood that the flowers are succeeded by porcelain-blue berries: but they have not appeared as yet in this country.

In fact, since the plant which flowered under Mr. Fortune's care in September, 1846, and which is here represented, no specimen of any beauty has appeared, and an idea has been entertained that the species is not worth cultivation because there has not yet been skill enough to manage it properly. What its precise treatment should be must be left to the determination of experiment. It is, however, to be inferred from what is known of Hong Kong, that the plant requires the climate of the tropics while growing, and that of Devonshire when at rest. In the rocky ravines of Hong Kong, it is deluged with torrents of water, and forced by a vehement heat into luxuriant growth; at that time it must become exposed to as much as 120° of direct sunheat, while ripening its wood; and afterwards it must endure a temperature of 40° until the rains and heats return, and once more force it into vigour. All this must be done with an abundance of air, for in its natural station it is continually exposed to violent gusts and storms of wind, which struggle with a burning sun. Manure, properly so called, it can have little or none; but must depend for its food upon what the air can bring it, and upon such saline matters as may be yielded to water by the decomposition of the granitic soil.

We should add, that the accompanying figure hardly does justice to the plant itself, the panicle of flowers, copied by our artist, having been in reality nearly a foot in diameter.

Another species, the skyblue Adamia (*A. cyanea*), is in cultivation, and has been figured in the Botanical Magazine. But it is far inferior to this in beauty; the flowers being whole-coloured, a bad violet, and not half the size. It would, however, redeem its character if it could be made to form its berries, which Dr. Wallich describes as appearing on the open rocky mountains of Nepal, in great profusion, of a deep blue colour, and rendering the plant "an object of great elegance."

It was to that species, named *cyanea* because of its blueness, that the denomination Adamia was first given by Dr. Wallich, in commemoration of the eminent services rendered to Indian science by his friend John Adam, Esq, formerly President of the Supreme Council of Calcutta. We have before us two more species of the same genus, one of which found in Java, by Mr. Lobb, is probably the *Cyanitis sylvatica* of Reinwardt, and must be more than a rival to the present plant. We believe, however, it never reached England alive.

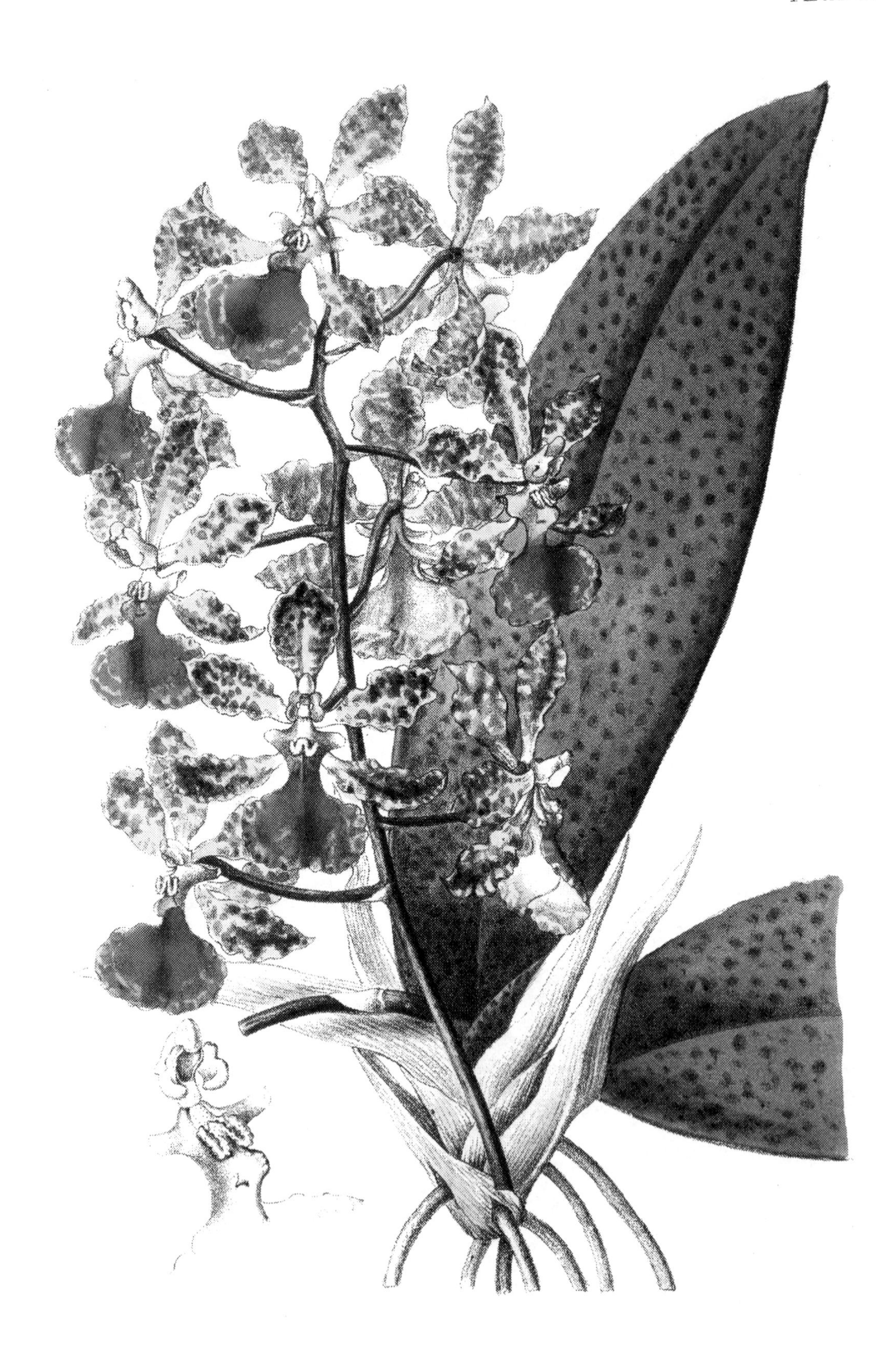

L. Constans Pinx & Lith.

Printed by Maclure, Macdonald & Macgregor Lith. London.

[PLATE 6.]

THE PURPLE-LIPPED ONCID.

(ONCIDIUM HÆMATOCHILUM.)

A Stove Epiphyte, from NEW GRENADA, *belonging to the Natural Order of* ORCHIDS.

Specific Character.

THE PURPLE-LIPPED ONCID.—(Sect. *Pluritubercudate.*) Bulbless. Leaves oblong, flat, thick, sharp-pointed, spotted, growing singly. Racemes compact, stiff. Sepals distinct, and the petals all of similar form, spathulate, wavy ; lip roundish, with auricles at the base ; the crest scarcely evident at the base, wavy in front like the letter W, thence raised into an eminence, with a toothlet on each side. Wings of the column rounded, curved downwards, somewhat lobed.

ONCIDIUM *HÆMATOCHILUM;* (sect. *Pluritubercudata*) ebulbe, foliis oblongis planis coriaceis acutis maculatis solitariis, racemis densis rigidis, sepalis liberis petalisque conformibus spathulatis undulatis, labello subrotundo basi auriculato, cristâ basi obsoletâ antice flexuosâ (literam W referente), inde in jugum productâ denticulo utrinque ; columnæ alis rotundatis decurvis sublobatis.

THE accompanying drawing was made in September, 1847, from a plant in the possession of Messrs. Loddiges, and we believe it is found in their list under the provisional name of *O. luridum purpuratum*. They had imported it from New Grenada; but it does not occur among any dried collections which we have examined from that country.

In foliage it resembles the Carthagena Oncid (O. carthaginense) and its allies; the leaves being hard, stiff, dull green, spotted with brown, and destitute of any evident pseudo-bulb. The flowers, too, grow in the same manner, but they are very different in details of structure, as well as in colour and size. The sepals and petals are a warm greenish yellow, strongly blotched with rich chesnut-brown. The lip, on the other hand, is of the richest crimson, except near the base, where it fades into bright rose-colour. The crest, by the minute peculiarity of which Oncids are often most certainly known, resembles the letter W, having in the rear a short, flattish, narrow space, and in front a well-defined projection, with a small tooth on each side.

By these circumstances it is readily distinguished from the neighbouring species, in none of which such an arrangement occurs, varied as are the forms assumed by the tubercles of their crest. In all the varieties of the Carthagena Oncid there is, for instance, a pair of strong warts in place of the small teeth, one on each side of the anterior elevation, and the W-like body is divided into two distinct Vs. In the sanguine Oncid the two posterior tubercles are more oblong, projecting with a furrow along the middle. In Professor Morren's new Rosette Oncid (*O. cosymbephorum*), nearly allied to this, there is quite a bunch of tubercles at the base of the lip.

Among Oncids this purple-lipped kind is one of the best, being inferior to none except Lance's, The contrast between the crimson of its lip, the greenish-yellow ground-colour of the petals, and their rich cinnamon spots, is of rare occurrence, and produces a charming effect.

Every one who has studied the genus Oncidium, or endeavoured to ascertain the names of his species, must have felt the task to be one of extreme difficulty, in some measure owing to the want of any sufficiently precise classification of the genus. What was sufficient when the number of species was small, became useless as they increased in number; and that which succeeded has proved insufficient in its turn. We have therefore endeavoured, upon a full review of the subject, to effect such a classification as may meet the exigencies of the case, now that the discovery of new species has much slackened, and that the main forms are probably ascertained.

In the first place, it is necessary to eliminate all those singular and little known species, of which *O. serratum* figured in another page, and Mr. Bateman's *O. microchilum* may be taken as examples. These have very distinct stalks to their sepals, and a lip so much smaller than the other parts, as in some cases nearly to escape observation. They constitute the true *Cyrtochilums* of Humboldt, but have nothing to separate them generically from Oncidium.

In all the other species the lip is the largest part of the flower.

Of these some have the leaves placed with their edges vertically, or "equitant;" others have the leaves tapering, like an onion; and the remainder have the ordinary flat leaves.

Among the herd of flat-leaved species some have the side sepals united, more or less, so as sometimes to give the flowers the appearance of having only four divisions instead of five; others, on the contrary, have five divisions, unmistakeably distinct. The first may be called *Tetrapetalous,* the second *Pentapetalous,* as we formerly proposed.

Some of the Tetrapetalous series have the true petals considerably larger than the sepals. In others, sepals and petals are of the same size.

Among the Pentapetalous set some have the lip entire, although in most it is distinctly eared. Some have it narrowest, some broadest at the base. For the separation of the narrow-based ear-lipped species into groups, there seems to be nothing more useful than the modifications of the crest. In one group the crest is a hairy cushion; in a second, it consists of a very few (not more than four) tubercles; in a third, the number of tubercles is greater; in a fourth, they are surrounded by minute warts.

In this way a dozen well-defined groups are obtained, under which about 150 species, of which the principal part are in gardens, may be readily arranged.

The fine species now figured belongs to the section having a pentapetalous structure with many tubercles on its crest. The remainder of the section is as follows:—

1. O. suave, *Lindl. in Bot. Reg.*, 1843. *misc.* 22.—Mexico.—Like O. reflexum, but the flowers are much smaller. Sepals and petals chocolate colour tipped with yellow; lip yellow with a cinnamon-brown middle. Has a slight agreeable odour.

2. O. Suttoni, *Bateman, in Bot. Reg.*, 1847. *misc.* 8.—Guatemala.—Leaves grassy. Flowers small, yellow and olive coloured; not worth cultivation.

3. O. tenue, *Lindl. in Journ. Hort. Soc.* iii. p. 76 *ic.*—Guatemala.—A species of little beauty, resembling *O. suave.* Flowers small, yellow, mottled with dull brown.

4. O. pentadactylon, *Lindl. in Ann. Nat. Hist.*, xv.—Peru.—Flowers small, in a large panicle—often altogether abortive ; not in cultivation, nor worth it.

5. O. maizæfolium, *Lindl. in Orchid. Linden.* No. 78.—New Grenada.—A mountain plant. Flowers bright yellow, spotted with red. Not in cultivation.

6. O. ramosum, *Lindl. in Bot. Reg.*, sub. fol. 1920. aliàs *O. Batemannianum*, Knowles and Westcott, Floral Cabinet, 3. 183. t. 137.—Brazil.—A fine species, with gay pale yellow flowers in a branched panicle as much as five feet high.

7. O. retusum, *Lindl. in Bot. Reg.*, sub t. 1920.—Peru.—A beautiful species, with deep chestnut and yellow flowers, and a yellow lip.

8. O. oblongatum, *Lindl. in Bot. Reg.*, 1844, *misc.* 11.—Mexico.—Like O. reflexum, but with coloured pseudobulbs and a speckled stem. Flowers very yellow, large, and handsome.

9. O. Barkeri, *Lindl. in Bot. Reg.*, 1841, *misc.* 174. *Sertum Orchid.*, t. 18.—Mexico.—A very handsome plant, with large yellow flowers with rich brown spots on the sepals and petals. Raceme simple.

10. O. unguiculatum, *Lindl. in Journ. Hort. Soc.*, i. 303, *ic.*—Mexico.—Near O. Barkeri, but stem erect, and branched, lip longer and narrower, and tubercles of the crest narrower. Lip bright yellow ; sepals and petals yellow, speckled with brown. Very handsome.

11. O. Pelicanum, *Martius, Bot. Reg.*, *misc.* 216., 1847, t. 70.—Mexico.—Very like O. reflexum, from which it differs in the tubercles being smooth, not downy, and the lateral lobes of the lip smaller in proportion to the intermediate segment.

12. O. reflexum, *Lindl. in Bot. Reg.*, sub. t. 1920.—Mexico.—A branched species, in the way of O. altissimum, but smaller. Flowers yellow, spotted with brown, except the lip.

13. O. nebulosum, *Lindl. in Bot. Reg.*, 1841, misc. 175 ; aliàs *O. Geertianum*, Morren in Ann. Gand. 1848, Feb.—Guatemala.—Flowers large, pale yellow, with faint spots of brown.

14. O. citrinum, *Lindl. in Bot. Reg.*, t. 1758.—Trinidad.—Flowers bright yellow, with faint traces only of greenish blotches.

15. O. leucochilum, *Bateman Orch. Mexic.*, t. 1 ; aliàs *O. digitatum*, Lindl. in Benth. plant. Hartweg. p. 94.—Mexico and Guatemala.—A charming species, with greenish flowers speckled with crimson, and a white lip fading to yellow.

16. O. sphacelatum, *Lindl. in Bot. Reg.*, 1842, t. 30—Mexico and Guatemala.—A fine handsome and branching species with yellow flowers spotted with rich brown. There are two varieties, of which the large flowered alone deserves cultivation.

17. O. altissimum, *Swartz, Bot. Reg.*, t. 1851.—West Indies.—Flowering stems sometimes 10-13 feet long. Flowers yellow and brown ; inferior to many others, notwithstanding its long panicles, which, however produce a striking effect when they have room to develope.

18. O. Baueri, *Lindl. Gen. and Sp. Orch.* 200., *Bot. Reg.* t. 1651.—Panama and Tropical America.—Much like the last, but the panicle more compound, and the column-wings truncate.

19. O. ensatum, *Lindl. in Bot. Reg.*, 1842, *misc.* 15.—Guatemala.—Also very like the last, but the leaves straight, long, and stiff, like sword-blades.

20. O. pictum, *Humb. Bonpl. and Kunth, nov. gen.* and *sp.* i., t. 81.—Popayan.—Like O. altissimum, but the panicle is more compact, the flowers larger and more yellow, and the edge of the leaf-sheaths very wavy.

21. O. sanguineum, *Lindl. Sertum*, t. 27 ; alias *O. Huntianum.* B. Mag., t. 3806 ; alias *O. roseum*, Lodd.; alias *O. Henchmanni*, Lodd.—La Guayra—A very variable plant near O. Carthaginense. Flowers small, blotched with crimson upon a straw-coloured ground.

22. O. hæmatochilum.—Of this plate.

23. O. cosymbephorum, *Morren, Annales de Gand.* t. 275—? —Flowers very pretty, bright rose colour, spotted with crimson and tipped with yellow. Lip cinnamon brown.

24. O. carthaginense, *Swartz*, alias *Epidendrum guttatum* Linn. ; aliàs *O. luridum*, Bot. Reg. t. 727 ; aliàs *O. intermedium*, Floral Cabinet, t. 60.—West Indies and tropical America—Another very variable plant, usually having dull olive brown speckled flowers ; but in the variety called *guttatum* they are rich brown and yellow, and very handsome. Other varieties are also known.

25. O. Lanceanum, *Lindl. in Bot. Reg.*, t. 1887.—Surinam—The finest of the section, with large deep brown speckled flowers and a rich violet lip. Fragant as Vanilla.

26. O. Cavendishianum, *Bateman Orch. Mex.*, t. 3 ; aliàs *O. pachyphyllum*, Bot. Mag. t. 3807.—Guatemala.—Leaves thick, fleshy, erect. Flowers large, bright yellow.

27. O. bicallosum, *Lindl. in Bot. Reg.*, t. 12, 1843.—Guatemala.—Very like the last, but flowers larger, slightly scented, with two great tubercles on its lip, besides smaller ones.

28. O. cultratum, *Lind. in Ann. Nat. Hist.*, xv.—Popayan.—A small, dwarf species, with not more than ten flowers in the panicle. Not in cultivation.

To this enumeration of the species in the Pluritubercúlate Section it may be useful to add a tabular view of the whole arrangement proposed in the beginning of this article.

I. MICROCHILA. Labellum nanum. I. CYRTOCHILUM H.B.K.

II. Macrochila. Labellum dilatatum.

A. Folia equitantia. II. EQUITANTIA.

B. Folia teretia. III. TERETIFOLIA.

C. Folia plana.

1. IV. TETRAPETALA MACROPETALA. Sepala lateralia connata. Petala multo majora.

2. V. TETRAPETALA MICROPETALA. Sepala lateralia connata. Petala sepalis subæqualia.

3. VI. PENTAPETALA MACROPETALA. Sepala lateralia libera. Petala multo majora.

4. Pentapetala micropetala. Sepala lateralia libera. Petala sepalis subæqualia.

* labellum indivisum ; (v. apice tantum lobatum; v. utrinque unidentatum) VII. INTEGRILABIA.

* * labellum auriculatum trilobum

= basi angustius, v. lobo terminali subæquali.

a. Cristâ pulvinatâ s. villosâ. VIII. PULVINATA.

b. Cristâ tuberculatâ (nec pulvinatâ)

‡ tuberculis 2—4. IX. PAUCITUBERCULATA.

‡ ‡ tuberculis 5—00, segregatis. X. PLURITUBERCULATA.

‡ ‡ ‡ tuberculis 5—10, verrucisq. circumstantibus. XI. VERRUCO-TUBERCULATA.

= basi manifestè latius. XII. BASILATA.

GLEANINGS AND ORIGINAL MEMORANDA.

33. CALANTHE SYLVATICA. *Lindley.* A beautiful terrestrial Stove Orchid, with long erect spikes of large flowers, at first white, but changing to bright yellow. Has flowered at Paris with M. Pescatore, from the Isles of France and Bourbon.

This is the most beautiful of all the species of Calanthe. To the foliage and general habit of the White Hellebore-leaved (*Calanthe veratrifolia*), it adds far finer flowers, which are at first pure white, but by degrees change to a clear bright yellow, very different from the livery of death. Thus, each spike of flowers resembles a massive plume, the upper part of which is snow-white, the lowest very yellow, while in the middle the one colour insensibly passes into the other through a tender cream-coloured tint.

34. ANGRÆCUM VIRENS. *Lindley in Botanical Register,* 1847, under t. 19. A showy white-flowered orchidaceous epiphyte, from Bourbon. Blossomed in January in the Garden of Plants, at Paris, under the care of Monsieur Houllet. (Figs. 9 & 10).

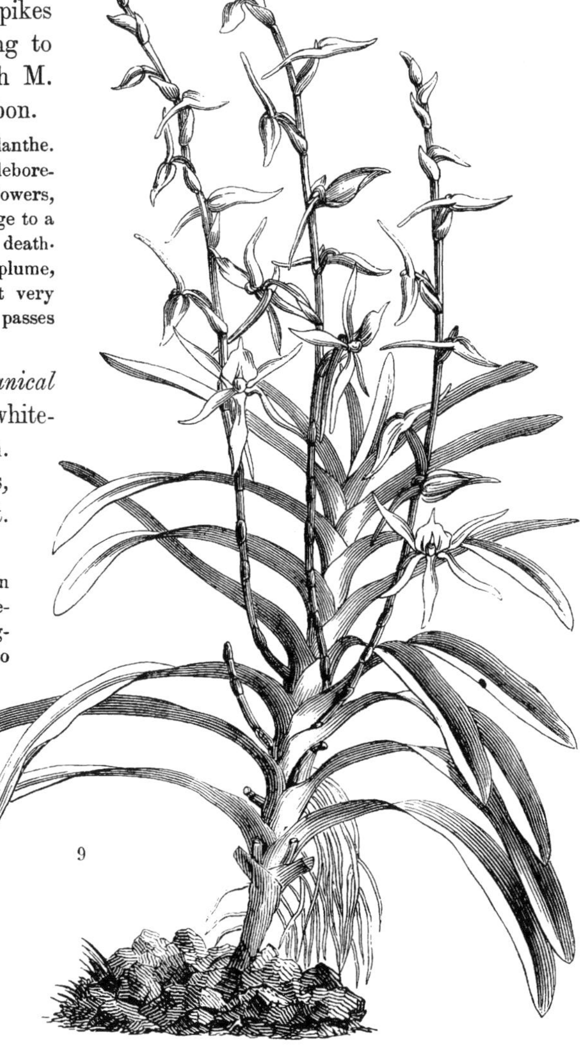

9

In the Garden of Plants, at Paris, were growing in January last two or three fine specimens of this remarkable plant, of which we had previously seen fragments only from the late Mr. George Loddiges, to whom it was said to have been sent from Serampore. The plants in question were as large as a full-grown Ivory Angurek (*Angræcum eburneum*); but their leaves were so flaccid and glaucous, as to render it evident that they belonged to some other species. From among them rose up several stately spikes of large unexpanded flowers, conspicuous for the dark-brown scales which supported them, the whole plant having the appearance represented in the accompanying fig. 9. Each spike was about two feet long.

At the time we saw them they were unexpanded, and led to hopes that they might show the species to be the little known superb Angurek of Dupetit Thouars (*Angræcum superbum*), the specimens having undoubtedly been received from Bourbon. Upon opening, they however proved to be what is now represented, each flower being of the size and form represented at figure 10. The sepals and petals, and the spur of the lip are greenish, and the lip itself, although white, is nevertheless conspicuously tinged with green in the middle ; not, however, to such a degree as in the plant which flowered with Mr. Loddiges, and which gave rise to the name which this plant bears. It is, however, a noble-looking plant, richly deserving a place among even the most select collections.

It may be useful to mention in this place, that the French collections contain some Bourbon and Isle of France Orchids, quite unknown among us. In addition to the subject of the last memorandum (No. 33), we observed the curious *Habenaria citrina*, *Eulophia scripta*, a showy species, *Bolbophyllum nutans*, and some other rarities, in the collection of M. Pescatore.

35. Passiflora belottii, *of the French Gardens*. A hybrid stove plant of uncertain origin ; apparently between P. cærulea and quadrangularis. Introduced by Messrs. Knight and Perry.

A robust shrub. Stems round. Leaves large, glabrous, deeply three-lobed, the lobes acuminate, or ovato-acuminate, entire. Flowers large and showy ; sepals flesh-coloured, tinged with green ; petals delicate light rose colour ; rays of the coronet blue, with indistinct purple transverse bars.—*Gardeners' Magazine of Botany*.

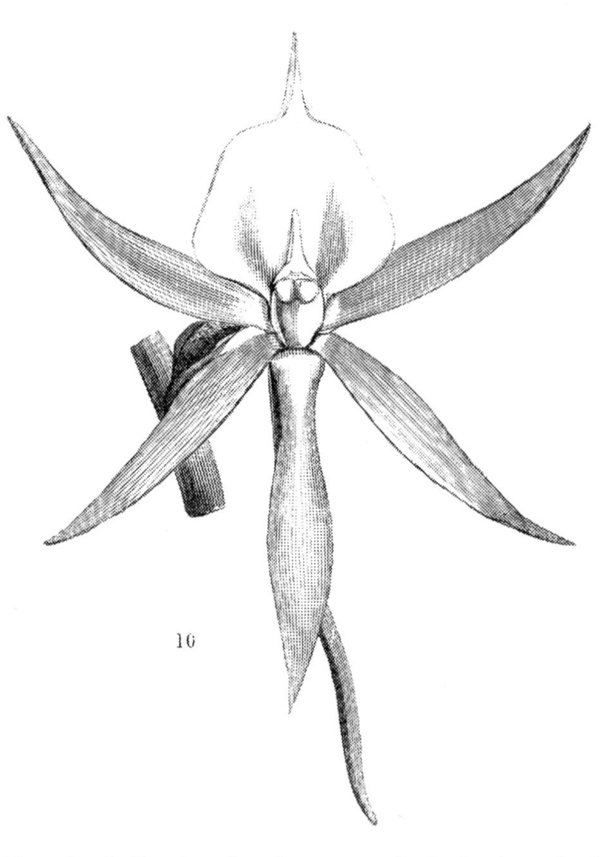

36. Metrosideros florida. *Smith*. (*aliàs* Melaleuca florida, *Forster ; aliàs* Leptospermum scandens, *Forster*). A beautiful greenhouse shrub, with rich crimson flowers, belonging to the order of Myrtleblooms (*Myrtaceæ*). Introduced to the Royal Botanic Gardens, Kew, from New Zealand. Flowers in May.

A shrub, about 5 feet high, everywhere glabrous, forming a compact mass, but every now and then sending out spreading branches, which indicate that under favourable circumstances it would be scandent. Leaves opposite, 1 inch or 1½ inch long, leathery, slightly glossy, distinctly and closely nerved on both sides ; dark-green above, pale beneath, where also the dotting is more distinct than on the upper side, but not visible to the naked eye. Corymbs terminal, almost sessile. Petals orbicular, concave, red, deciduous, longer than the calycine lobes. Stamens numerous, at first involute, then spreading, four times as long as the petals, red. A fine glossy-leaved evergreen shrub, forming a handsome bush, having much resemblance to the Myrtle. Although a native of New Zealand, the climate of which is said to be similar to that of Great Britain, yet we find it not sufficiently hardy to bear the open air in this country, during the low temperature of some of our winters, especially such as are sometimes experienced in the eastern and midland districts. The climate of the coasts of Devon and Cornwall, and the south and west of Ireland would probably be suitable for the plant in the open air. Its habit shows it to love moisture, and although with us it grows freely, treated as a greenhouse plant, in a pot or tub in loam, yet in its own country it assumes a very different habit, being epiphytal, climbing up and extending itself on trees to a great height, becoming fixed by its aerial roots and branches, which interlace with the trees on which it grows, forming dense leafy masses, similar to Ivy in this country, but of a much gayer appearance when in flower. We find it disposed to throw out roots on the main branches ; it therefore readily increases by cuttings treated in the usual way.—*Botanical Magazine*, t. 4471. We doubt, however, whether the plant thus described is the real M. florida, or Raka-pika of New Zealand, said to have obovate leaves, and yellowish petals somewhat cut. It looks very like a smooth state of Metrosideros robusta, the Rata of the New Zealanders.

37. Echites peltata. *Velloso.* A fine climbing stove plant of the order of Dogbanes, (*Apocynaceæ*), imported from Brazil by Mons. H. Galeotti, and flowered with M. Van Houtte of Ghent. Leaves large, thick, massive. Flowers large, bright yellow, clustered. (Fig. 11.)

A native of hedges near Rio Janeiro, where it grows to a considerable length. Leaves broad, rounded at the end, but with a point there, when young, covered with rusty down ; when full grown, 5 to 6 inches long, and 3½ to 7½ broad. The flowers grow in clusters of six or eight, with short downy stalks. The corolla, which is a clear bright—but not dark—yellow, is rather more than 2 inches long, twice contracted in the tube, and with five very much imbricated, broad somewhat crisp segments ; the tube is white (but is coloured yellow in the plate). It requires a damp stove, strong loam mixed with white sand, and a thorough drainage.—*Van Houtte's Flore*, t. 390.

38. Clematis indivisa ; *variety* lobata, *Hooker.* A beautiful greenhouse climbing plant from New Zealand. Flowers large, pure white, with crimson anthers. Flowers in April. (Fig. 12.)

In its native country it quite festoons the trees with its dense foliage and large panicles of flowers. A climber, with ternate leaves, and firm, leathery leaflets, slightly downy, and coarsely lobed, or almost pinnatifid. The panicles are often a foot long ; those in gardens have only hitherto produced small flowers, which measure full 2½ inches across ; whether fragrant or not is not stated.—*Botanical Magazine*, t. 4398.

39. Linum grandiflorum. *Desfontaines.* A hardy annual from Algiers, with brilliant crimson flowers. In the French Gardens, flowers from July to October. (Fig. 13.)

A glaucous erect annual, branching upwards. Ordinary leaves narrow, obtuse, closely packed ; those of the stem ovate, acute, or acuminate, with some delicate fringes on the edge. Flowers of the colour of Portulaca Gilliesii, more than an inch across, with five whitish spaces in the eye. It flowers abundantly and in succession, and, being a dwarf plant, it answers remarkably well for borders.—*Revue Horticole*, vol. ii., p. 404.

40. Eriocnema marmoratum. *Naudin.* A soft, herbaceous, stemless, stove-plant, from Brazil, belonging to the Melastomads. Leaves green, striped with white. Flowers rose-coloured, produced with Mons. Morel of Paris. (Fig. 14).

Possibly only an annual. Stem very short, fleshy, resembling a tuber. Leaves

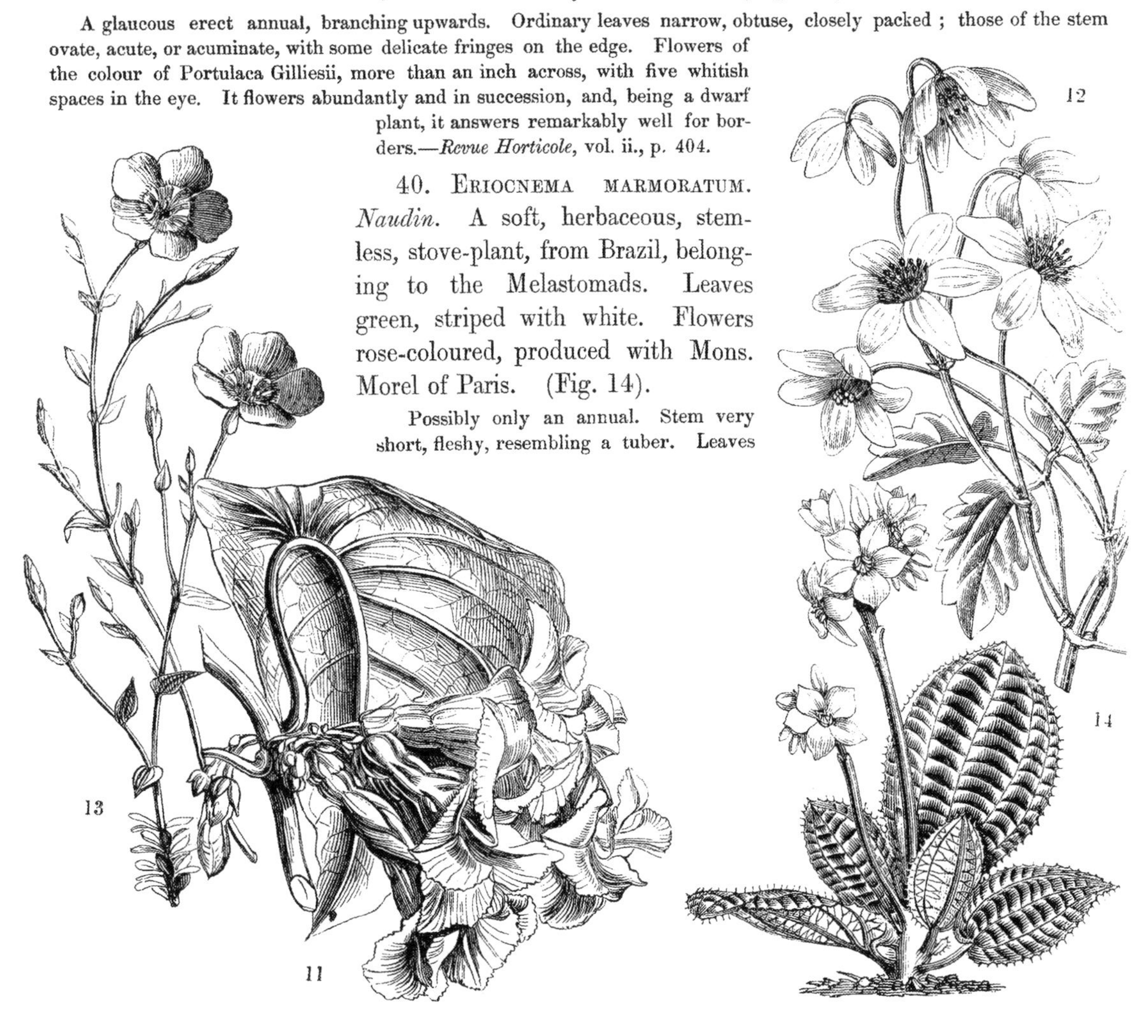

hairy, oval, 5-ribbed, stalked, oblong, heart-shaped, on the upper side bright green, beautifully marked with brown stains and broken streaks of white, on the under side rich purple. Flowering branches or scapes about 5 inches high, terminated by a bent short spike of rich rose-coloured blossoms, about as large as in Cyclamen coum, with five petals.—*Revue Horticole*, vol. ii., p. 381, fig. 20.

41. Eriocnema æneum. *Naudin*. A stove-plant, also from Brazil, with much the appearance of the last, except that the leaves are deep bronze colour. Also with M. Morel.

Flowers somewhat larger than in the last, and perfectly scorpioid, as in a Forget-me-not. Scape shorter, and more velvety. Leaves greenish brown, almost black, shining, with quite a metallic lustre. These two plants are very delicate. They are grown in peat, but require to be kept continually shaded and damp.—*Revue Horticole*. The species seem to demand the same treatment as the gay-leaved sylvan Orchids from the tropics, such as the Physaures, Anœctochiles, and the like.

42. Oncidium serratum. *Lindley*. A very striking, orchidaceous, half-twining epiphyte from Peru. Flowers large, brownish-olive, and brilliant yellow, produced with M. Pescatore of Paris. (Fig. 15.) Rather more than twice the natural size.

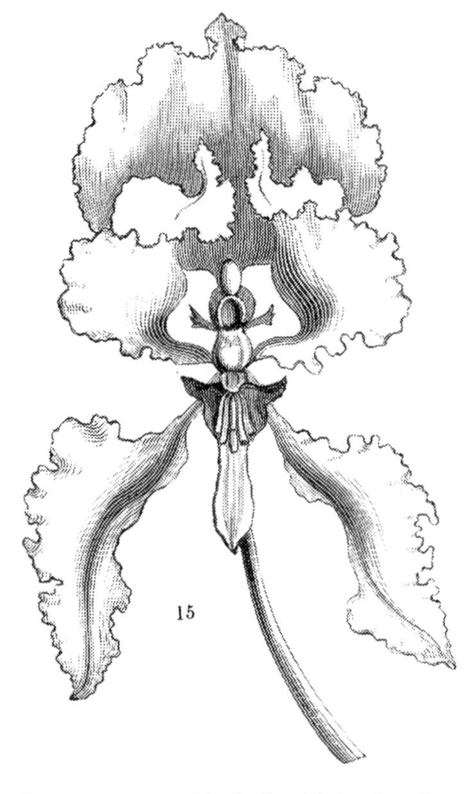
15

Till we received a flower of this curious species from M. Pescatore, it was only known to us from a rude copy of an old Spanish drawing, sent from Peru by the late Mr. Mathews, and preserved in Sir Wm. Hooker's Herbarium. The plant has oblong, smooth, terete pseudo-bulbs, each having two broad sword-shaped leaves at the point, and several others below the pseudo-bulbs. The flower-stem was nine feet long, partly twining, with five or six lateral branches, each carrying from four to six flowers near the extremity. These flowers have the very singular form shown in the annexed figure, which is about twice as large as they are represented in the Spanish drawing above alluded to, and perhaps four times as large as they were with M. Pescatore. The colour is said to be cinnamon-brown in Peru, with bright yellow tips to the upper divisions. In the fresh flower they had the colour of Oncidium luridum, only brighter; but the yellow on the upper half of the delicately fringed and crisped petals was clear and brilliant. If flowered in the summer, the species would no doubt be much finer: as it is, we must regard it as one of the most remarkable of the short-lipped Oncids.

43. Calceolaria flexuosa. *Ruiz and Pavon*. A greenhouse shrub, belonging to the Linariads. Introduced by Messrs. Veitch and Co. Flowers yellow. From Peru.

Stems hairy, flexuose. Leaves cordate, ovate, much wrinkled, coarsely crenate, whitish beneath, with numerous hairs. Flowers yellow, in large terminal panicles. Corolla with a broadly ovate slipper, not longer than the leafy calyx. Probably a fine plant for large beds.—*Gardeners' Magazine of Botany*. We should doubt its value as an ornamental species; its habit is coarse; the corolla is quite overpowered by a great leafy calyx; its habit is evidently that of a prostrate, not erect, plant, and we may observe, that in a wild state its flowers become so small, and the foliage so shabby, as to render it in that state a mere weed. Cultivation may, however, improve it. We trust that Messrs. Veitch have also raised Lobb's No. 344, the finest Calceolaria yet known.

44. Lardizabala biternata. *Ruiz et Pavon*. A hardy evergreen climbing shrub from Chili, belonging to the order of Lardizabalads. Leaves in threes, prickly at the edge. Flowers dark purple, in close drooping racemes, appearing in December. Introduced by Messrs. Veitch and Co. (Fig. 18.)

A climbing shrub, with terete, twisted branches. Leaves, especially in the flowering branches, generally simple, ternate, but sometimes bi and triternate; leaflets rather thick, evergreen, ovate, here and there almost spinously twisted,

dark green above, paler and reticulated beneath. Flowers in close drooping spikes, of numerous, rather large, deep purplish chocolate-coloured flowers. The calyx of the male of six rhombeo-ovate, spreading, fleshy, nearly equal sepals. Petals six, spreading, lanceolate, or almost subulate, white, mealy, membranaceous. Stamens six, united into a column and bearing six spreading, oblong, slightly incurved, apiculated, two-celled anthers, opening at the back. A native of woods in the south of Chili, and perfectly hardy. A plant in this garden (Kew) has withstood the cold of the last three winters without injury, and Mr. Veitch reports that in his nursery there is a specimen 12 feet high, growing against a wall. It is a beautiful evergreen creeper, with dark green foliage, and well adapted for covering high walls. It is a rapid grower, and apparently not particular as to situation; but from its habit, we infer that shady places suit it best. —*Botanical Magazine*, t. 4501.

45. Tropæolum Deckerianum. *Moritz.* A downy, handsome, twining, greenhouse perennial, with blue, green, and scarlet flowers. Apparently very pretty. Introduced from Venezuela to the Botanic Garden, Berlin. (Fig. 16.)

Roots fibrous. Stems grey, downy, climbing and rooting; with blunt, peltate, sinuated ovate leaves. The flowers, which grow singly have a scarlet spur 2 inches long, tipped with green; green hairy sepals; five intensely blue, wedge-shaped, toothed, short petals; and stamens of the same colour. It may be grown out of doors in summer, or may be kept in a pot and trained like other small species of the genus. Propagated by cuttings, or by seeds. *Van Houtte's Flore des Serres*, t. 490. A very great acquisition, remarkable for the singular intermixture of green, scarlet, and blue in its flowers.

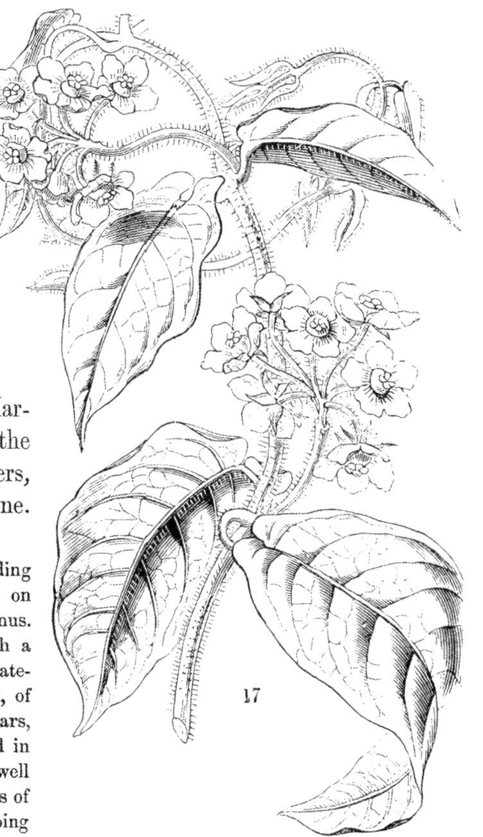

46. Gonolobus Martianus. *Hooker.* (*aliàs* Fischeria Martiana, *Decaisne.*) A Brazilian stove twiner belonging to the Asclepiads, with many-flowered umbels of greenish-white flowers, possessing little beauty. Flowers at Kew in May and June. (Fig. 17.)

Climbing, much branched; branches densely clothed with spreading hairs, which become reddish in drying. Leaves oblong-ovate, hairy on both sides, almost velvety, mucronate, cordate, with a deep but closed sinus. Flowers in many-flowered umbels with hairy pedicels, white, with a deep-green radiating ring at the base; lobes spreading, ovate-rotundate-obtuse, longitudinally plaited in the middle. A soft-wooded plant, of rapid and extensive growth, well adapted to cover trellis-work, pillars, &c. Where it is required to cover a great space, it should be planted in a mixture of loam and peat, about eighteen inches in depth, and well drained. It may also be grown in a pot, and trained up the rafters of the house, or on a wire trellis fixed to the pot; and by occasionally stopping the leading shoots it may be made to flower abundantly.—*Bot. Mag.* t. 4472.

47. Maranta? ornata. *Linden.* Var. 1. albo-lineata; var. 2. roseo-lineata. Two charming stove plants from Columbia, introduced by M. Linden. Flowers unknown. Leaves rich deep green, striped in one variety with clear white, in another with clear pink.

Until these have flowered their real genus cannot be satisfactorily determined. In the meanwhile, their foliage forms a most beautiful object among other vegetation; their green is of the rich deep tone of *Calathea zebrina*, while their stems and under side have the same rich stain of purple. In addition, they are brilliantly banded by well defined oblique streaks, of a clear delicate pink colour in one variety, and of yellowish white in the other. They require a rich well-worked, mixed soil, frequent watering while growing, a shady place in the stove, and a diligent care to keep "scales" off them. Easily propagated by division.—*Van Houtte's Flore*, tt. 413 and 414. Both these exquisite plants were exhibited before the Horticultural Society at one of their meetings at Chiswick in 1849, ? on which occasion they received a medal. ? ? ?

48. Chorozema cordatum. *Lindley.* (*aliàs* C. flava, *Henfrey.*) A yellow variety of this well-known little greenhouse shrub has been imported by Messrs. Henderson, of the Wellington Nursery, and published in the *Gardeners' Magazine of Botany* as a new species. Except colour, which is variable in its wild state, there is nothing essential by which it can be distinguished.

49. Berberis undulata. *Lindley.* An evergreen shrub, apparently hardy, imported by Messrs. Veitch and Son, from the mountains of Peru, where it grows at the elevation of 12,000 feet. Has not yet flowered in this country.

In a young state, as now with Messrs. Veitch, this has slender branches, and weak palmated spines. The leaves are dull green, scarcely glaucous, oblong, tapering to the base, remarkably wavy, and furnished with a few spiny distinct teeth, without any distinct trace of netted veins. The flowers have not yet appeared. In a wild state, it is a stout stiff bush, with 3-parted or 5-parted spines, sometimes as much as $1\frac{1}{4}$ inch long. The leaves are thick, narrower than in the cultivated plant, but still preserve their undulated appearance. The flowers appear in small, roundish, nearly sessile racemes, which are scarcely so long as the leaves.—*Journ. Hort. Soc.*, Vol. v. p. 7.

50. Erica elegantissima. *Gardeners' Magazine of Botany.* A pretty hybrid, said to have been raised between E. hiemalis and E. Hartnelli. Flowers tubular, deep rose, with a white flat border.

51. Æschynanthus Javanicus. *Hort.* A most beautiful stove epiphyte introduced by Messrs. Rollisson, from Java, with close racemes of bright red ascending flowers, each more than 2 inches long, with a starry yellow throat. Belongs to the order of Gesnerads.

At first sight this bears much resemblance to the Æ. pulcher. The plant is more compact, the leaves smaller, the flowers all over down as well as the pedicels, the calyx truly cylindrical (not swollen below), the limb spreading, the corolla more slender and graceful, the stamens exserted. Leaves opposite, oval or ovate, sometimes approaching to oblong, between coriaceous and fleshy, obscurely angular and toothed, the veins sunk in the substance of the leaf. Corymbs terminal, of many large, handsome, richly-coloured flowers. Calyx large, greatly wider than the tube of the corolla it includes, downy, dark green, red-brown above; the tube cylindrical, faintly striated, the five lobes of the limb spreading horizontally. Corolla bright red, about thrice the length of the limb, the tube slender, funnel-shaped, downy, laterally compressed, with a prominence under the throat, mouth oblique, limb of four nearly equal, spreading, large ovate lobes, the upper one notched, the rest entire and streaked and blotched with yellow. Stamens all exserted, especially the upper ones.—*Botanical Magazine*, t. 4503.

52. Theresia persica. *C. Koch.* A hardy Liliaceous plant from Mount Ararat, where it is found at the elevation of 4000 feet. It is said to have the flowers of the same form as in Fritillaria, but the habit of a Lily.

This is described as having a bell-shaped, hexapetaloid flower, with oblong coloured sepals, provided with a nectariferous cavity in the inside; six hypogynous stamens included within the flower; oval anthers, deeply pierced below to receive the filament; a 5-celled, many-seeded, 5-angular, columnar ovary; with a linear, entire style, and a scarcely distinguishable stigma. The bulbs are said to be like those of the Crown Imperial. It does not appear from the *Annales de Gand*, whence this account is taken, whether the plant is in the Belgian gardens or not; its presence in a

work treating on Garden plants, leads to the supposi-
tion that it is in culti- vation, though Professor
Morren does not say where.

53. STANHOPEA CIRRHATA. *Lindley.* A
stove orchidaceous epiphyte, from Nicaragua, introduced by Mr. Skinner. Has
not yet flowered. (Fig. 19.)

A few pseudo-bulbs of this remarkable plant
were sold at one of Mr. Skinner's sales, having
been collected in Nica- ragua by Mr. Warcziewitz.
A couple of specimens in spirits enable me to
define it. Among Stan- hopeas it is unique, for the
flowers being absolutely solitary, not
in spathaceous spikes, and for the column
being wing- less, and extended into a
pair of feelers like some Odontoglossums.
Its lateral
horns, too,
are extremely

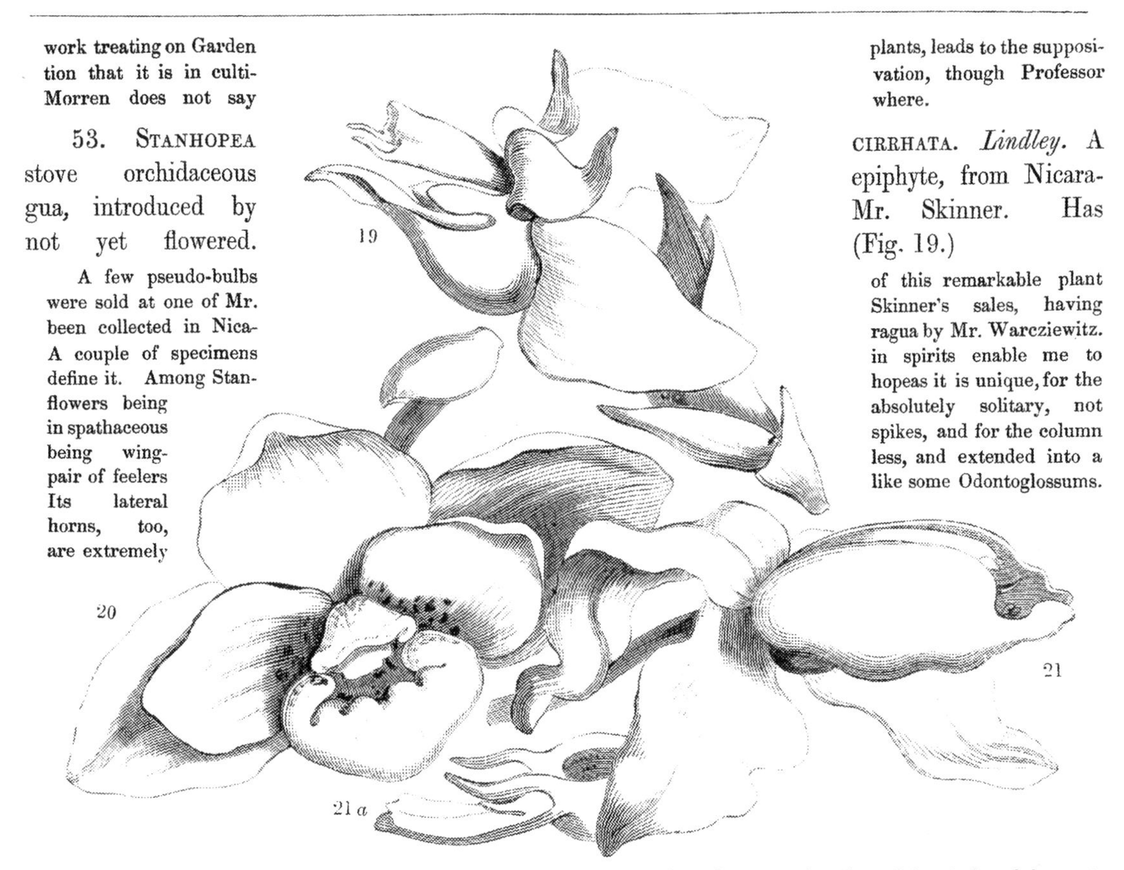

short and fleshy. Its colours are unknown, but it does not promise to be a showy species of much horticultural interest. —*Journ. Hort. Soc.*, Vol. v. p. 37.

54. STANHOPEA ECORNUTA. *C. Lemaire.* A stove orchidaceous epiphyte, from Central America, whence it was sent to Mr. Van Houtte by Mr. Warcziewitz. Flowers white, with the base of the lip yellow. (Fig. 20).

This extremely curious plant differs from all the previously known Stanhopeas, in having a lip wholly free from horns, and without any break in its middle. It may be regarded as a species with the hypochil (or lower half of the lip) alone present. This body is described as being "ovate, obsoletely triangular at the end, very short. It has much the form of a slipper, extremely fleshy, of a very bright yellow-orange colour, passing towards the point into pure white, and mottled on its sides with handsome purple blotches. Four little tumours, two near the articulation with the column, and two near the point, indicate four abortive horns." The flowers are otherwise pure white, with a few purple spots near the base of the petals, which are short, firm, concave, and not reflexed, as in most other Stanhopeas. "The column is very short, very fleshy, compressed, rounded above, winged at the sides, channelled in front." "The flowers, which grow in pairs, are about 4½ inches across, and have very short bracts."—*Van Houtte's Flore*, t. 181. Can it be a monster of Stanhopea tricornis ?

55. STANHOPEA TRICORNIS. *Lindley.* An orchidaceous epiphyte, from Western Peru; plants of which were dispersed at one of Mr. Skinner's sales. Has not yet flowered. (Fig. 21.)

A very curious thing. The figure of the lip is most remarkable, there being a third horn at the base of the middle lobe of the lip in addition to the two always present at the side. In a figure sent home by Mr. Warcziewitz the petals are represented to be pink and the rest of the flower white ; the petals moreover are very fleshy, firm, and apparently incapable of rolling back as in the rest of the genus.—*Journ. of Hort. Soc.* iv. Fig. 21 *a* represents a portion of the lip.

56. ACINETA CHRYSANTHA. *Lindley.* (*aliàs* Neippergia chrysantha, *Morren.*) A stove epiphyte, supposed to be from Mexico, exhibited at Ghent, by M. Auguste Mechelynck, in September, 1849.

Flowers the size of A. Barkeri, in erect racemes, of a bright golden yellow colour. Very handsome. Natural order of Orchids.

This noble looking plant has exactly the habit of the other Acinetes, except that the raceme grows erect, to the height of a foot or so, instead of being pendulous. It is loaded closely with golden yellow blossoms, each more than 17½ inches wide, very like those of A. Barkeri, except in colour. The lip appears to be white, and the column crimson. At night the flowers have a sweet aromatic odour ; by day they are scentless. From the other Acinetes it is distinguished especially by the presence of a long, blunt, papillose horn arising from the hypochil. *Annales de Gand*, t. 282. We do not perceive any ground for separating this plant from Acinete, the horn upon which Professor Morren relies, being equally present upon both Barker's and Humboldt's Acinete, although of a different form. Nor do we feel certain that the erect position of the flowering raceme is habitual with this plant, for, according to the drawing, while one raceme rises upright, another is bent downwards in the same manner as in the Acinetes. Annexed to the article which describes this plant, M. Morren makes the following startling announcement : "I shall prove in another place that *Anguloa*, *Lycaste*, or *Maxillaria*, are simply *isophorous* forms of the same organisation, that is to say, that one may be transformed into another, so that the same plant will produce one year the flower of Anguloa, and another that of Lycaste. This strange fact I have witnessed, and, connecting it with other analogous facts, well ascertained to exist in the Vegetable Kingdom, I think of soon bringing forward a general theory of isophorism in plants, a doctrine exactly analogous to that of isomerism, now perfectly established in chemistry and mineralogy. I suspect that this Neippergia is also an isophorous form, that is to say, transformable into another genus."

57. Cuphea purpurea. *Lemaire.* A very pretty hybrid perennial, obtained by M. Delache, of St. Omer, between C. miniata ♀ and C. viscosissima ♂. Flowers large bright rose-colour, handsome.

To the habit and foliage of *C. miniata*, and its two large upper petals, it adds the four small petals of *C. viscosissima*, but has little of its viscidity. The colour of the flowers is a fine bright rose, slightly shaded with violet, a charming tint, which cannot be given by art. It requires the same treatment as other Cupheas.—*Van Houtte's Flore*, t. 412. Seems to be a good bedding-out plant.

58. Warrea candida. (*aliàs* Huntleya candida, *Hort.*) An orchidaceous epiphyte from Bahia, with handsome purple and white flowers. Introduced by M. Morel of Paris, flowered with M. Pescatore in Feb., 1850. (Fig. 22 magnified).

W. candida ; foliis latoligulatis apice recurvis, floribus 2-3, sepalis petalisque ovalibus acutissimis, labello subquadrato apice angustiore retuso basi saccato angulato inflexo carnosissimo dente crasso tridentato in medio et altero simplici acuminato utrinque plicisque 3 parvis in faciem superiorem.

The accompanying figure represents a flower of this plant about four times the natural size. M. Pescatore, from whom we received it by post, states that he bought it from M. Morel, under the name of Huntleya Meleagris. M. Morel informs us that he imported it in 1848 from Bahia, his collector having found it about 150 leagues in the interior of that province. According to M. Luddemann, the director of M. Pescatore's garden at La Celle, the species is handsomer than *Huntleya violacea*. The flower is pure white, the centre of the lip purple, towards the edge blue-violet, at the base white, streaked with red. The plant is of small stature, the full-grown leaves not being more than 9 inches long. The flowers grow three together, in the same manner as in the Huntleyas. It seems to be a nice plant, in the way of *Warrea Wailesiana*.

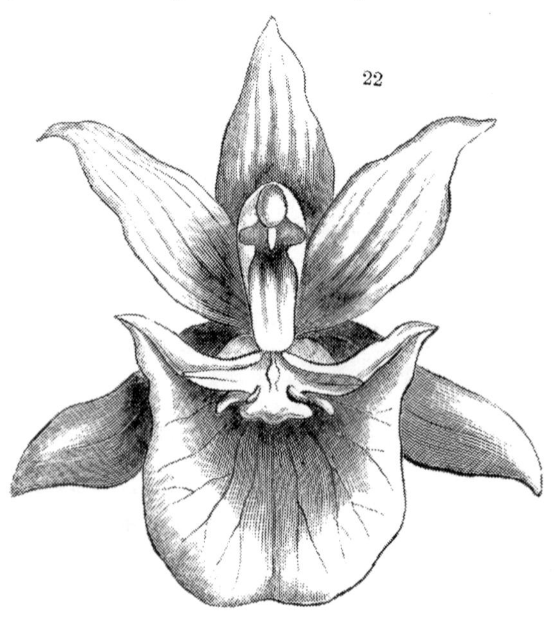

J. Constans, Pinx & Zinc.

Printed by C.F. Cheffins, London.

[PLATE 7.]

THE CEYLON RHODODENDRON.

(RHODODENDRON ROLLISSONII.)

A half-hardy Tree, from the MOUNTAINS OF CEYLON, *belonging to the Natural Order of* HEATHWORTS.

Specific Character.

THE CEYLON RHODODENDRON.—A small tree. Leaves short, oblong, acute, obtuse, or even heart-shaped at the base, wavy, very rugose and convex, revolute at the edge, covered beneath with close pale brown wool. Flowers in small heads. Flowerstalks woolly. Calyx obsolete. Corolla campanulate, slightly spotted. Ovary many-celled.

RHODODENDRON *ROLLISSONII.*—Arboreum; foliis brevibus oblongis acutis basi obtusis cordatisve undulatis rugosissimis convexis margine revolutis subtus tomento denso fulvo tectis, umbellis densifloris, pedunculis tomentosis, calyce obsoleto, corolla campanulata parcè punctata, ovario multiloculari.

Rhododendron Rollissonii: *Botanical Register, t.* 25, 1843, *aliàs* R. zeylanicum *of the Gardens.*

THE following notice of this plant appeared in the *Gardeners' Chronicle* for March 9, 1850:—

"This is now in great beauty in the open border, and proves to be a very fine thing, far surpassing, in my opinion, the old Rhododendron arboreum, or any of the numerous hybrid varieties that have originated from it. The rugged corky bark, and rough, wrinkled thick leaves, revolute at the margin, and clothed underneath with a somewhat rusty-coloured pubescence, give a peculiar character to the plant, by which it may be easily recognised. The head of flowers is round and compact, like that of R. arboreum, but the colour is much richer, being a deep blood red, with a few dark spots at the bottom of the tube. The plant we have under the name of R. Rollissonii I consider to be identical with R. zeylanicum, although the former has not yet flowered with us. Both have been growing for several years in the open air, and although considered as being rather tender, they have been found sufficiently hardy to withstand 10 degrees of frost (the greatest cold experienced here last winter) without injury."

This letter was written by Mr. W. B. Booth, gardener to Sir Charles Lemon, Bart., M.P., at Carclew, near Penrhyn, in Cornwall, whence also the specimens were received from which the accompanying drawing was made.

We are glad to reproduce a figure of this plant,—firstly, for the sake of making a highly interesting species better known; and, secondly, for the sake of removing the error of supposing that what is called *R. Rollissonii* is some hybrid form. It is nothing whatever except the wild Tree-Rhododendron of the Cingalese Hills. As far as our information now goes, it seems to be found nowhere else.

As a species, this differs manifestly from the other Indian Tree-Rhododendrons in its very peculiar leaves, which, instead of being long and narrow, and more or less flat, are broad and short, very obtuse, and even heart-shaped at the base, wavy, excessively wrinkled, and remarkably rolled back at their edge. The hairiness of their under-side is like neither the coarse brown shagginess of the Cinnamon Rhododendron, nor the close silvery surface of the Scarlet Tree-Rhododendron, nor the short pale-brown starry pile of the Campanulate Rhododendron. On the contrary, the fur, although copious, is of a pale-brown dull colour, and so close that it would not be taken for hairiness without a minute inspection.

There are now in general cultivation five very distinct races of Indian Rhododendrons, concerning which a few observations require to be made.

Firstly, we have the Old Scarlet Tree-Rhododendron (*R. arboreum*) with rich blood-red flowers, and long flat leaves, silvery underneath. Whether there is really any white variety of this, is uncertain.

Next, there is the Cinnamon Tree-Rhododendron (*R. cinnamomeum*), so well known by its long, flat, deep green, wrinkled, narrow leaves, covered beneath with a coarse, shaggy, rusty wool. This, originally published in 1824 by Dr. Wallich, and afterwards in 1837, as a variety of the Scarlet Tree-Rhododendron, in the Botanical Register, t. 1982, is chiefly known as a white-flowered plant. Nevertheless it varies to Rose colour, as is proved by the Neilgherry Rhododendron (*R. Nilagherieum*) which is figured in the Botanical Magazine, t. 4381; and which is absolutely identical, except in colour. We ought to state that this last was introduced by Messrs. Loddiges, and not by Lucombe and Pince of Exeter, to whom belongs no other credit than that of flowering it. Whether the *R. nobile* of Wallich, which we have not seen, is this or the Ceylon Tree-Rhododendron, is uncertain.

Then, there is the Bearded Tree-Rhododendron (*R. barbatum*), little known at present, but long since dispersed by Messrs. Loddiges, and which is remarkable for the coarse stiff hairs of the leaf-stalks.

After this species follows the Campanulate Rhododendron (*R. campanulatum*) with its broad flat leaves, cordate at the base, and short stellate rusty down; and finally we have

The Ceylon Rhododendron (*R. Rollissonii*), the subject of this article.

We are the more anxious to make this clear, because the wondrous discoveries of Dr. Hooker, and the new things come or coming from the islands of India, will render the Garden Botany of Asiatic Rhododendrons very difficult a few years hence. Nor can we say that it appears to be clearly understood even now.

PLATE 8.

L. Constans, Pinx & Zinc

Printed by C.F. Cheffins, London.

[PLATE 8.]

THE TETRANDROUS BORONIA.

(BORONIA TETRANDRA.)

A Greenhouse Shrub, from NEW HOLLAND, *belonging to the Natural Order of* RUEWORTS.

Specific Character.

THE TETRANDROUS BORONIA. A smooth shrub. Leaves pinnated; leaflets in three or four pairs, with an odd one, linear, blunt. Peduncles three-parted shorter than the leaves. Stamens 8, four being smaller than the rest.

BORONIA *TETRANDRA*; lævis, foliis pinnatis, foliolis 3-4-jugis cum impari linearibus obtusis, pedunculis trifidis foliis brevioribus, staminibus 4 minoribus.

Boronia tetrandra: *Labillardière, Stirpes Nov. Holl.*, i., p. 98, t. 125; *aliàs* B. microphylla *of the Nurseries; aliàs* B. pilosa *Labillardière.*

BY what strange blunder this plant, having eight stamens, gained the reputation and name of having four, we are unable to say. It owed its name to the French botanist Labillardière, who, it is to be inferred, did not possess very accurate powers of observation. English nurserymen, in return, have given it another name, which belongs to a totally different species, and which is just as inapplicable.

The accompanying drawing was made in the Garden of the Horticultural Society a month since, and is thus noticed in the Society's Journal.

"This little shrub is not unlike a dwarf Boronia pinnata; but it has a less number of leaflets, and seldom produces more than 1 flower at a time in each axil. These are pale pink, rather large, and very pretty. The leaflets are usually 7, but occur to the number of 5, or even 9; they are narrow, blunt, and smell rather agreeably. The whole plant is destitute of down or hairs."

The native country of the plant is the south-east of Australia (where it seems to be common, especially in Van Dieman's Land), and where numerous forms, or supposed forms of it have been detected by Mr. Ronald Gunn. If all these supposed forms really belong to one and the same species, we must confess that the tetrandrous Boronia is as variable a plant as we know of. We would recommend Nurserymen and others corresponding with Van Dieman's Land, to procure seeds of as many of the forms as possible, for some of them seem much better adapted to cultivation than even this. Dr. Hooker distinguishes five in particular, viz.:

1. *Floribunda.* This has linear stalked leaflets in three or four pairs, hairy branches, and very numerous lateral and terminal flowers.

2. *Terminiflora;* with linear stalked leaflets, broader than in No. 1, a more erect habit, and flowering invariably in terminal clusters.

3. *Grandiflora.* Here the leaflets are longer, ovate-lanceolate, and only in two pairs, the flowers much larger, and the branches nearly smooth. This variety is said to smell like Tansy or Rue.

4. *Laricifolia.* An upright twiggy branched form, with leaves in distant fascicles, the leaflets in three or four pairs, nearly sessile and pressed close to the stem, and clusters of small terminal flowers.

5. *Pilosa.* A fine leaved hairy form, not much different from No. 2. This has been considered a distinct species by systematic Botanists, who call it Boronia pilosa.

To none of them does the garden plant properly belong. It is most like No. 3, but the leaflets are almost invariably in three or four pairs.

Although inferior to the pinnated Boronia (*B. pinnata*) this is by no means an uninteresting species, its flowers being produced abundantly, and having a delicate blush colour like that of an apple blossom.

As to the *B. microphylla,* whose name has been ignorantly applied to this plant, we need only say that it bears it the least possible resemblance.

I. Constans Pinx. & Zinc.

Printed by C.F. Cheffins, London.

[PLATE 9.]

THE LONG-TAILED LADY'S-SLIPPER.

(CYPRIPEDIUM CAUDATUM.)

A Greenhouse herbaceous plant, from PERU, *belonging to the Natural Order of* ORCHIDS.

Specific Character.

THE LONG-TAILED LADY'S-SLIPPER. — Stemless. Leaves distichous, sword-shaped, leathery, smooth, spotless. Scape erect, bearing several flowers, longer than the leaves. Bracts like spathes, as long as the ovary. Sepals ovate-lanceolate, gracefully curved. Petals extended into very long pendent wavy linear tails. Lip oblong, glandular on the edge, near the base. Sterile stamen broader than long, 2-lobed, with bristles on the ends of its lobes.

CYPRIPEDIUM *CAUDATUM.*—Acaule; foliis distichis ensiformibus coriaceis glabris immaculatis scapo stricto plurifloro brevioribus, bracteis spathaceis ovarii longitudine, sepalis ovato-lanceolatis arcuatis, petalis in caudas longissimas pendulas flexuosas lineares productis, labello oblongo margine versus basin glanduloso-serrato, stamine sterili transverso bilobo apicibus setosis.

Cypripedium caudatum: *Lindley, Genera and Species of Orchidaceous Plants, p.* 531.

THIS extraordinary plant was for many years known only by a few fragments preserved in Herbaria. At last the collector Hartweg met with it in wet, marshy places near the hamlet of Nanegal, in the province of Quito; but he did not send it home. Subsequently, the collectors of Messrs. Veitch, of Exeter, and of Mr. Linden, fell in with it; and to the latter is, we believe, owing its introduction to Europe in a living state.

For the opportunity of figuring it we have to acknowledge our obligations to Mrs. Lawrence, who first succeeded in bringing it into flower, and who exhibited it to the Horticultural Society in March last. Since that time a weaker specimen has blossomed with Mr. C. B. Warner.

The accompanying plate is a faithful representation of the plant as it flowered at Ealing Park, but is far from giving an adequate idea of the natural beauty of the species. The great sheathing bracts, which in South America are as large as those of a Heliconia, were mere abortions; and we learn from drawings brought home by Mr. Warczewitz that the flowers are very much larger and finer-coloured in its native swamps. The stains on the lip, for instance, are numerous, and of a rich warm brown, giving quite another appearance to the flowers. On one of Hartweg's dried specimens are remains of six flowers of this sort, placed at the end of a scape more than two feet high.

The petals are the extraordinary part of the species. In most Lady's-slipper flowers they are short, and little distinguishable from the sepals; but here they extend into the most curious narrow tails, which hang down and wave in the wind, in a manner of which we have in gardens no other such example, not even in the genus of Strophanths. What adds to the curiosity of these singular appendages is the fact, first remarked by Mrs. Lawrence, that they are quite short when the flower begins to open, and that they acquire length day by day, at a rate which would enable an attentive observer to see them grow. This lady has favoured us with some measurements made by herself, from which we learn that—

When the flower first opened, the petals were	$\frac{3}{4}$ of an inch long.
During the second day they grew	$3\frac{3}{4}$ of an inch.
On the third day they advanced	4 inches more.
The growth of the fourth day amounted to	$4\frac{1}{2}$ inches.
And on the fifth day they still extended	$5\frac{1}{2}$ inches.

At this time the growth is supposed to have ceased, the petals having in four days lengthened $17\frac{3}{4}$ inches, and being $18\frac{1}{2}$ inches long when full grown.

Another example of this tendency to lengthen the petals into tails, but in a less degree, occurs in the "sedgy Lady's-slipper," mentioned further on. And a third case is found in the strange genus Uropedium, in which not only do the petals turn to tails, eight or ten inches long, but their example is followed by even the lip, which for this purpose flattens itself, entirely unfolds, and pushes itself out into a long and narrow tongue. It may be useful to state that this Uroped, which is not yet in cultivation, has the habit of the "bannered Lady's-slipper," and was found wild by Linden, growing in the soil of little woods in the savannah which occurs on the high part of the Cordillera that looks down upon the vast forests of the Lake of Maracaybo. Its elevation above the sea was 8,500 feet, in the territory of the Chiguará Indians, where the specimens now before us were gathered in flower, in June, 1843.

The reason of this marvellous structure seems to deserve inquiry at the hands of some proficient in the doctrine of final causes. There is evidently a tendency towards it in other Orchids, as, for example, in Brassias, some Oncids, the genus Cirrhopetalum, and the long-tongued Habenarias.

The long-tailed Lady's-slipper belongs to a section of the genus which is distinctly characterised by having no foliage on the sides of the stem, instead of which a number of thick narrow leaves spring up from its very base, and allow the flowering stem to rise freely into the air.* They all inhabit tropical countries, but are generally found at considerable elevations above the sea. As most of them are in cultivation, the following enumeration may be useful:—

1. The Handsome Lady's-slipper. (*C. venustum*, Wallich.)

From the mountains of Sylhet, and the Khasiya hills of Continental India. We have not seen this from the Malay Islands.

Leaves spotted with deep green and purple, almost as long as the scape. Lip and sepals veined with green. Petals stained with purple, and fringed with long hairs.

2. The Java Lady's-slipper. (*C. javanicum*, Reinwardt ined.)

Found wild in Java. (Not in cultivation ?)

Leaves speckled with green, and much shorter than the scape. Sepals veined with green. Petals

* The stemless Lady-slipper (*C. acaule*) has the leafless scape of this division, together with the broad, thin-ribbed leaves of the other, and serves to connect the two. It is here intentionally passed by.

distinctly spotted with purple on a green ground, tipped with pink, and fringed with long hairs. Lip deep olive-green, not veiny.—Dr. Blume refers this to *C. venustum*, and perhaps with reason; but a drawing before us from Dr. Reinwardt, and a dried specimen brought home by Lobb (No. 304), suggest the propriety of further examination. The short comparative memoranda given above, sufficiently show that if the same species, it is a well-marked variety.

3. The Bearded Lady's-slipper. (*C. barbatum*, Lindley.)

On Mount Ophir, where it was found by Mr. Griffith.

Like No. 1, but the upper edge of the petals is marked with purple glands, and all the parts of the flower are much stained with rich purple.

4. The Purple-stained Lady's-slipper. (*C. purpuratum*, Lindley.)

Grows wild in wet mossy crevices near the summit of Mount Ophir.

Also in the way of No. 1. But the dorsal sepal is convex, white with purple veins, and all the other parts are deeply stained with purple. The leaves are much shorter and more oblong than in any of the preceding.

5. Low's Lady's-slipper. (*C. Lowei*, Lindley.)

In Borneo and Sarawak.

Remarkable for the extension of the petals into two long spathulate bodies blotched with purple. When wild it has 8-10 flowers on a scape.

6. The Glandular Lady's-slipper. (*C. glanduliflorum*, Blume.)

New Guinea, on old decaying trunks of trees. (Not in cultivation.)

Leaves like those of No. 7. Flowers large, about 2 or 3 on a scape, with long twisted petals, bearing hairy glands on their edge; and a large pale pink lip, which bears within it a pair of long reversed horns.

7. The Bannered Lady's-slipper. (*C. insigne*, Wallich.)

Mountains of Sylhet and Khasiya.

Leaves narrow, not stained. Flowers large, with an orange-coloured lip, a broad dorsal greenish sepal, edged with white, and long spreading flat greenish petals.

8. Lindley's Lady's-slipper. (*C. Lindleyanum*, Schomburgk.)

Damp meadows of Guayana, among Sundews, Sunjars (*Heliamphoras*), and similar plants. (Not in cultivation.)

A stout, hard leaved plant, with a stem 2 feet high, covered with rusty down. Flowers brown, in a one-sided panicle, having coarse spathaceous bracts at their base. Lip small, oblong, green. A very curious, but not handsome plant.

9. The Sedgy Lady's-slipper. (*C. caricinum; foliis angustissimis coriaceis acutis unicostatis scapi tomentosi longitudine, racemo plurifloro, bracteis ovatis spathaceis glabris ovario glabro brevioribus, sepalis lateralibus connatis labelli longitudine, petalis in caudam acuminatis.*)

Found in Bolivia by Bridges. (Not in cultivation.)

The flowers in our possession are mere fragments, but they suffice to show that the species is perfectly distinct from all others. The leaves are about a foot long, and $\frac{1}{4}$ inch wide, but they appear as if narrower in consequence of their edges being rolled back.

10. THE LONG-TAILED LADY'S-SLIPPER. (*C. caudatum*, Lindley.)
Mountains of Peru.

The subject of this Plate. The following woodcut gives some idea of the appearance of the plant in a wild state.

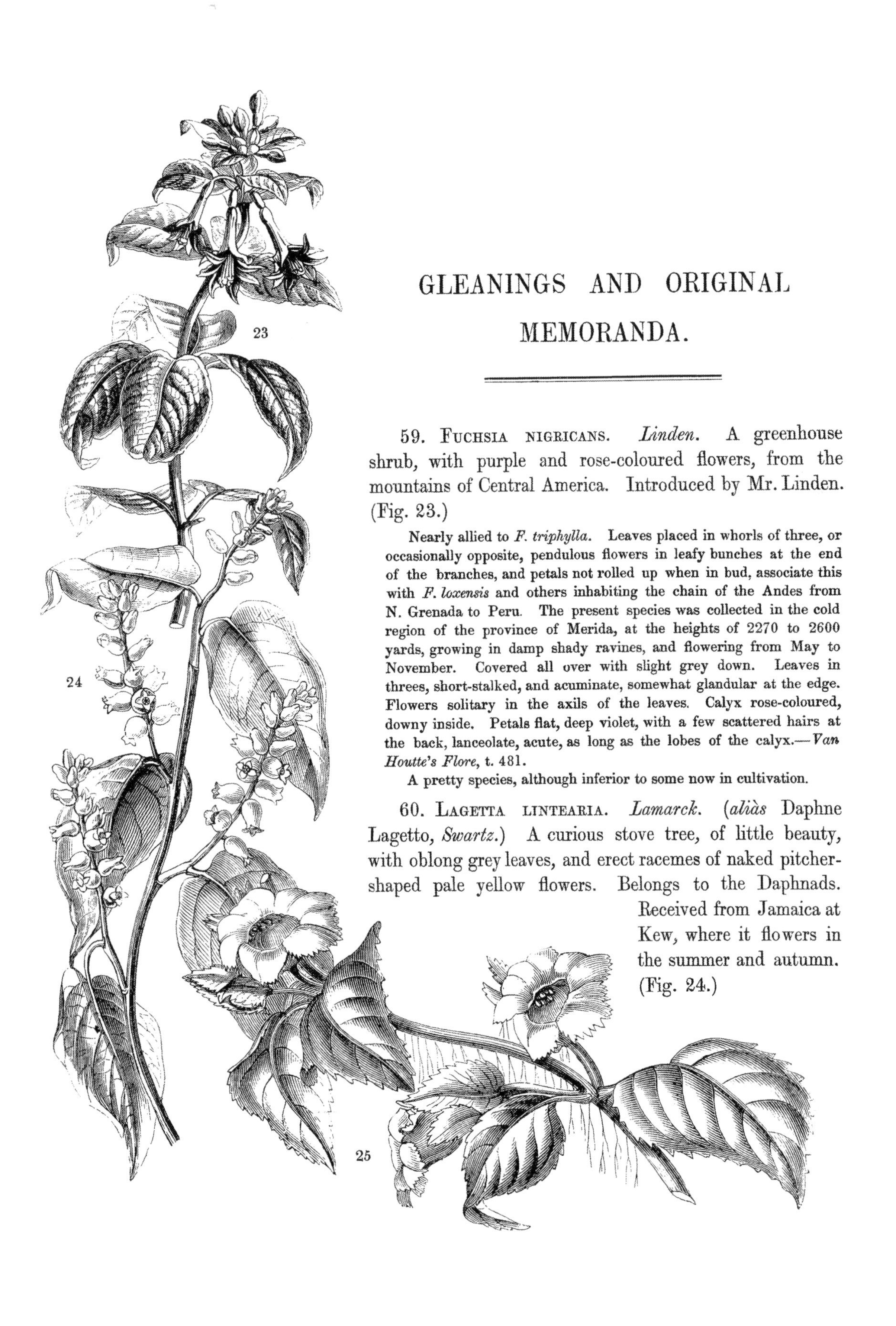

GLEANINGS AND ORIGINAL MEMORANDA.

59. Fuchsia nigricans. *Linden.* A greenhouse shrub, with purple and rose-coloured flowers, from the mountains of Central America. Introduced by Mr. Linden. (Fig. 23.)

Nearly allied to *F. triphylla*. Leaves placed in whorls of three, or occasionally opposite, pendulous flowers in leafy bunches at the end of the branches, and petals not rolled up when in bud, associate this with *F. loxensis* and others inhabiting the chain of the Andes from N. Grenada to Peru. The present species was collected in the cold region of the province of Merida, at the heights of 2270 to 2600 yards, growing in damp shady ravines, and flowering from May to November. Covered all over with slight grey down. Leaves in threes, short-stalked, and acuminate, somewhat glandular at the edge. Flowers solitary in the axils of the leaves. Calyx rose-coloured, downy inside. Petals flat, deep violet, with a few scattered hairs at the back, lanceolate, acute, as long as the lobes of the calyx.—*Van Houtte's Flore*, t. 481.

A pretty species, although inferior to some now in cultivation.

60. Lagetta lintearia. *Lamarck.* (*aliàs* Daphne Lagetto, *Swartz.*) A curious stove tree, of little beauty, with oblong grey leaves, and erect racemes of naked pitcher-shaped pale yellow flowers. Belongs to the Daphnads. Received from Jamaica at Kew, where it flowers in the summer and autumn. (Fig. 24.)

The liber or inner bark of this tree consists of layers of reticulated fibre, exactly resembling well-prepared lace ; and its nature is best exhibited by taking a truncheon from a branch, tearing down the bark, and separating it by the hand into as many layers as that portion of the tree is years' old. "The ladies of Jamaica," Dr. Lunan observes, "are extremely dextrous in making caps, ruffles, and complete suits of lace with it. In order to bleach it, after being drawn out as much as it will bear, they expose it (stretched) to the sunshine, and sprinkle it frequently with water. It bears washing extremely well with common soap, or the 'curatoe' soap, and acquires a degree of whiteness equal to the best artificial lace. The wild negroes have made apparel with it of a very durable nature, but the common use to which it is applied is ropemaking." A tree from 20 to 30 feet high, with branches too straggling and foliage too thin to form a striking object, though really of a good size, glossy and handsome when in flower. Leaves alternate, on rather short petioles, which are jointed on the branch, hence the leaves readily fall off in drying ; they are heartshaped-ovate, acute, reticulated, palish green. Flowers pure white, or, in bud, greenish-white, arranged in spikes which are solitary and terminal on a main branch, or on short side branches. In growing it at Kew we have made use of good yellow loam, mixed with a little leaf-mould and sand. In this it has attained the height of 8 feet, and continues in a perfectly healthy state.—*Botanical Magazine*, t. 4502.

61. Drymonia cristata. *Miquel.* A creeping, downy, fleshy-leaved, hothouse Gesneriad, with large lacerated green flowers. Native of Dutch Guiana. Bloomed at Ghent in October, 1848, with M. Van Houtte. (Fig. 25.)

Stems round, rooting from any part of their surface. Leaves coarsely toothed. Flowers solitary, axillary, with great leafy calyxes nearly as long as the pale green uneven corolla. Described as handsome, on account of its long creeping branches and broad deep-green foliage, and as suitable for mixing with Epiphytes in an Orchid house.—*Van Houtte's Flore*, t. 388. Seems to be very near Drymonia bicolor.

62. Abies Jezoënsis. *Siebold.* A magnificent evergreen coniferous tree from Japan. Introduced by Messrs. Standish and Co. Leaves of a brilliant green. (Fig. 26.)

According to Siebold, the Jezo Spruce is so called because it grows on the islands Jezo and Krafto, in the empire of Japan, whence it has been introduced into the gardens of the wealthy inhabitants of Jedo. He describes it as a large tree, with a soft light wood employed by the Japanese for arrows, and in the construction of domestic utensils. The leaves are said to remain for seven years upon the branches. The cones were unknown to him. He only saw the tree in flower in the month of June.

The plant now introduced by Messrs. Standish and Co. has leaves of the most brilliant green on both sides, placed when young in two rows, about 1¼ inch long, and a line and a half wide, thin and soft when young, stiff when old, and terminated gradually by a very distinct spine, which is the end of the midrib. The branches when very young are covered with a rusty down ; when old they become smooth. The cones are narrow, tapering, rather more than 6 inches long, with broad convex loose rounded scales, which do not readily separate from their axis, and have at their base a short roundish slightly serrated bract, which is just visible at the point of intersection of the lateral scales. Although the cones of the Jezo Spruce are unknown, we can hardly doubt that this is the plant intended by Siebold ; at least we observe nothing at variance with his figure and description, except that he describes the young branches of that species as being smooth ; in the plant before us they are covered with short down, but they become smooth with age ; and as he describes those which he saw as having a yellowish rusty coat, the apparent difference is reduced to little. Probably perfectly hardy, but that is not as yet ascertained.

63. Oncidium trilingue. A remarkable half climbing Orchid from Peru, with large brown and yellow flowers thinly arranged upon a racemose panicle. Introduced by Sir Philip de Malpas Grey Egerton, Bart., M.P. Blossomed at Oulton Park, in April, 1850.

O. trilingue, (Microchila) foliis .... racemo subvolubili basi paniculato, floribus raris. bracteis oblongis spathaceis ovario quadruplo brevioribus, sepalis lateralibus unguiculatis basi connatis lanceolatis undulatis elongatis dorsali subrotundo-ovato crispo ungue auriculato columnæ longitudine, petalis lanceolatis revolutis valde crispis, labelli pugioniformis crispi revoluti auriculis grossè dentatis carnosis ascendentibus cristâ maximâ valdè convexâ a fronte trilingui a tergo bituberculato laminâ tenui interjectâ denticulo carnoso utrinque, columnæ glabræ alis parvis setaceis.

This remarkable species belongs to the same natural division of the genus as the *O. serratum* mentioned and figured at p. 28, No. 42, in a previous number of this work. It is, however, perfectly distinct from it and all others known to us. Its flowers are of a deep chocolate brown, the petals and crest of the lip being edged and spotted with bright yellow. Of the crest the structure is so singular and complicated that it is difficult to describe ; in this, however, it is remarkable that in front of a large quasi-rocky elevation there project three flat yellow tongues which are quite peculiar to the species. Before it flowered the plant was supposed to be O. macranthum, which is a very much finer thing, and, if drawings can be trusted, must be one of the best of all Oncids.

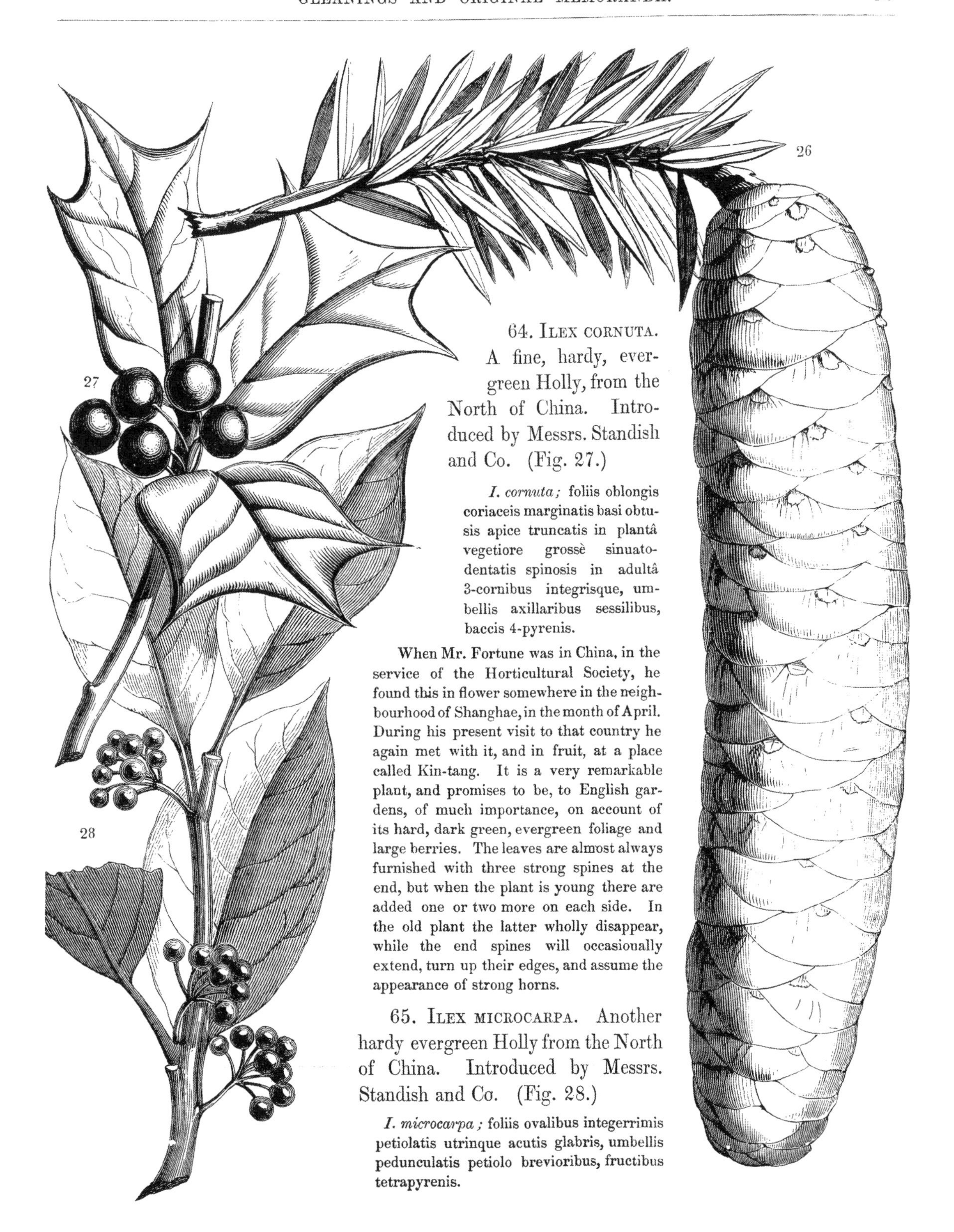

64. ILEX CORNUTA. A fine, hardy, evergreen Holly, from the North of China. Introduced by Messrs. Standish and Co. (Fig. 27.)

I. cornuta; foliis oblongis coriaceis marginatis basi obtusis apice truncatis in plantâ vegetiore grossè sinuato-dentatis spinosis in adultâ 3-cornibus integrisque, umbellis axillaribus sessilibus, baccis 4-pyrenis.

When Mr. Fortune was in China, in the service of the Horticultural Society, he found this in flower somewhere in the neighbourhood of Shanghae, in the month of April. During his present visit to that country he again met with it, and in fruit, at a place called Kin-tang. It is a very remarkable plant, and promises to be, to English gardens, of much importance, on account of its hard, dark green, evergreen foliage and large berries. The leaves are almost always furnished with three strong spines at the end, but when the plant is young there are added one or two more on each side. In the old plant the latter wholly disappear, while the end spines will occasionally extend, turn up their edges, and assume the appearance of strong horns.

65. ILEX MICROCARPA. Another hardy evergreen Holly from the North of China. Introduced by Messrs. Standish and Co. (Fig. 28.)

I. microcarpa; foliis ovalibus integerrimis petiolatis utrinque acutis glabris, umbellis pedunculatis petiolo brevioribus, fructibus tetrapyrenis.

Concerning this new shrub we have no information beyond the statement that it was found at Tein-tung. The aspect of the plant is not unlike that of an evergreen oak, but the leaves are perfectly smooth on each side. The berries when ripe are very small, and appear to be unusually pulpy, for, on drying, they shrivel up, and leave the ribs of the 4 stones which they enclose quite apparent. It seems allied to Thunberg's Ilex rotunda.

66. CATTLEYA SPECTABILIS, of which there is a figure in the *Florist* of April (vol. iii. p. 92.), is only a finely blown specimen of *C. pumila,* and thus adds another to our list of *aliàses* at p. 6.

67. TROPÆOLUM WAGNERIANUM. *Karsten.*

Judging from a coloured print circulated by Mr. F. A. Haage, Jun., of Erfurt, we should say that this is scarcely more than a variety of the *Tropæolum* figured at p. 9 of this volume; differing in little except the form of the leaves which are represented to be hastate, and in the colour of the petals which appear to be dark violet instead of blue.

68. HELICONIA ANGUSTIFOLIA. *Hooker.* A noble hothouse herbaceous plant from Brazil, with large crimson spathes, and snow-white flowers. Blossomed at Kew in January, 1846. Belongs to the order of Musads.

A very handsome and rather dwarf species, introduced to Liverpool from Brazil. Its beautiful bright red spathes, deep orange-coloured ovaries, and white sepals tipped with green, have a very handsome effect. The flower-stem is sheathed by the bases of the long petioles, and the principal leaf is 1½ foot long and about 3 inches wide, with a stout rib and parallel oblique veins, narrowed to a point at both ends, and glabrous, except that the rib beneath the very long taper petioles and cylindrical sheaths (at least in their upper part), is clothed with a scattered pulverulent or scurfy down. The rachis is a span and more long, deep red, bearing at distances of an inch or more, six or seven bright red spathes, the lowest one 6 inches long, the rest gradually shorter and less acuminated. This belongs to a genus of tropical plants inhabiting moist places, conspicuous by their fine broad leaves and showy flowers; forming, with allied genera, dense thickets in their native localities. The present may be considered a dwarf species of the genus, as it does not attain more than between three and four feet in height. It requires to be grown in a large pot, in light loam, supplying it freely with water during summer.—*Botanical Magazine,* t. 4475.

69. GARRYA ELLIPTICA. *Douglas.* The Female. A hardy evergreen shrub, from North Western America. Introduced by the Horticultural Society. Belongs to the order of Garryads.

Hitherto the male only of this fine Evergreen bush has been known in our Gardens; in which its good foliage and long massive tails of yellowish catkins, appearing in the earliest days of spring, have deservedly rendered it a universal favourite. The female, which in foliage is like the male, has flowered now for the first time in Europe, and proves to be as destitute of beauty as the male is conspicuous for it. The catkins are short, green, and, at a little distance from the bush, are not to be observed. To Botanical Gardens the plant is an acquisition, as it is to Horticulture, inasmuch as it will probably now ripen fruit, and thus afford a ready means of propagation. It is possible, also, that the deep purple berries, with long clusters of which the plant is loaded in North-West America, may prove ornamental; but of that we can at present have no certain knowledge.—*Journ. Hort. Soc.,* Vol. v. p. 137.

70. TRICHOPILIA SUAVIS. A delicious Orchid, of which a figure will appear in an early number of this work.

T. suavis; pseudobulbis tenuibus obcordatis, foliis latis oblongis coriaceis, pedunculis bifloris, petalis linearibus rectiusculis, labello maximo bilobo undulato crispo basi arctè convoluto sursùm abruptè ventricoso.

71. TUPA CRASSICAULIS. *Hooker.* (*aliàs* Siphocampylus canus, of the *Belgian Gardens.*) A Brazilian Greenhouse Lobeliad, of little interest, with long serrated leaves, hoary underneath, and dull yellowish red flowers. Blooms at Kew in summer and autumn. Introduced by M. Makoy, of Liége.

Our plants are nearly three feet high, and exhibit a stout but woolly or cobwebby stem, leafy at the top, something after the manner of the *Daphne Laureola.* Leaves soft, four to six inches long, patent or deflexed, lanceolate or broad-lanceolate, acute, serrated, tapering at the base into a short foot-stalk, dark green and slightly downy above, tomentose and hoary beneath. Peduncles solitary, one to two inches long, woolly. Calyx woolly, the limb of five acuminated spreading segments. Corolla yellowish, or greenish red, at length quite red; tube two inches long, nearly straight, laterally compressed; limb two-lipped, lips long, superior one inclined upwards, bifid, segments linear-acuminate; lower lip deflexed, trifid, segments linear-lanceolate.—*Botanical Magazine,* t. 4505.

72. Cycnoches barbatum. *Lindley.* A very pretty orchidaceous epiphyte from Costa Rica, with long drooping hairy racemes of yellow flowers spotted with brown, and with a shaggy lip. Flowered with Mrs. Lawrence.

A singular and handsome plant, which appears almost to connect *Gongora* with this very sportive genus. A young plant scarcely exhibits a pseudo-bulb at all, only several imbricating, leafy scales terminated by an oblong-oval, acuminated, plaited leaf. When the leaf is fully developed the almost naked pseudo-bulb appears, ovate, compressed, green, smooth, with the withered scales at the base. Scape from the base of the pseudo-bulb, a foot long, dark purple, pubescent or hairy, jointed, sheathed with scales at the joints ; this is terminated by a drooping many-flowered raceme, a foot long, of which the rachis and pedicel-like ovaries are dark purple, and hairy. Flowers moderately large, at first sight a good deal resembling those of *Gongora maculata,* but larger. Lip very hairy or bearded, hanging down, white tinged with yellow, and elegantly spotted with deep blood colour.—*Botanical Magazine*, t. 4479.

73. Griffinia Liboniana. *De Jonghe.* An unimportant hothouse Amaryllid from Brazil, with pale blue flowers. Introduced by M. de Jonghe of Brussels. Flowers in March.

A bulb, with narrow, oblong, flaccid leaves, which much resemble those of a Drimia, being mottled with pallid blotches upon a dark green ground. The scape is about 6 inches high, and round. The flowers are small, pale ultra-marine with very narrow segments, whitish on the lower half. Stamens very short, and delicate. It is no doubt a very distinct species of the genus, the narrow, unstalked, blotched leaves being quite peculiar to it. But it is not likely to possess any interest as an object of beauty.—*Journ. Hort. Soc.*, Vol. v. p. 137.

29

74. Catasetum Warczewitzii. (*aliàs* Warczewitzia, *Skinner.*) A most fragrant terrestrial Orchid from Panama. Introduced by Mr. Skinner. Flowers pale green. Discovered by Warczewitz; blossomed at Penllergare in April with J. D. Llewelyn, Esq. (Fig. 29.)

C. *Warczewitzii,* (Monachanthus) racemis brevibus densis pendulis, sepalis petalisque subrotundo-ovatis patulis incurvis, labelli galeâ anticâ basi compressâ apice ventricosâ lobis membranaceis planis lateralibus parvis serratis intermedio bilobo laciniis divaricatis fimbriatis, columnâ muticâ.

This has found its way into cultivation under the name of Warczewitzia, Mr. Skinner having supposed the genus to be new, and desiring to give it to the bold and indefatigable naturalist who discovered it. We quite agree with Mr. Skinner that if patience, and unwearied industry, courage that never quails before danger, and enthusiasm which despises difficulty, should give a naturalist a claim to a genus, Mr. Warczewitz most eminently has one. But he must wait for another opportunity, the plant that was given him being undoubtedly a Catasetum, and nearly related to *C. discolor.* As a species it is perfectly distinct from all others; the flowers which grow in a close pendulous raceme, consist of roundish ovate sepals and petals, and a helmetted lip which spreads into a thin 3-lobed limb, the middle lobe of which divides into two diverging fringed halves. They are pale green, with bright emerald green veins, and though not gaudy are extremely pretty. Their charm consists, however, in their delicious fragrance, which is quite equal to that of Aërides odoratum.

75. Achimenes Jaureguia. *Wcz.*

This appears from a figure received from Mr. Haage of Erfurt, to be only a white flowered variety of *A. longiflora.*

76. ONCIDIUM LONGIPES. A little unimportant Orchid from Brazil (?) with yellow and brown flowers. Received from M. Morel of Paris in April.

O. longipes; (TETRAPETALA MACROPETALA) pseudobulbis ovalibus diphyllis, foliis angustis tenuibus, scapo bifloro foliis æquali, pedunculis elongatis, sepalis lateralibus elongatis pendulis basi connatis dorsali breviore latiore refracto, petalis oblongis planis, labelli lobis lateralibus parvis obtusis intermedio transverso apiculato sinu convexo serrato, cristâ pubescente depressâ basi simplici truncatâ papillâ utrinque adpressâ apice 3-lobâ, columnæ alis minimis sinuatis.

The habit is plainly that of *O. uniflorum*, but the sepals and petals are deeply stained with dull brown. Having been sent to Messrs. Loddiges some years since by M. Morel, it probably exists in our collections; but it is not worth cultivating except by mere Botanists.

77. BORONIA SPATHULATA. *Lindley.* (*aliàs* B. mollina, of Gardens.) An evergreen Swan River shrub, with a heavy unpleasant odour, and small pink flowers. Flowered with Mr. J. G. Henderson in March. Belongs to the Rueworts. (Fig. 30.)

An erect shrub, of little beauty, with compressed branches, which are rather rough when young. Leaves dull olive-green, simple, veinless, smooth, short and roundish-obovate on the early branches, becoming narrower and spathulate on the later. Flowers pink, small, in small terminal cymes, inconspicuous; their stalks are defended by coarse glands. Even in its native country, after having been burnt down, and reduced in stature to 9 inches or a foot, this can be a plant of very small interest. When extended by cultivation into long straggling branches sparingly covered with leaves, it is quite destitute of interest for gardens, and must be regarded as the worst of the Boronias.—*Journ. Hort. Soc.*, Vol. v. p. 142.

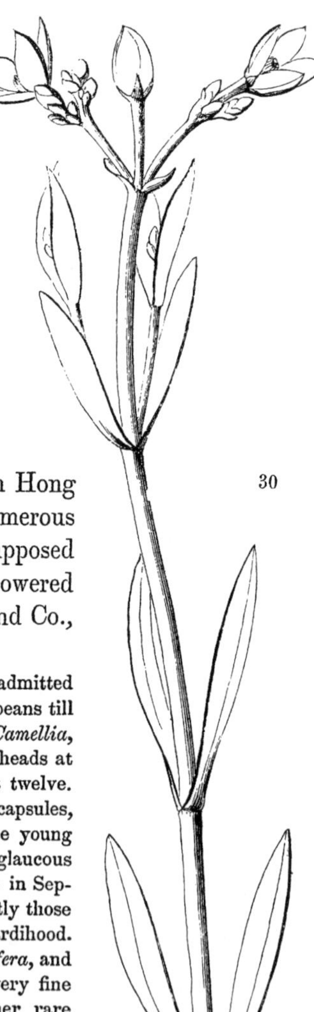
30

78. RHODOLEIA CHAMPIONI. *Hooker.* A greenhouse shrub from Hong Kong, of exquisite beauty, with heads of flowers surrounded by numerous large closely packed floral leaves, of a brilliant deep rose colour. Supposed to belong to the order of Witch Hazels (Hamamelidaceæ). Has not flowered in England. Living plants have been received by Messrs. Standish and Co., of Bagshot.

Captain Champion, writing from Hong Kong, December, 1849, says, "This is admitted by all here to be the handsomest of Hong Kong flowering trees, and new to Europeans till I discovered it in February last. It is a small tree, but would probably, like the *Camellia*, blossom as a shrub profusely, each branch bearing six to eight flowers. Flower-heads at its extremity, and these 2½ inches in diameter. Outer leaflets of involucre about twelve. Inner leaflets of involucre, rose-coloured, about eighteen. Fruit of five radiating capsules, each about the size of a small hazel-nut, birostrate, two-celled, many-seeded; in the young state crowned by two long filiform styles. Leaves long, petiolated, bright green, glaucous beneath. Flowers in February, and the fruit only attains its full size and ripens in September, splitting, when ripe, from the apex downwards. Conditions of growth exactly those of *Camellia Japonica*, I should say, and the tree of about the same degree of hardihood. There was a tree of *Camellia Japonica* in flower in the same wood, also *C. oleifera*, and another probably new species, together with Dr. Siebold's *Benthamia*, a new and very fine *Pergularia*, an *Ornus*, six or seven Oaks, a Chestnut, a *Liquidambar*, and other rare trees."—*Botanical Magazine*, t. 4509.

The account given in the *Botanical Magazine* of this extraordinary genus, is not sufficient to enable us to offer any opinion upon its affinity; but it appears to be the finest flowering shrub that has reached England since the arrival of the *Camellia* itself. Mr. Bentham compares it to *Sedgwickia*, an Asiatic genus unknown in Gardens; and it must be confessed that in the scaly buds of the two there is a very striking resemblance. We should however observe that the leaves on the live plants received at the nursery of Messrs. Standish of Bagshot, have not at all the texture or appearance of those of *Sedgwickia*, but in those respects are similar to *Viburnum Tinus*.

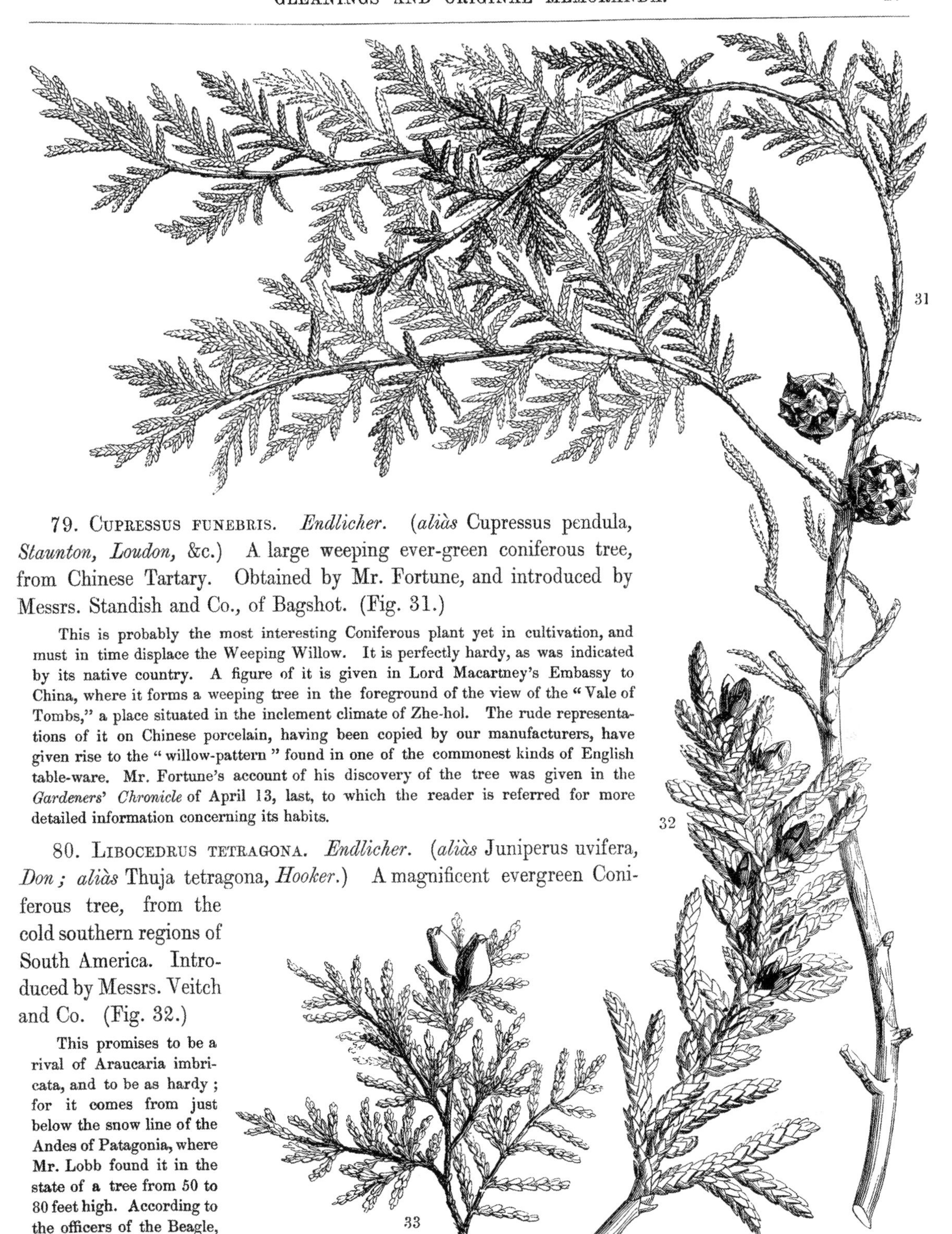

79. Cupressus funebris. *Endlicher.* (*aliàs* Cupressus pendula, *Staunton, Loudon,* &c.) A large weeping ever-green coniferous tree, from Chinese Tartary. Obtained by Mr. Fortune, and introduced by Messrs. Standish and Co., of Bagshot. (Fig. 31.)

This is probably the most interesting Coniferous plant yet in cultivation, and must in time displace the Weeping Willow. It is perfectly hardy, as was indicated by its native country. A figure of it is given in Lord Macartney's Embassy to China, where it forms a weeping tree in the foreground of the view of the "Vale of Tombs," a place situated in the inclement climate of Zhe-hol. The rude representations of it on Chinese porcelain, having been copied by our manufacturers, have given rise to the "willow-pattern" found in one of the commonest kinds of English table-ware. Mr. Fortune's account of his discovery of the tree was given in the *Gardeners' Chronicle* of April 13, last, to which the reader is referred for more detailed information concerning its habits.

80. Libocedrus tetragona. *Endlicher.* (*aliàs* Juniperus uvifera, *Don; aliàs* Thuja tetragona, *Hooker.*) A magnificent evergreen Coniferous tree, from the cold southern regions of South America. Introduced by Messrs. Veitch and Co. (Fig. 32.)

This promises to be a rival of Araucaria imbricata, and to be as hardy; for it comes from just below the snow line of the Andes of Patagonia, where Mr. Lobb found it in the state of a tree from 50 to 80 feet high. According to the officers of the Beagle, it is one of the trees called by the Spaniards, Alerçe ; but this is doubted by Dr. Hooker. The young branches are covered with small thick dark green scales or leaves, so placed as to constitute a four-sided arrangement, and being much larger than is usual among the scale-leaved Conifers, produce a massive appearance, which is quite peculiar to the species. The cones are small bodies, consisting of two opposite pairs of scales, each having a long horn at its back, and the exterior pair not being half the length of the inner. These scales appear to be whitish inside, and inclose four winged seeds, which stand in pairs at the base of the larger scales; the smaller scales are seedless. These scales, of two different sizes, are placed in what botanists call a valvate position; that is to say, they all touch at the edge without overlapping any interior scale; and in this resides the distinctive character of the Libocedars. In the Arbor-vitæs (Thuja), on the contrary, the outer scales of the cones are all alike in size, and always inclose two or more smaller scales. In other words, the cones of a Libocedar are much more simple in their structure than those of an Arbor-vitæ, in which we have the first distinct commencement of the spiral arrangement found in the higher branches of the Coniferous order.

81. LIBOCEDRUS CHILENSIS. *Endlicher*. (*aliàs* Thuja chilensis, *Don; aliàs* Thuja andina, *Pöppig*.) From Chili. A noble evergreen, with the habit of an Arbor Vitæ. Imported by Messrs. Low and Co. Natural order Conifers. (Fig. 33.)

A fine evergreen tree. Mr. Bridges says that it is from 65 to 80 feet high ; Sir W. Hooker, that it is a tree from 30 to 40 feet high, of great beauty, and well worthy of being introduced into our gardens. Pöppig relates that it resembles the American Arbor Vitæ, but is less robust, sometimes branching from the base, and gaining the habit of a Cypress, but in other cases forming a conical head. "The trunk," he adds, "of this last variety is simple as high as the middle, straight, taper, clothed with a rough cracked bark of a brownish ash-colour, knotty, scarcely more than a foot thick, with a yellowish, resinous, hard strong scented (*olente*) wood." Whether it will bear the climate of Great Britain without protection is at present uncertain. The young branches of this tree, when they are visible, are compressed, obovate between the nodes, and bright green, with glaucous furrows ; they are, however, for the most part, hidden by the leaves. The latter, which are compressed, blunt, and keeled, are glaucous at the sides, but bright green at the back and edges ; they stand in two pairs crosswise, the lower pair being much larger than the upper pair, which resembles two tubercles. These leaves evidently represent the type of the cones, which are drooping, short-stalked, about half an inch long, and consist of four woody scales, also standing crosswise, in two very unequal pairs. These scales are applied face to face, and have a sharp tubercle on the outside below the point. The two larger scales have each two seeds at their base ; the two smaller are seedless. The four seeds stand erect in the cones, with unequal-sided wings.—*Journ. Hort. Soc.*, Vol. v. p. 35. It is stated in that work that the plant had been also introduced by Messrs. Standish & Co. This, however, proves to have been an accidental error, Mr. Low having been the sole importer.

82. DENDROBIUM PALPEBRÆ. *Lindley*. From the East Indies. A handsome stove epiphyte belonging to the natural order of Orchids, flowering in November. Flowers white, with a lip stained with yellow. Introduced by Messrs. Veitch and Co.

A charming species, in the way of *D. densiflorum*, with the perfume of distant hawthorn. Its stems are more slender than those of the species just named; the flowers in loose racemes and white, with a deep yellow stain at the base of the lip, which is not only covered with soft down, but is fringed near the base with long hairs, like eyelashes. These elevated lines pass along the middle, and terminate near the base in a 3-lobed tubercle, for the purpose of receiving which the base of the column is hollowed out into an oblong cavity. It was received from Messrs. Veitch, in November, 1849.—*Journ. Hort. Soc.*, Vol. v. p. 33.

83. ACHIMENES ESCHERIANA. *Lemaire*. A hybrid between A. rosea ♀ and longiflora ♂. Said to be handsome.

Raised by M. Regel, of Zurich. It has the habit of *A. rosea*, but is rather stronger. The flowers are intermediate in size between the two parents ; the limb is a rich crimson, spotted with bluish violet when going off ; the orifice is golden yellow, dotted with purple, as in the mother.—*Van Houtte's Flore*, 1848, p. 405 *d*.

I. Constans. Pinx & Zinc

Printed by C.F.Cheffins, London.

[Plate 10.]

DAMPIER'S CLIANTH.

(CLIANTHUS DAMPIERI.)

A Greenhouse perennial trailer, from New Holland, *belonging to the Order of* Leguminous Plants.

Specific Character.

DAMPIER'S CLIANTH.—Herbaceous, shaggy, decumbent. Leaflets opposite, very seldom alternate, obovate-oblong. Stipules cut or toothed. Peduncles bearing a kind of umbel at the point, shorter than the leaves. Calyx 5-cleft, with acuminate segments, and acute re-entering angles. Ovary shaggy.

CLIANTHUS *DAMPIERI.*—Herbaceus, villosus, decumbens; foliis oppositis rarissime alternis obovato oblongis, stipulis laciniatis v. dentatis, pedunculis apice subumbellatis foliis brevioribus, calycibus 5-fidis laciniis acuminatis sinubus acutis, ovariis villosis.

Clianthus Dampieri, *Cunningham in Hort. Soc. Trans.* II. *series* i. 522. *R. Brown, in Sturt's Narrative* (1849) II. 71; *aliàs* Clianthus Oxleyi, *Cunningham; aliàs* Donia speciosa, *Don* (according to Brown); *aliàs* Kennedya speciosa, *of Cunningham.*

This beautiful plant has been raised from New Holland seeds, by Messrs. Veitch of Exeter, under the name of Kennedya speciosa; and received the large silver medal of the Horticultural Society when exhibited in Regent Street, in April last; an honour never conferred upon any new plants, except such as are of surpassing value as objects of cultivation.

It formed a stout decumbent herbaceous perennial, of a pallid aspect, covered with long hairs. The pinnated leaves were in about five pairs, with an odd one; the leaflets being oblong, or slightly obovate, opposite in most cases, and furnished with a pair of coarsely toothed or slashed stipules. From the axils of these leaves, and shorter than they, arise angular peduncles, having on the end four or five quasi-umbellate flowers of the most brilliant colour. Their calyx is tubular, shaggy, with five acuminate lobes, and acute re-entering angles. The standard is ovate, oblong, acuminate, bright scarlet, with a deep purple stain at the base, which is convex and shining; the keel is acuminate, scarlet, and very like that of the Crimson Clianth (*Clianthus puniceus*), as are the wings, which are also scarlet. The ovary and stamens appear not to be different from the organs belonging to the last-mentioned species.

Dr. Brown, who seems to have studied this plant, speaks of it thus in the Appendix to *Captain Sturt's Narrative of an Expedition into Central Australia*:—

"In July, 1817, Mr. Allan Cunningham, who accompanied Mr. Oxley in his first expedition into the western interior of New South Wales, found his Clianthus Oxleyi on the western shore of Regent's Lake, on the River Lachlan. The same plant was observed on the Gawler Range, not far from the head of Spencer's Gulf, by Mr. Eyre in 1839, and more recently by Capt. Sturt, on his Barrier Range, near the Darling. I have examined specimens from all these localities, and am satisfied that they belong to one and the same species.

"In March (not May) 1818, Mr. Cunningham, who accompanied Capt. King in his voyages of survey of the coasts of New Holland, found on one of the islands of Dampier's Archipelago, a plant which he then regarded as identical with that of Regent's Lake. This appears from the following passages of his MS. Journal:—

"'I was not a little surprised to find Kennedya speciosa, (his original name for Clianthus Oxleyi) a plant discovered in July, 1817, on sterile bleak open flats, near Regent's Lake, on the River Lachlan, in lat. 33° 13′ S., and long. 146° 40′ E. It is not common; I could see only three plants, of which one was in flower. This island is the Isle Malus of the French.' Mr. Cunningham was not then aware of the figure and description in Dampier above referred to, which, however, in his communication to the Horticultural Society in 1834, he quotes for the plant of the Isle Malus, then regarded by him as a distinct species from Clianthus Oxleyi of the River Lachlan. To this opinion he was probably in part led by the article 'Donia, or Clianthus,' in *Don's System of Gardening and Botany*, vol. II. p. 468., in which a third species of the genus is introduced, founded on a specimen in Mr. Lambert's Herbarium, said to have been discovered at Curlew River, by Capt. King. This species named Clianthus Dampieri, by Cunningham, he characterises as having leaves of a slightly different form, but its principal distinction is in its having racemes instead of umbels; at the same time he confidently refers to Dampier's figure and description, both of which prove the flowers to be umbellate, as he describes those of his Clianthus Oxleyi to be. But as the flowers in this last plant are never strictly umbellate, and as I have met with specimens in which they are rather corymbose, I have no hesitation in referring Dampier's specimen, which many years ago I examined at Oxford, as well as Cunningham's, to Clianthus Dampieri. This specimen, however, cannot now be found in his Herbarium, as Mr. Heward, to whom he bequeathed his collections, informs me; nor can I trace Mr. Lambert's plant, his Herbarium having been dispersed.

"Since the preceding observations were written, I have seen, in Sir William Hooker's Herbarium, two specimens of a Clianthus, found by Mr. Bynoe, on the north-west coast of Australia, in the voyage of the Beagle. These specimens, I have no doubt, are identical with Dampier's plant, and they agree both in the form of leaves and in their subumbellate inflorescence, with the plant of the Lachlan, Darling, and the Gawler Range. From the form of the half-ripe pods of one of these specimens, I am inclined to believe that this plant, at present referred to Clianthus, will, when its ripe pods are known, prove to be sufficiently different from the original New Zealand species, to form a distinct genus; to which, if such should be the case, the generic name *Eremocharis* may be given, as it is one of the greatest ornaments of the desert regions of the interior of Australia, as well as of the sterile islands of the north-west coast."

It is possible that this may be intended to cover some further meaning than can be assigned to the words as they would be interpreted by ordinary readers. We can only remark that we find in this plant no indication of a genus different from Clianthus; in fact, we see less to separate it from the Clianths than is to be found in Endlicher's Streblorhize (Clianthus carneus.) At all events, it is much to be regretted that naturalists should thoughtlessly encumber books with names of which there is no present or probable want. It is early enough to add to the chaos of Botanical nomenclature when a clear case of scientific necessity can be made out.

The plant will prove of the easiest cultivation, demanding no more care than is given to Pelargonium, the habits of which it probably possesses.

As the work from which the preceding remarks of Dr. Brown have been extracted is not likely to be in the hands of many of our readers, we fill a vacant space with his remarks upon two other Leguminous plants from New Holland, which this eminent botanist supposes to constitute new genera.

CLIDANTHERA.

Calyx 5-fidus. *Petala* longitudine subæqualia. *Stamina* diadelpha: *antheræ* uniformes; loculis apice confluentibus, valvula contraria ab apice ad basin separanti dehiscentes! *Ovarium* monospermum. *Stylus* subulatus. *Stigma* obtusum. *Legumen* ovatum, lenticulari-compressum, echinatum.

Herba, v. Suffrutex, *glabra, glandulosa; ramulis angulatis.* Folia *cum impari pinnata; foliolis oppositis, subtus glandulosis.* Stipulæ *parvæ, basi petioli adnatæ.* Flores *spicati, parvi, albicantes.*

Subgenus forsan Psoraleæ, cui habitu simile, foliis calycibusque pariter glandulosis; diversum dehiscentia insolita antherarum!

6. Clidanthera *psoralioides.*

Suffrutex bipedalis in paludosis. *D. Sturt.*

Herba, vel suffrutex, erecta, bipedalis, glabriuscula. Ramuli angulati. Folia cum impari pinnata, 4-5-juga; foliola opposita, lanceolata, subtus glandulis crebris parvis manifestis, marginibus scabris. Spicæ densæ, multifloræ. Calyx 5-fidus, parum inæqualis, acutus, extus glandulis dense conspersus. Corolla: *Vexillum* lamina oblonga subconduplicata nec explanata, basi simplici absque auriculis; ungue abbreviato. *Alæ* vexillo paullo breviores, carinam æquantes, laminis oblongis, auriculo baseos brevi. *Carinæ petala* alis conformes. Stamina diadelpha, simplex et novemfidum; antheræ subrotundæ v. renifornes, valvula ventrali anthera dimidio minore subrotunda. Ovarium hispidum ovulo reniformi. Legumen basi calyce subemarcido cinctum, echinatum. Semen reniforme, absque strophiola; integumento duplici. Embryo viridis; cotyledones obovatæ, accumbentes.

PENTADYNAMIS.

Calyx 5-fidus subæqualis. *Vexillum* explanatum, callo baseos laminæ in unguem decurrenti. *Carina* obtusa, basin versus gibba, longitudine alarum. *Stamina* diadelpha; *antheris* 5 majoribus linearibus, reliquis ovatis. *Ovarium* polyspermum. *Stylus* e basi arcuata porrectus, postice barbatus. *Legumen* compressum.

Herba (Suffrutex sec. D. Sturt), bipedalis sericeo-incana; caule angulato erecto. *Folio* ternata; foliolis sessilibus, linearibus, obtusis. *Flores* racemosi, flavi.

9. Pentadynamis *incana.*

"On sand-hills with Crotalaria Sturtii." *D. Sturt.*

Herba erecta, ramosa, sericeo-incana. Folia alterna, ternata; petiolo elongato, teretiusculo, foliolo terminali longiore vix unciali. Racemi multiflori, erecti; pedicelli subæquantes calycem. Bracteolæ subulatæ, infra apicem pedicelli, basin calycis attingentes. Calyx 5-fidus; laciniis acutis tubum æquantibus. Corolla flava, calyce plus duplo longior. Vexillum explanatum, basi absque auriculis sed callo in unguem decurrenti ibique barbato auctum. Carina infra medium gibba pro receptione baseos styli. Staminum antheræ majores lineares, basi vel juxta basin affixæ; 5 minores ovatæ, incumbentes. Ovarium lineare, pubescens. Stigma terminale, obtusum. Legumen immaturum incanum, stylo e basi arcuata porrecto terminatum, calyce subemarcido subtensum.

In the collection of the plants of his last expedition, presented to the British Museum by Sir Thomas Mitchell, there is a plant which seems to belong to the genus Pentadynamis, which is probably, therefore, one of the species of Vigna, described by Mr. Bentham.

PLATE 11.

I. Constans. Pinx & Zinc

Printed by C.F. Cheffins, London

[PLATE 11.]

THE SWEET TRICHOPIL.

(TRICHOPILIA SUAVIS.)

A stove Epiphyte, from CENTRAL AMERICA, *belonging to the Natural Order of* ORCHIDS.

Specific Character.

THE SWEET TRICHOPIL.—Pseudo-bulbs thin, oblong, obcordate, one-leaved. Leaves broad, oblong, wavy, leathery, nearly sessile. Peduncles about 2-flowered Petals linear, nearly straight. Lip very large, 3-lobed. wavy, crisp, closely rolled up at the base, suddenly inflated upwards. Hood of the column 3-lobed, with all the lobes fringed, the middle one being the narrowest.

TRICHOPILIA *SUAVIS.*—Pseudo-bulbis tenuibus oblongis obcordatis monophyllis, foliis latis oblongis undulatis coriaceis subsessilibus, pedunculis sub-bifloris, petalis linearibus rectiusculis, labello maximo bilobo undulato crispo basi arctè convoluto sursum abruptè ventricoso cuculli trilobi laciniis omnibus fimbriatis intermediâ angustiore.

Trichopilia suavis : *Suprà p.* 44, *no.* 70.

AMONG the Vandeous Orchids, that is to say among the Orchids having waxy pollen-masses on a well-defined gland, and usually with a caudicle in addition, stands conspicuous, a group which we have elsewhere named BRASSIDS, comprehending the genera Oncidium, Odontoglossum, Brassia, Cymbidium, and many more. (See *Vegetable Kingdom,* p. 181.)

It is among these genera that the genus Trichopil is stationed, and well defined by its four pollen-masses at the end of a long wedge-shaped caudicle, its convolute free lip, and the remarkable hood of the column, divided, in the species hitherto seen, into three unequal lobes. Helcia, which is nearest it, has a flat lip with a distinct fleshy hypochil, and a fringed, not hooded, anther-lid.

It is not improbable that many more Trichopils lurk in the forests of Central America than we have any actual knowledge of. To the Cork-screw Trichopil (*Tr. tortilis*), so named on account of the spiral form of the petals, a second species, from Mexico, was some years since added by Messrs. Richard and Galeotti, with narrow stem-like pseudo-bulbs, and large solitary yellow flowers, under the name of *Tr. Galeottiana.* The plant now figured forms a third ; and a fourth, still unnamed, has flowered with Sir Philip Egerton.

For the opportunity of publishing a coloured plate of this, the Sweet Trichopil, we are indebted to R. S. Holford, Esq., whose specimens reached us a few weeks since in admirable condition. It had also been flowered about the same time by Mrs. Lawrence and Mr. Loddiges. Its broad thin pseudo-bulbs and large leathery leaves will distinguish it when not in flower, and have led to the confusion of it with the large-flowered Tooth-tongue, *Odontoglossum grande.* The flowers emit the most delicate odour of Hawthorn. They are, when well grown, full five inches in diameter, delicate in texture, nearly white, with a few slight stains of red on the sepals and petals, and a great convolute lip richly spotted with clear rose, which, it seems, becomes, in the bright natural climate of the species, a rich and brilliant red.

The cultivation of the plant is exactly that of Lycaste Skinneri, and similar terrestrial Orchids. This has been well described in the *Journal of the Horticultural Society,* vol. v. p. 14.

"It should be recollected that no plants can exist for any very great length of time without rest, and that rest is induced in a tropical climate by drought, in the same way as low temperature in our own country suspends vital energy: therefore Orchids must be subjected to the usual seasonable changes of rest and activity. Rest is induced by withholding moisture from their roots, and partly from the air, and this state of things may be considered to represent their winter. Spring should be imitated by gradually reviving vital energy by increase of moisture, first to the atmosphere, and afterwards to the roots or soil, accompanied by a proportionate increase of temperature: this period of their growth should be very slow. Summer must be represented by a greater increase of both heat and moisture; partial shade should also be resorted to, to bring the energy of the plant into full force. And lastly, an autumn must be created to bring about maturity, by gradually reducing the quantity of both heat and moisture, until the plants are again brought to a fit state for repose. The first and last stages should be of but short duration, and require caution, otherwise much mischief may be done to the plants.

"By growing Orchids in the mean instead of the maximum of heat and moisture, they will not make such rapid growth; but they will become more robust and healthy, and be less liable to receive injury from sudden transitions, either of heat, drought, or moisture, in the atmosphere.

"The temperature of the house can only with certainty be kept regular by night, particularly in summer; therefore the fire should never raise the heat of the principal house higher than 60°, and about five degrees less should be maintained where the plants are in a less excitable state: but as the days lengthen, so the temperature may rise; yet it should if possible never range higher than 75° by night in summer; it will occasionally, however, be higher in very warm weather, and should be counteracted as much as possible by evaporation and ventilation by night, and by both, as well as by shading, by day."

T. Constans. Pinx. & Zinc.

Printed by C.F. Cheffins. London.

[PLATE 12.]

THE MAGNIFICENT MEDINILL.

(MEDINILLA MAGNIFICA.)

An evergreen stove Shrub from JAVA, *belonging to the Natural Order of* MELASTOMADS.

Specific Character.

THE MAGNIFICENT MEDINILL.—An evergreen erect bush, perfectly smooth in every part, with compressed 4-winged branches, setose at the nodes. Leaves opposite, leathery, obovate-oblong, cordate, somewhat stem-clasping, suddenly pointed, triple-nerved below the middle, and with pinnate ribs at the base. Panicles terminal, long, pendulous, with whorled branches. Bracts very large, bright rose-colour, in whorls of 4, many-nerved, deciduous. Flowers decandrous.

MEDINILLA *MAGNIFICA.* — (Sect. *Sarcoplacuntia*) ramis compressis tetrapteris ad nodos setosis, foliis oppositis coriaceis glabris sessilibus obovato-oblongis cordatis subamplexicaulibus cuspidatis infra medium triplinerviis pone basin pinnato-costatis, paniculis terminalibus elongatis pendulis, ramis verticillatis, bracteis maximis coloratis quaternatis multinerviis deciduis, floribus decandris.

Medinilla bracteata of *the Gardens*, but *not of Blume.*

THE genus Medinill, founded originally by M. Gaudichaud, upon a shrub from the Marianne Islands, has become known in Gardens by the introduction of the Showy and the Red-leaved species (*M. speciosa* and *erythrophylla*); the former, a plant of striking beauty; the latter, much less remarkable in appearance. These two may be taken as good examples of the genus generally, some of which are among the handsomest shrubs of the Malay Archipelago, while others would be passed by without notice. Many species have been made known by Dr. Blume, and other Dutch naturalists. They seem all to inhabit the islands of Asia within the tropic, and to require a damp forest climate. Blume says that he has seen some of them climbing up the trunks of trees to the height of from 60 to 80 feet. He adds that they have a mucilaginous bark, which, stripped of its epiderm, is employed by the Malays for poultices, in dislocations and tumours, and that the subacid leaves are, in Celebes, boiled with fish.

The species now before us was imported from Java by Messrs. Veitch, and gained one of the large medals of the Horticultural Society, early in the present spring. By some error it was called *Medinilla bracteata,* a name to which it has not the slightest claim; the plant once so called by

Dr. Blume, and now before us, not being even a member of the genus, but having been separated by the learned Dutchman himself as a DACTYLIOTE. (*Museum Bot. Lugd. Bat., p.* 18.) It is a poor insignificant thing, not worth cultivation. This, on the contrary, is one of the most noble-looking plants in India. Its massive leaves are nearly a foot long, and 4 or 5 inches broad, of a firm leathery texture, and of the richest green. From the ends of the branches hang down panicles, from 15 to 18 inches long, of rich glossy rose-coloured flowers, with purple petals and large many-ribbed bracts of the richest and clearest pink. Of the effect thus produced, the accompanying figure gives a correct, and in no degree exaggerated, illustration; it however only shows the lower part of a panicle—all that the page can be made to contain.

It is strange that so noble a form of vegetation should have escaped the acute eyes of the Dutch botanists; and yet we must conclude that it has done so, for no trace of it appears among the five or six-and-twenty species they have published. It certainly belongs to the section to which Blume gives the name of *Sarcoplacuntia*, well characterised by a short truncate calyx and fleshy placentæ; in fact is very nearly allied to the Showy Medinill (*M. speciosa*) itself. That such a plant as this should have remained unnoticed in an island so much explored as Java, is one of the best illustrations that could be produced of the inexhaustible richness of vegetation in the Malay forests.

What the true cultivation of this Medinill should be, can hardly be said to have been ascertained. Messrs. Veitch, we believe, have treated it as a hardy stove or warm green-house plant. According to Dr. Blume the species are mostly mountain plants (*Rumphia*, vol. i. pp. 11. &c.), and Reinwardt places Melastomads generally in such places. Speaking of the forests above 3000 feet in elevation above the sea, the latter author says:—"The singular Pitcher-plant here, hangs down from the lofty branches, and the broad and elegantly divided fronds of a beautiful Fern, the Dipteris, rise upon their slender stems. This elevated situation is more particularly characterised by the different kinds of laurels which here predominate. Java is especially rich in laurels, as well as in figs; these, with some Eugenias and other Myrtaceous plants, with a very large Gardenia, perpetually in flower, cover everywhere the highest spots in the mountains of India, associated with tall *Melastomas*, Rhododendrons, Magnolias filling the air with their fragrant perfume, and several sorts of oak. Intermixed with these, Orchids constantly prevail, and in great variety. It is only where the forest of laurels ceases, and the summit of the mountains becomes narrower and can no longer retain a covering of vegetable mould, when the air becomes more rarefied and colder, at an elevation of more than 7000 feet, that the appearance of the forest trees changes."—(*Journal of the Horticultural Society*, vol. iv. p. 232.)

Hence we may infer that the climate which suits the Pitcher-plant and the Java Rhododendron, will also be that adapted to the Medinills.

GLEANINGS AND ORIGINAL MEMORANDA.

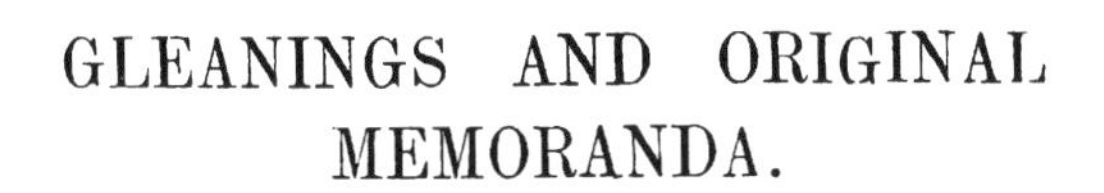

84. Acacia macradenia. *Bentham.* A fine New Holland greenhouse bush, with hard green smooth curved phyllodes eight or nine inches long, and innumerable zig-zag spikes of yellow heads. Flowers in March. (Fig. 33.)

An unknown Correspondent signing himself T. D., Pembroke, sent us living specimens of this in March last, intimating that it had been raised from South Australian seeds received from Drummond in 1847. It is certainly the same plant as was found by Sir Thomas Mitchell in his journey to the Victoria River, occupying the beds of rivers, and

forming bushes ten or twelve feet high. It is very handsome where there is room for it, its long narrow sabre-shaped phyllodes (leaves) having a bright colour and firm texture, and bending downward gracefully from singularly flexuose branches. The inflorescence is similarly zig-zag, much shorter than the leaves, and often forms an entangled mass of branches each of which is terminated by a yellow head about as large as the seed of the Sweet Pea.

85. Cephalotaxus Fortuni. *Hooker.* A fine, and probably hardy coniferous shrub, with long, narrow, deep-green distichous leaves; from the north of China. Introduced by Messrs. Standish of Bagshot. (Fig. 34.)

In the absence of a well-grown plant, little or nothing can be said of this tree, save that it is stated by Mr. Fortune to grow to a height of from 40 to 60 feet. Its branches are probably spreading or drooping, obscurely streaked or furrowed, distichous, pale brown, slender. Leaves quite distichous, alternate or opposite, close together, 3 to 4 inches long, linear, tapering a little at the base, much and gradually acuminate, one-nerved, dark full green above, paler beneath. A plant in the Bagshot Nursery stood in the open air during the last winter, without being in the least injured. As it increases from cuttings as readily as the common yew, and grows freely, we may expect to see this rare tree soon become common.—*Botanical Magazine*, t. 4499.

86. Galanthus plicatus. *Bieberstein.* A charming hardy bulb, from the Caucasus. Flowered in the Garden of the Horticultural Society in March 1850.

This beautiful Snowdrop, although long cultivated in gardens, is hardly known to the public. There appears to be no doubt as to its specific difference from the common species, its leaves being very much broader, and, as it were, plaited, not flat, its flowers being larger, and the green on the petals far more conspicuous. In a horticultural point of view it is a much finer thing than the old Snowdrop, just as hardy, and as easily managed.—*Journ. Hort. Soc.*, Vol. v. p. 138. *With a figure.*

87. Cereus Tweediei. *Hooker.* An erect, round-stemmed, furrowed Cactus, covered with stiff spines, from among which arise handsome curved narrow orange tubular flowers, each almost 3 inches long. From Buenos Ayres by Messrs. Lee and Co. Flowered at Kew, in September, 1849.

About 1 foot to 1½ foot high, and 1 inch in diameter, of a very glaucous green hue, simple, but increasing readily by offsets at the base. The shape is cylindrical, very slightly tapering upwards, numbered with many, about sixteen, moderately deep furrows perfectly straight, the ridges obtuse and even (not tubercled). Spine-tufts on the ridges close together, oval, with brown wool. Spines many in each tuft, four or five stouter than the rest, white, blotched with brown ; of the stout ones three or four (half to three-quarters of an inch long) are nearly erect ; a solitary stout one together with the other lesser ones, which are white, generally, all point downwards. Flowers rich orange-crimson, numerous, from the side of the stem, 3 inches long, curved upwards, the mouth oblique. Calyx-tube funnel-shaped, the scales remote, subulate, oppressed, lower ones ciliated with white hairs. Petals small, scarcely longer than the teeth of the calyx, acute. Stamens lying against the upper side of the tube, and there much longer than the flower ; lower ones scarcely protruded. Anthers deep purple.—*Botanical Magazine*, t. 4498. Will probably be a good breeder.

88. Juniperus sphærica. An evergreen tree from the north of China. Introduced by Messrs. Standish and Noble. (Fig. 35.)

J. sphærica ; arborea, foliis omnibus squamæformibus quadrifariis obtusis dorso foveâ circulari notatis, ramulis gracilibus tetragonis obtusis, galbulis sphæricis glaucis breviter pedunculatis.

Found in the north of China by Mr. Fortune, who describes it as a tree 30 to 50 feet in height. The young branches are four-cornered, blunt, and usually more slender than in the accompanying figure. All the leaves are minute, scaly, with a circular pit at their back. The fruit is quite round, about as large as the ball of a pocket pistol. The species differs from J. chinensis apparently, in not having any acicular leaves, and very decidedly in the size and form of its fruit, which is twice as large as in that species, and not at all depressed at the end, but very regularly spherical.

89. Quercus inversa. An evergreen Oak, from the north of China. Imported by Messrs. Standish and Noble. (Fig. 36.)

Q. inversa ; sempervirens, ramis tomentosis, foliis coriaceis obovatis petiolatis cuspidatis obtusis nunc apice serratis supra glaberrimis subtus glauco-tomentosis, glandibus spicatis obovatis cupulâ brevi tomentosâ squamulosâ multò longioribus.

From specimens of this fine oak sent to Messrs. Standish and Noble by Mr. Fortune, we presume that it forms a tree with the habit of the Evergreen Oak. The leaves are deep green, covered with a short glaucous down on the underside, but quite smooth and shining on the upper ; they are always contracted into a short blunt cusp at the point, where they are also sometimes serrated. The cups of the acorns are much like those of the Evergreen Oak, but the acorns are wider at the upper than at the lower end. The male flowers are produced at the ends of the same branches as carry the females, but are much more compactly arranged, forming long downy tails. The inflorescence consists of many such branches produced at the points of the shoots. The female flowers are tolerably regularly sessile in threes.

90. Quercus sclerophylla. An evergreen Oak from the north of China, sent by Mr. Fortune to Messrs. Standish and Noble. (Fig. 37.)

Q. sclerophylla; sempervirens, ramis glabris, foliis petiolatis coriaceis glabris acuminatis obtusis ultra medium grosse serratis supra lævibus subtus glauco-pubescentibus, glandibus spicatis pubescentibus sphæricis paulo ultrà cupulam protrusis, cupulis tomentosis squamis elevatis quasi tuberculatis.

A much finer oak than the last, with a very peculiar aspect. Some of the leaves are six inches long and nearly three inches broad ; their texture is that of a Spanish Chestnut, but thicker ; their colour rich bright green on the upper side, and glaucous with fine down on

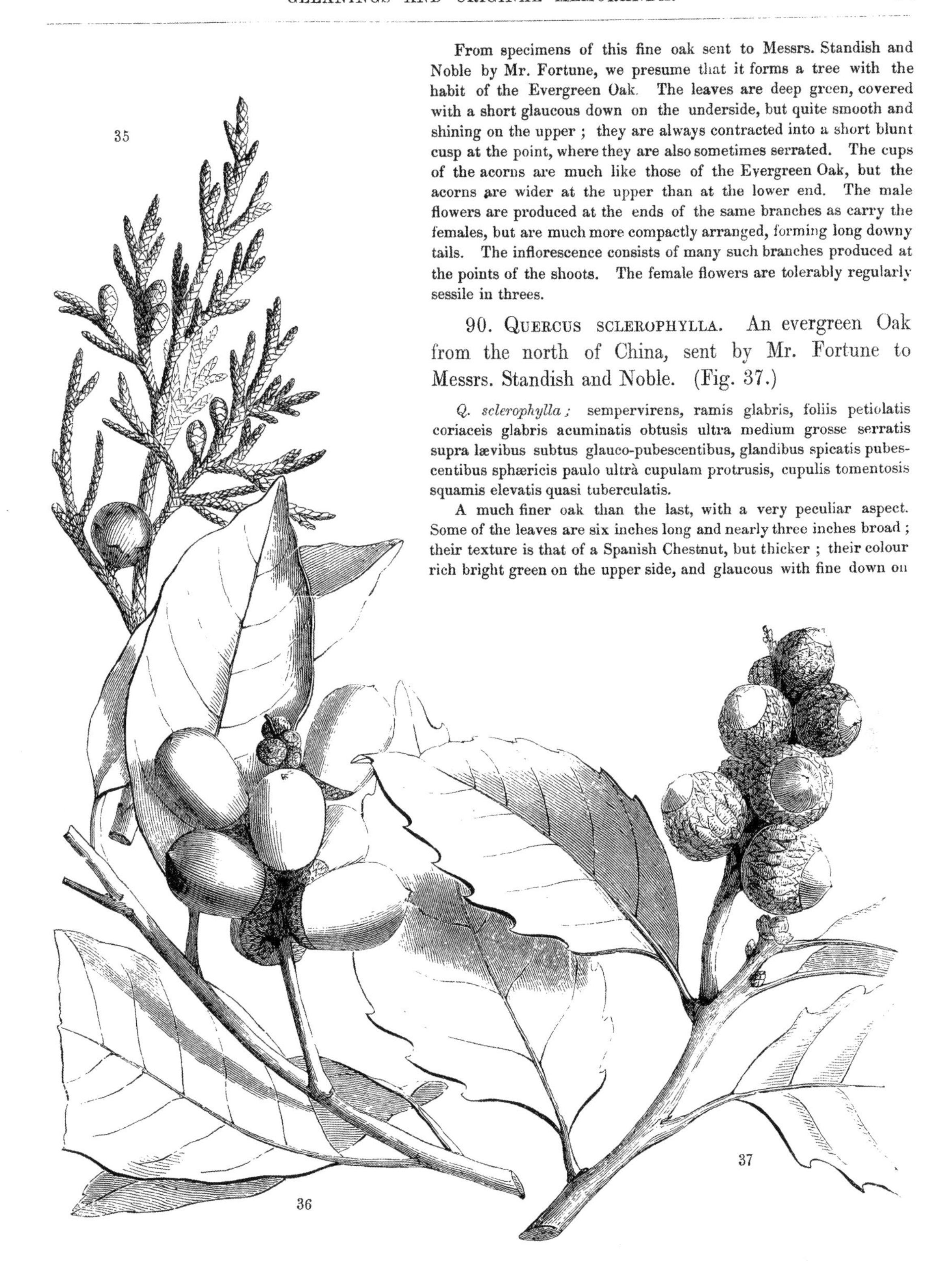

the under side. The spikes of the fruit are 3 or 4 inches long, very compact, with small downy acorns almost enclosed within very tomentose cups, the scales of which are large, distinct, and so much elevated as almost to give the cups the appearance of being covered with soft warts. A very fine thing.

91. LÆLIA GRANDIS. An Orchidaceous Epiphyte, with very large nankeen-coloured flowers. A native of Bahia. Flowered in May with M. Morel, of Paris. (Fig. 38.)

L. grandis; caule clavato monophyllo, folio coriaceo basi latius pedunculo bifloro basi spathaceo longiore, floribus subhorizontalibus, sepalis lanceolatis reflexis, petalis late-lanceolatis denticulato-crispis convexis labello parallelis et paulo longioribus, labello membranaceo venoso nudo undulato trilobo: laciniis lateralibus circa columnam convolutis et multo longioribus.

The accompanying Figure, the natural size, was taken from a flower received from Mons. Morel, along with a sketch of the leaf and stem. It is a plant with all the habit of a Cattleya, but the pollen-masses are 8, not 4. The stem appears to narrow to the base, as in Cattleya maxima; the leaf is represented as being firm, stiff, and rather broader at the base than the point. The flowers grow in pairs, on a peduncle issuing from a spathe, as in the species just named. The sepals and petals are nankeen-coloured; the lip white, washed with rose at the base in the inside, with purple veins, and a pure white border. The nearest affinity of this curious thing is with the Lælias Perrinii et majalis.

92. CYANOTIS VITTATA. *Lindley.* (*aliàs* Tradescantia zebrina, of *Gardens.*) A trailing Mexican stove perennial belonging to the Natural Order of Spiderworts (Commelinaceæ.)

It has handsome striped purple and grey leaves; but its flowers are insignificant. The stems, which are much branched, lie prostrate, or hang down from the shelf on which the plant is placed, and are of a deep rich purple; the leaves have the same colour, but are striped with a greenish grey, and when fresh are exceedingly pretty; on which account the plant is a favourite for covering rough unsightly places in hothouses. The flowers are violet-coloured; they appear for a long time, one after the other, from within a couple of terminal bracts, or spathes, of which one is shaped like the ordinary leaves, except being sessile; the other is much shorter, and boat-shaped. The stamens bear a tuft of jointed hairs in the middle, protrude beyond the tube of the corolla; the anthers are transversely linear, or almost crescent-shaped, with a small cell on each horn.—*Journ. Hort. Soc.*, Vol v. p. 139.

93. CUPANIA CUNNINGHAMI. *Hooker.* (*aliàs* Stadtmannia australis, *Cunn.*) A stove tree, belonging to the Natural order of Soapworts (*Sapindaceæ*), with panicles of small green flowers. Introduced, in 1825, to the Royal Botanic Gardens, Kew, from Australia. Flowers in the spring.

Most visitors to the great stove of the Royal Gardens, Kew, are attracted to a lofty shrub or tree among the Palms, exceeding many of them in height, with large pinnated leaves, and the young branches clothed with rusty down; it flowers in the spring, and is succeeded by large clusters of orange-coloured downy fruit. This is the plant here mentioned. It is a native of New Holland, on the north-east coast, near the tropic, and was discovered by Allan Cunningham, who speaks of it in his notes as a "tree 30—40 feet high, found in dark woods at Five Islands district, and on the banks of Hastings at Port Macquarrie, and Brisbane in Moreton Bay." It is a noble plant, with handsome foliage and fruit, but rather insignificant flowers, and of too lofty growth for ordinary cultivation.—*Botanical Magazine*, t. 4470.

94. SYMPLOCOS JAPONICA. *De Candolle.* (*aliàs* S. lucida, *Zuccarini.*) A hardy (?) evergreen shrub, from Japan, belonging to the natural order of Sapotads. Flowers in small, pale yellow clusters. Introduced by Messrs. Standish and Noble. (Fig. 39.)

This is said to grow in Japan to the size of the European ash-tree, with a close head: or to become a coppice-bush 20 feet high. The Japanese call it *Furoggi*. Thunberg took it for a myrtle! It is much used by the Japanese for decorating the shrines of their Idols, for which its evergreen habit renders it suitable. Whether or not it is hardy is uncertain; according to Siebold, it grows naturally in the southern provinces of Japan along with true Laurels, Terebinths, Magnolias, and Buckthorns.

95. RHAPONTICUM ACAULE. *De Candolle.* (*aliàs* Cynara acaulis, *Linnæus*; *aliàs* Cynara humilis, *Jussieu*; *aliàs* Serratula acaulis, *De Candolle*;

aliàs Cestrinus carthamoides, *Cassini.*) A fragrant tap-rooted perennial; native of Barbary, belonging to the Cynaraceous division of Composites, and said to have been introduced in the year 1799; now lost. (Fig. 40.)

The author of a Diary of a Tour in Barbary, as quoted in the *Gardener's Chronicle*, speaks thus of the present plant:— "The air was filled with the aroma of a multitude of TOFFS, which the Bedouin children had gathered for us. I know no European flower which I could put in comparison, as regards odour, with this seemingly insignificant Thistle; and here in Tunis, where kind Nature seems to have created it in such abundance, in order to overpower the pestiferous exhalations of the town, I have become too fond of it not to say a few words about it. One or two days after our arrival in Tunis, F— brought me a very ugly flower, a sort of vegetable polypus, as it were, which had neither leaves nor stalk, nor, as I supposed, smell. For want of a stalk it was stuck on the end of a small twig. Almost offended at the imputation against my taste, implied by F—'s offering me so ugly a thing, I paid no attention to his present, but let it lie on the chimney-piece. Often, however, as I passed the spot I perceived a delicious odour, and in vain inquired where were the concealed beds of Violets or Mignonette from which it proceeded. Neither F— nor T— could give me any information on the matter. The perfume, meanwhile, grew stronger and stronger every day, and with it grew my amazement at the phenomenon. It was my despised Thistle which diffused its incomparable fragrance over the whole room. I found it limp and faded lying under a heap of newspapers; I took it up, and pulled out the pointed twig that had been thrust into its tender heart, entreated its forgiveness for having so mistaken its worth, laid it into a saucer of water, and behold, it did forgive me; for its shrivelled florets expanded themselves again, and sent forth their fragrance more abundantly than ever. It is now the season when they are in bloom, and they stretch their heads by hundreds out of the earth; for they grow so close to the ground that one must actually dig them out, to get the flower entire. The exquisite perfume of this Thistle is universally acknowledged, for many fragrant essences are prepared from it." This is evidently the plant described by Desfontaines, under the name of Cynara acaulis, and we reproduce his figure of it, in the hope that it will lead to its re-introduction. He says that it is called TAFGA, that its heads are yellow, that its flowers smell like the Farnese Acacia, the sweetest of Italian plants, that its root is eatable, and that the Moors employ the plant to keep moths off their clothes (ad vermes vestimentis fugandos).

96. CALLIANDRA BREVIPES. *Bentham.* A stove shrub from Brazil, with clusters of pink mimosa-like flowers, appearing in October. Belongs to the Leguminous Order. Sent to Kew by M. Van Houtte.

A branching shrub, 4 to 5 feet high. Leaves double, each portion oblong, very closely pinnated with small linear-oblong, acute leaflets, and these generally drooping. Heads of flowers on short peduncles from the axils of the leaves, few in each head. Corolla, yellow, four-cleft. Stamens six times as long as the corolla, very slender, pale red or rose-colour. A pretty shrub which grows luxuriantly in the warm stove, if potted in light loam mixed with leaf mould. Being a dry, fibrous-rooted plant, it requires to be freely supplied with water. With a little attention to tying up and pruning, it may be made a compact, handsome bush. When in flower it is highly ornamental, its bright red tufts contrasting strongly with the delicate green foliage. It is readily increased by cuttings, which should be planted under a bell-glass and placed in bottom heat.—*Botanical Magazine*, t. 4500.

97. GALPHIMIA GLAUCA. *Cavanilles.* A Mexican hothouse Malpighiad, with handsome glaucous foliage, and an abundance of gay yellow blossoms. Introduced by the Horticultural Society. Flowers in the autumn and early winter.

A beautiful shrub, easily kept in the form of a bush. The leaves are a deep bluish green, ovate, obtuse, glaucous on the underside, and furnished with a pair of glands on the edge near the base. The flowers, which are golden yellow, appear in close terminal racemes, between 3 and 4 inches long in strong plants. Each has five distinct petals, with almost exactly the form of a trowel. Grows freely in a mixture of loam and sandy peat, and is easily increased by cuttings of the half-ripened young shoots. It requires to be kept rather dry for a few months, and afterwards, during the growing season, to be freely supplied with moisture both to the roots and in the atmosphere.—*Journ. Hort. Soc.*, Vol. v. p. 139. *With a figure.*

98. TERNSTRÖMIA SYLVATICA. *Chamisso and Schlechtendahl.* An evergreen Mexican greenhouse shrub, of no beauty, with greenish sweet-scented blossoms. Belongs to the Natural Order of Theads. Flowered at the Apothecaries' Garden, Chelsea, in February.

Not unlike a Sweet Bay, but more spreading. Leaves narrow, oblong, bluntly acuminate, deep green on the upper, very pale on the under side, perfectly smooth. The flowers grow singly on short curved stalks, and are quite hidden among the leaves. They are of the pale, dull, greenish purple of Magnolia fuscata, and quite destitute of beauty. When fresh gathered they have a very agreeable hawthorn-like scent.—*Journ. Hort. Soc.*, Vol, v. p. 141.

99. Dendrobium crepidatum. A beautiful species from the Indian Archipelago, with slender erect stems, and pink white and yellow flowers. Blossomed with R. S. Holford, Esq., March 1850. (Fig. 45.)

D. crepidatum ; caulibus teretibus erectis, foliis . . . , floribus geminis, sepalis petalisque oblongis obtusis firmis, labello oblongo integro subsinuato obtuso lateribus erectis intus levissimè pubescente basi utrinque plicato-venoso, cornu brevi obtuso.

Although we have an imperfect knowledge of this extremely pretty plant, we are able to state that it is perfectly distinct from all others. It first came to us in a letter from Mr. Bassett, Gardener to R. S. Holford, Esq., and afterwards the whole plant was transmitted by that gentleman's orders. The leaves however were absent, and the flowers much faded. It has slender erect stems ; and the flowers, which are white, tipped with delicate pink, but deep yellow in the middle of the lip, appear in pairs as in *D. Pierardi* and its allies. They have a very firm texture, more like that of a Lycaste than a Dendrobe, and are about as large as those of *D. aduncum*. The lip has a peculiar form very much like that of the old-fashioned slippers, which, without a hollow for the foot, were merely latched round the instep.

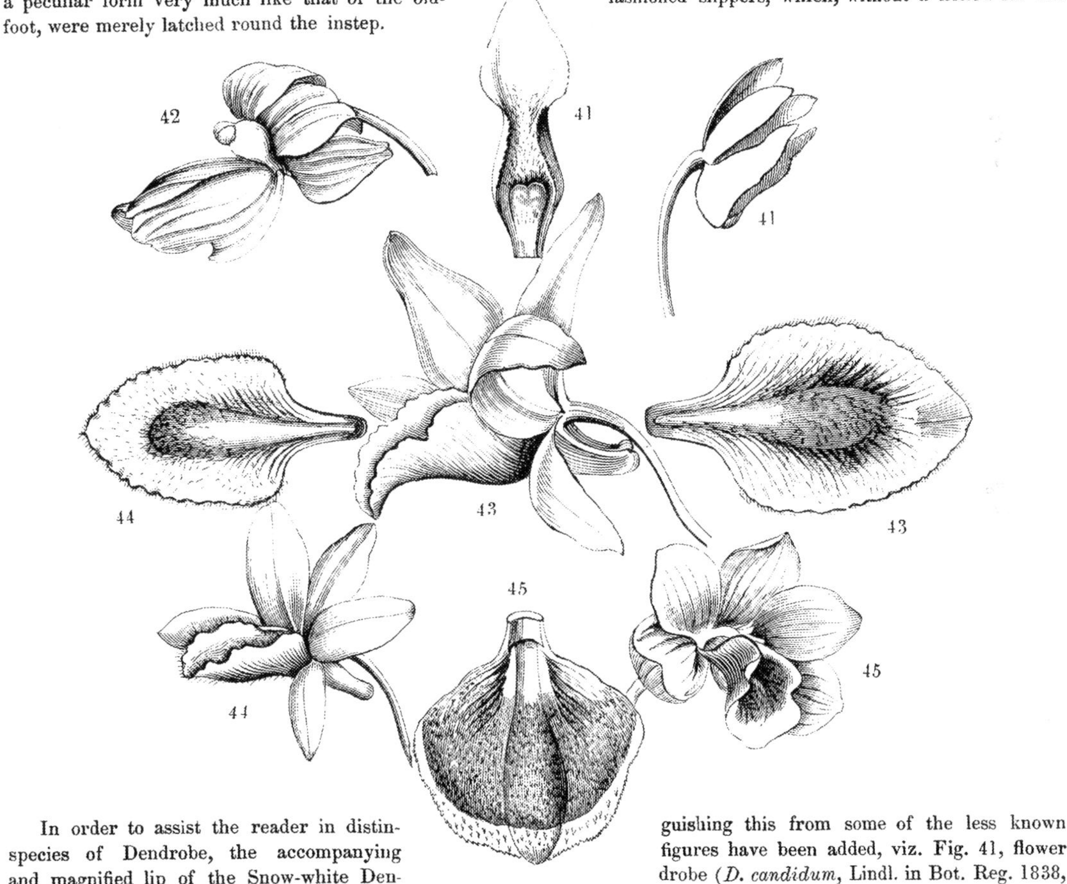

In order to assist the reader in distinguishing this from some of the less known species of Dendrobe, the accompanying figures have been added, viz. Fig. 41, flower and magnified lip of the Snow-white Dendrobe (*D. candidum*, Lindl. in Bot. Reg. 1838, misc. no. 54) ; Fig. 42, a flower of the Revolute Dendrobe (*D. revolutum*, Lindl. in Bot. Reg. 1840, misc. 110) ; Fig. 43, Flower and magnified lip of the Green-centred Dendrobe (*D. mesochlorum*, Lindl. in Bot. Reg. 1847, t. 36) ; Fig. 44, Flower and magnified lip of the Egerton Dendrobe, (*D. Egertoniæ*, Lindl. in Bot. Reg. 1847, t. 36.)

100. Brachysema aphyllum. *Hooker*. An ugly leguminous bush from Swan River, with winged leafless stems, and crimson flowers, not in cultivation but figured from New Holland materials. —*Bot. Mag.*, t. 4481.

101. Isoloma breviflora. (*aliàs* Gesnera breviflora, *Lindley ; aliàs* Gesneria Seemanni, *Hooker.*)

A fine hot-house Gesneraceous plant, with long whorled shaggy racemes of scarlet spotted flowers. Native of Panama. Blossoms in October at Kew.

A very handsome, copious-flowering, and bright coloured species, approaching nearest to *G. longifolia*, but differing much in the form of the leaves and in the limb of the corolla. It was discovered by Mr. Seeman, at Panama. Stem two feet or more high, simple, rather stout below, nearly terete, villous with spreading hairs, as is almost every part of the plant. Leaves opposite and ternate, the lower ones large, broadly ovate, or sub-obovate, on rather long petioles, coarsely serrate, acute, rather than acuminate ; upper ones gradually smaller and more tapering to a point, all obtuse at the base. From the whorls of the upper floral leaves, the hairy peduncles appear fasciculato-verticillate, longer than the petioles, and the uppermost ones longer even than the leaves, single-flowered. Calyx shallow, cup-shaped, with five nearly regular, acute, spreading lobes. Corolla very villous, bright brick red, a little inclined to orange. Tube nearly cylindrical, short, tapering, orange at the base ; the limb of five, nearly equal, rounded segments, spotted with deeper red, and clothed with glandular hairs. Ovary roundish ovate, very villous, having at the base four conspicuous, hypogynous, broad glands, of which one is bifid. The rhizome of Gesneraceous plants is either in the form of a thick, fleshy round tuber, or consists of a number of fleshy scales, compactly seated on an elongated axis, and, therefore, analogous to an underground surculose stem. The rhizome of this species belongs to the latter form, resembling that of *Gloxinia* and *Achimenes*, and requiring the same kind of treatment. It will thrive in a mixture of light loam and leaf mould ; and, in order to start the roots, they should be placed in bottom-heat in a warm stove, taking care not to give much water till they have made some progress in growth. If, during the summer, they happen to be placed in a position fully exposed to the south, they will require to be shaded during the middle of the day.—*Botanical Magazine*, t. 4504. The plant here spoken of under the name of *Gesneria Seemanni*, is only a well-grown specimen of the *Gesnera breviflora*, described in the *Journal of the Horticultural Society*, vol. iii., p. 165, (April, 1848.) It is one of the Isolomes which M. Decaisne has, with much reason, elevated to the rank of a genus, as had Regel, before him, under the name of *Kohleria*. Other Isolomes, are *G. longifolia*, *Bot. Reg.*, t. 40, 1842 ; *G. Hondensis*, *Bot. Mag.*, t. 4217 ; *G. trifolia*, *ib.*, t. 4342 ; *G. mollis* ; *G. lasiantha*, Zuccarini ; *G. tubiflora*, Cav. ; and, perhaps, *G. verticillata*, Cav. ; as M. Decaisne has pointed out in the *Revue Horticole*, 3rd. Ser., vol. ii., p. 465.

102. Clerodendron Bethuneanum. *Lowe.* A fine stove Verbenaceous shrub, with the appearance of *C. Kæmpferi*. Flowers crimson, in large panicles, produced in September, 1849, with Lucombe and Co. A native of Borneo.

Each flower of this plant is exceedingly beautiful in itself ; peduncles, pedicels, bracts, calyx, corolla, the very long and graceful stamens, all are of the deepest crimson, while the two side lobes of the corolla have a purple spot near the base, and the upper lobe has a much larger white spot. The species has been named after Capt. Bethune, R.N. who brought it and several other fine plants from Borneo. When its flowering season is past, it does not lose all its charms, for the crimson bracts and calyces remain, and the latter contain each a four-seeded berry of the richest blue colour. Although in its native country attaining a height of ten feet, it is one of those plants that flower readily when but of small size, and confined in a pot.—*Botanical Magazine*, t. 4485.

103. Tabernæmontana longiflora. *Bentham.* A stove shrub of the order of Dogbanes (Apocynaceæ) with long white fragrant flowers and a green tube. Blossomed with Lucombe and Co. A native of Sierra Leone.

The shrub has close-placed, ample dark green foliage, and remarkably large white or pale cream-coloured flowers, diffusing a delicious aromatic fragrance, resembling that of cloves. Dr. Vogel, who found the plant at Sierra Leone, speaks of the shrub as very handsome, with the aspect of a Citrus, and yielding a milky juice. Leaves elliptical, large, with a short point, and a short but dilated petiole, the veins diverging almost horizontally from the mid-rib. Peduncles erect, stout, each bearing about three large white flowers. Calyx lobes broadly oval, obtuse : at their base is a circle of minute glandular scales. Corolla with the tube twisted, 4 inches in length, swollen below the middle ; limb of five waved or reflexed ligulate lobes. This shrub requires a warm stove. It will thrive in a mixture of loam and peat soil, if placed so as to have the benefit of bottom-heat, and watered and syringed freely during the summer ; but care should be taken that at no time (especially during its season of rest) the mould becomes saturated, for the soft and slightly succulent roots are apt to suffer if kept in too wet a state, while the plant indicates a cessation of growth.—*Botanical Magazine*, t. 4484.

PLATE 13.

J. Constans, Pinx & Zinc.

Printed by C.F. Cheffins, London.

[PLATE 13.]

DOUBLE CHINESE PEACH TREES.

(AMYGDALUS PERSICA; FLORE SEMIPLENO.)

Hardy shrubs from CHINA, *with the habit of the Common Peach.*

THE Chinese and Japanese have long been known to possess several fine double varieties of the common Peach-tree. Such plants appear in their rude drawings, among their embroidery, and upon their paper hangings. Travellers talk of the exquisite beauty of these things when tortured into dwarfness. They are probably intended by Kæmpfer under the name of *Prunus flore rubro,* and *Prunus flore pleno,* of which last he says: "This is cultivated because of the beauty and abundance of its flowers. The older and more distorted or deformed it is, the more is it prized." Thunberg speaks also of a single white and a double red variety, adding that the Peach is cultivated everywhere in gardens, because of the *beauty of its flowers.*

Among the valuable and authentic Chinese drawings in the possession of the Horticultural Society, no doubt the finest collection in Europe, the following varieties may be readily distinguished:—

1. Large semi-double Crimson; with flowers as large as a Sasanqua Camellia; very handsome, petals acute.
2. Large semi-double Rose; like the first, but the colour not deeper than that of a China Rose.
3. Large semi-double Red; with flowers as large and deep red as No. 1, but with blunt petals, somewhat irregularly lobed.
4. Small semi-double Red; like the next, but of a deep rich rose colour; very pretty.
5. Small semi-double White, with very round petals, not much longer than the stamens.

When Mr. Fortune was sent to China by the Horticultural Society, he was particularly instructed to procure these things; and the result has been the acquisition of the two beautiful varieties now represented; namely, A SEMI-DOUBLE CRIMSON, which is probably the first of the foregoing list, and a SEMI-DOUBLE WHITE, which is not found there. These have now flowered in the Garden of the Society, and prove to be great acquisitions. They have, in all respects, the habit of the common Peach tree, except that they are more excitable, in which respect they approach the Almond; and consequently they are better suited for forcing or for flowering under glass, than in the open air; because, although hardy, they suffer from wet cold nights, which brown their flowers and ruin their gay appearance. It is not improbable, however, that seedlings may be in time produced from them in which this precociousness will disappear; for, being semi-double, it is to be expected that they will occasionally ripen fruit.

That semi-double Peaches will fruit has been pointed out by Monsieur Jacques, in the Journal of the Horticultural Society of Paris; and this writer adds the curious fact that the seedlings come true from seed. His experiment is thus detailed: "In the autumn of 1845 I put in sand twelve stones of double Peach trees, and I planted them in March, 1846. By the end of May five only came up, and by the end of the year were from 16 to 18 inches high. In the spring of the following year I pinched off some of the lower branches, and the plants continued to grow at the same rate. Political events in the beginning of 1848 prevented my transplanting them; they, therefore, went on growing in the seed-bed. In the course of that year they became a yard and half and two yards high, and were pretty well covered with branches from top to bottom. On the 5th of April, 1849, four out of these five plants were covered with flowers all along the branches, and at almost every bud; and the whole of the flowers appear to be the same as those of the common budded double Peach trees. Another interesting fact is, that this result had not to be waited for, for these shrubs were in full flower by the time they were three years old."

PLATE 14.

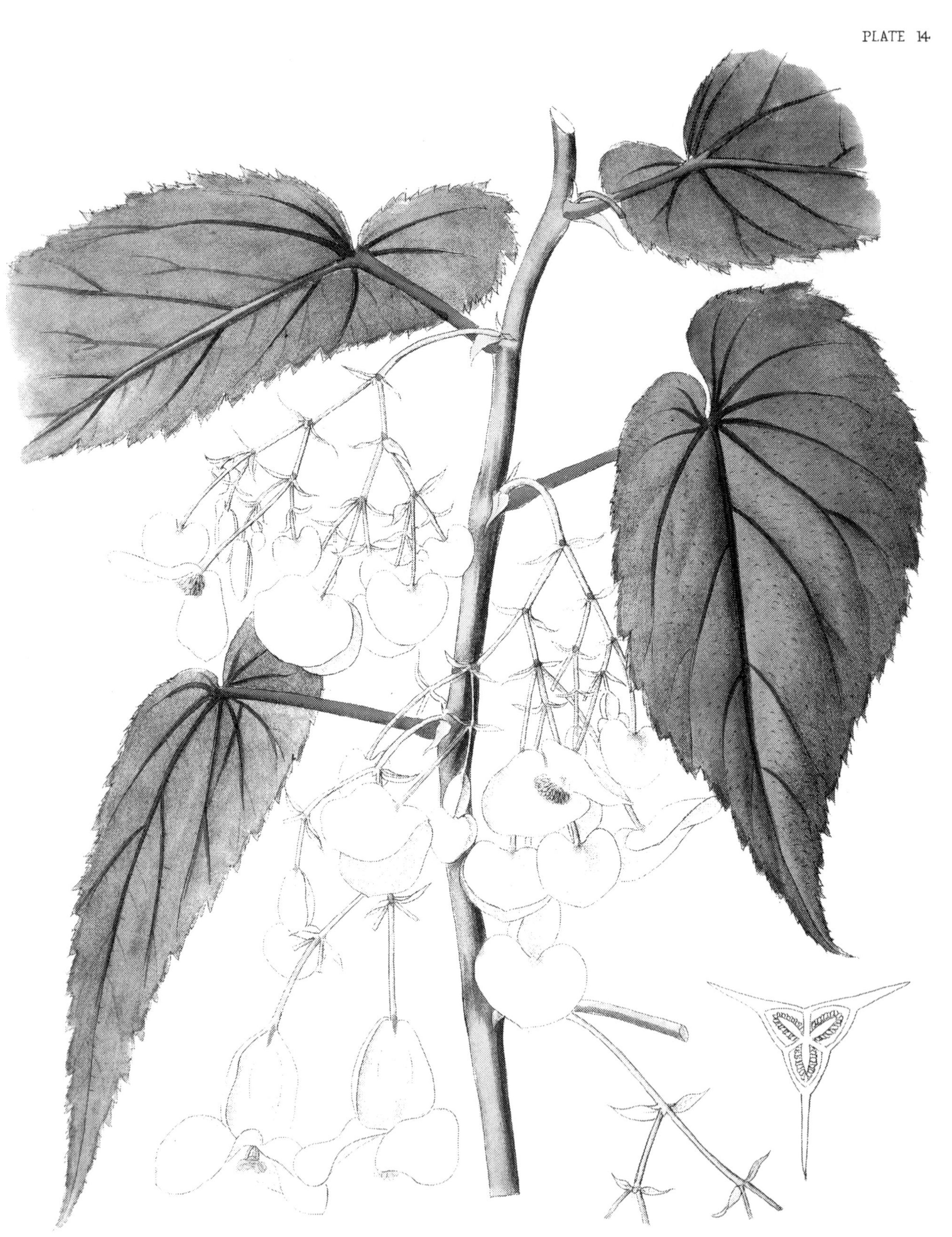

I. Constans, Pinx. & Zinc.

Printed by C.F. Cheffins, Lond

[PLATE 14.]

THE TWO-PETALLED BEGONIA.

(BEGONIA DIPETALA.)

A hothouse herbaceous plant from the EAST INDIES, *belonging to the order of* BEGONIADS.

Specific Character.

TWO-PETALLED BEGONIA.—Single-stemmed, erect. Stem and petioles quite smooth. Leaves obliquely cordate, acuminate, doubly serrated, ciliated with soft spines, hairy upon the upper side, nearly smooth on the under; not spotted. Flowers in loose few-flowered pendulous cymes. Petals 2, roundish, heart-shaped. Wings of fruit rounded, nearly equal.

BEGONIA *DIPETALA*.—Simplicicaulis, erecta, caule petiolisque glabris, foliis oblique cordatis acuminatis duplicato-spinuloso-serratis supra pilosis immaculatis subtus glabris, cymis paucifloris laxis pendulis, petalis 2 subrotundis cordatis, capsulæ alis rotundatis subæqualibus.

Begonia dipetala: *Graham in Botanical Magazine*, t. 2849. *Loddiges' Botanical Cabinet*, t. 1730.

THE genus Begonia is now taking in gardens the place which it deserves, for it is certainly one of the richest in brilliant colours, or variety of form; and in the hands of good managers it is one of the most easily cultivated of all known genera. The blossoms too appear for the most part during the winter months, and keep well when cut for the decoration of sitting rooms.

But it must be confessed that among the many species now in cultivation, a large number are very incorrectly named, so that the whole business of arranging the genus, and reducing it to order, has still to be undertaken. The first step to a proper arrangement is the determination of what really constitutes a Begonia, for the genus has now become almost as full of diverse forms as the old Linnæan Orchis. As a first step to this it appears necessary to take into account the placentation, limiting the name Begonia to those, which, like that before us, have simple placentæ, and putting aside those with double placentæ under the separate genus Diploclinium. (*See Vegetable Kingdom*, p. 319.)

Among the species which will have to be referred to Begonia proper, if it is thus limited, is the present, which, although long since introduced, is by no means so well known, or so well figured as it deserves to be. The original figure in the Botanical Magazine does it little justice, and represents its leaves as being covered with the grey blotches which are so striking in some other species. Loddiges says that these blotches come only in young plants, and disappear on the old ones. But we have never been able to find them at all in the two-petalled Begonia; on the contrary, the foliage has always that peculiar even tint represented in the accompanying plate. In fact the leaves are very nearly the same as in the pimpled Begonia (*B. papillosa*: Graham in Bot. Mag. t. 2846*), which differs in little except the leaves having shorter and hairy petioles, and in there being four petals instead of two.

The native country of this species is said to be Bombay; but we have seen no wild specimens of the plant. In gardens it flowers all the year round, and must be regarded as one of the most delicate and beautiful.

The spots on the leaves of some Begonias, and which have been said to exist here also, are caused by the presence of a stratum of air beneath the epiderm or skin; where the spots are missing, the green cells of the parenchym grow to the ends of those of the epiderm, no air intervening. When examined with the microscope the cells of the colourless skin look exactly like empty honeycomb placed on the surface of the leaf, while that part of the skin which is green has no such appearance. If the spotted leaves are boiled, the spots swell up by the distension of the air beneath them, and then look exactly like brown blisters, the green being changed to brown by the act of boiling. This would therefore seem to be an organic peculiarity of a very different degree of importance from mere peculiarity of colour, and one not likely to disappear. The history of the structure and its use is unknown. It is remarkable that it occurs only on the upper side of the leaves of Begonias, where there are no breathing pores (stomates), and never on the under side, whose stomates are large, active and abundant.

* This is a very different plant from *B. papillosa* of the Botanical Register.

J. Constans, Pinx & Zinc

Printed by C.F. Cheffins, London

[PLATE 15.]

THE CERVANTES ODONTOGLOT.

(ODONTOGLOSSUM CERVANTESII.)

A Greenhouse Orchid, from MEXICO.

Specific Character.

THE CERVANTES ODONTOGLOT.—Pseudobulbs ovate, angular. Leaves solitary, oblong, narrowed into a channelled footstalk. Scape few-flowered. Bracts and sheaths membranous, acute, equitant, long. Sepals membranous, oblong-lanceolate, acute. Petals broader, somewhat unguiculate. Lip slightly cordate, ovate, acute, with a fleshy, cup-shaped, downy stalk, having in front a double tooth, and in advance of that a pair of long hairy processes. Column downy, with rounded ears.

ODONTOGLOSSUM *CERVANTESII;* pseudobulbis ovatis angulatis, foliis solitariis oblongis in petiolum canaliculatum angustatis, scapo paucifloro, bracteis vaginisque membranaceis acutissimis equitantibus elongatis, sepalis membranaceis oblongo-lanceolatis acutis, petalis latioribus subunguiculatis acutis, labello subcordato-ovato acuto unguiculato, ungue carnoso cyathiformi pubescente antice bidentato medio tuberculato processubus 2 elongatis pilosis ante cyathum, columnæ pubescentis auriculis rotundatis.

Odontoglossum Cervantesii, *La Llave and Lexarza, Orch. Mex.* 2, 34 ; *Botanical Register*, 1845, t. 36.

THERE is probably not a group of Orchids the species of which are more generally beautiful than the white-lipped Odontoglots, of which this is one. They all agree in having the same habit, the same large, semi-transparent flowers, the same long membranous bracts, and the same delicacy of tint, varied by blotches of deep purple, or brown, or cinnamon.

Of these one of the rarest is the subject of the present plate, of which we received a specimen from Mr. Loddiges in the spring of this year. Its natural locality is among the mountains in the west of Mexico, whence we believe it was first brought by the late Mr. Barker's collector. In general it has a pale tinge of pink; when wild it is said to be snow-white; but in the state now represented it had gained a very distinct rose-colour, which greatly augmented its beauty.

In many respects it is nearly related to the membranous Odontoglot (*O. membranaceum*), from which it differs in the following particulars: its flowers are more pink, and rather smaller, and the lip is by no means spotted at the base; its petals are much more acute; its lip is very slightly heart-shaped, and quite acute at the point; the two front teeth of the lip are very much longer and more hairy; and the concavity at the base of the lip has a much larger central tubercle.

In addition to those two species the gardens now contain the following, which approach them very nearly, and constitute the nucleus of the white-lipped group, viz.:—

O. maxillare. Flowers white; the base of the sepals, petals, and lip equally stained with crimson, and a very large yellow appendage.

O. rubescens. Flowers lilac; the sepals narrow, and spotted with crimson all over; the petals broad, and a little spotted near the base; the lip with no spots at all.

O. Rossii (aliàs *O. Ehrenbergii;* aliàs *O. acuminatum*). Flowers not half as large as the last; sepals green, spotted with crimson; petals and lip pure white, the former only spotted with crimson at the base.

O. stellatum. Flowers much smaller than in the last; both sepals and petals green and spotted; lip lilac in the middle, white at the edge, and strongly toothed.

There are also some other species of the groups still to introduce from the west of Mexico, which are even finer than those now enumerated.

It does not much signify in what kind of material this is grown, provided only that it be of such a nature as to detain damp, while water passes off freely and air replaces it. Fibrous peat and decayed leaves are among the best substances; the management of such plants is more important. On this head Mr. Gordon's directions are among the best we have.

"Injury is often effected by a sudden rise of temperature by fire-heat in winter, while little or none is caused if the rise is occasioned by sun-heat; care should therefore be taken to guard against a rise of temperature by fire-heat, particularly in midwinter; rather suffer a depression of a few degrees of heat in very severe weather than use over-strong fires, which will over-dry the atmosphere, and, on the other hand, create too much moisture if water is supplied. Moisture, however, is by no means injurious to Orchids, provided they can part with it freely, but they are impatient of stagnant damp.

"When in a dormant state they should receive no more moisture than is sufficient to prevent their leaves from shrivelling; hence many of the more tender kinds do much better on blocks of wood suspended from the roof, where they can part with the superabundant moisture freely, than in pots. Nature herself indeed sets us an example to follow in regard to moisture, for we find, where the atmosphere is saturated with moisture (and a truly moist atmosphere cannot exist without a corresponding amount of heat), that the Orchids climb the loftiest trees; but, as the climate becomes drier, so they descend, until at last they are to be found growing upon the surface of the ground or upon rocks in shady places.

GLEANINGS AND ORIGINAL MEMORANDA.

104. Roupellia grata. *Hooker*. A hothouse climbing plant from Sierra Leone, with large coarse white flowers. Belongs to the Dogbanes (*Apocynaceæ*). Introduced by Mr. Whitfield. (Fig. 46).

This plant produces what is called "Cream-fruit" in Sierra Leone; a name that has probably arisen from its yielding an abundance of cream-like juice when wounded. We should, however, be unwilling to put such a dainty in the mouth; for it can hardly be destitute of the acridity for which its race is notorious. In the *Botanical Magazine* it

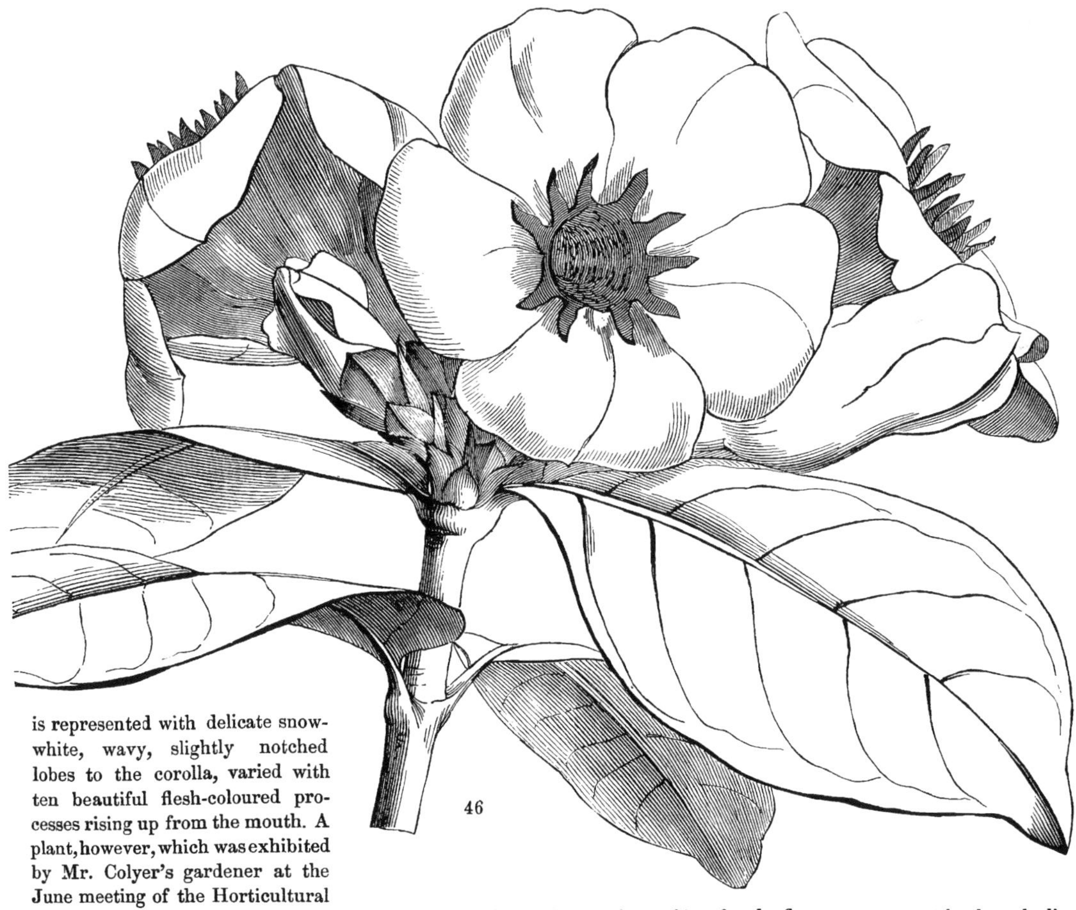
46

is represented with delicate snow-white, wavy, slightly notched lobes to the corolla, varied with ten beautiful flesh-coloured processes rising up from the mouth. A plant, however, which was exhibited by Mr. Colyer's gardener at the June meeting of the Horticultural Society by no means justified the flattering account that had been given of it; for the flowers were great leathery bodies, not white, but dirty, like half-soiled kid gloves; while the delicate flesh-coloured teeth proved to be ten huge, ugly, brown tusks. It is difficult to imagine a flower with a more uninviting appearance. As to the fragrance attributed to it, we perceived nothing more than a sickly or at least by no means agreeable odour. When compared with a Stephanote, or a Beaumontia, it shrinks into insignificance, notwithstanding the large size of the flowers. The following account of its habits is given in the *Botanical Magazine*, t. 4466 :—"This handsome, climbing, shrubby plant, requires to be grown in a warm and moist hothouse. It is of free growth, and being a smooth clean-leaved plant, not subject to insects, is well adapted for a trellis, or to train up a pillar or rafter ; and it will also form a bushy plant grown in a pot, if supported by a wire trellis, or by neat stakes. Good fresh loam with a little leaf mould will suit it. As it is a fast grower, it requires water freely during summer ; but care must be taken that the soil does not become stagnant. It is propagated by cuttings, which strike root readily when placed under a bell-glass, and the pot plunged in bottom heat. It appears to be a shy flowerer ; for although we have known it in cultivation for several years, we have not heard of its producing flowers, except in the collection above mentioned."

105. PENTSTEMON AZUREUS. *Bentham.* A hardy herbaceous plant from California. Flowers bright blue, very handsome. Belongs to the order of Linariads (*Scrophulariaceæ*). Introduced by the Horticultural Society.

A smooth, glaucous, erect perennial, about 2 feet high. Leaves linear-lanceolate, quite entire upon the stem, but near the root oblong and slightly heart-shaped at the base. Flowering racemes about a foot long or rather less, slightly downy, with one short peduncle in the axil of each opposite bract, bearing from 1 to 2 flowers. The latter are rather more than an inch long, clear violet blue, much deeper in the limb than on the tube. This hardy perennial is stated by Mr. Bentham to have been gathered in the dry river beds of the Valley of the Sacramento. Hartweg wrote on his seed papers that it was a mountain plant. It is very handsome as a border flower, but as its narrow foliage is not good, it is best grown among other species, such as Pelargoniums, &c.—*Journ. Hort. Soc.*

106. BEGONIA CINNABARINA. *Hooker.* A very handsome Bolivian greenhouse (?) plant with large nodding scarlet flowers. Introduced by Messrs. Henderson of Pine-Apple Place.

Extremely handsome ; the contrast between the green stem and darker green leaves, with the deep bright red of the long and stout peduncles and stipules, together with the red or rather deep large cinnabar-coloured flowers, is very striking, and renders this the most desirable of all the species for cultivation : add to which, it blooms very freely in an ordinary stove (I suspect it would do so in a greenhouse) and continues long in flower. Stem erect but zigzag, stout, succulent, pale green, slightly downy, as are the leaves and petioles. Leaves on rather short, stout, terete, green petioles, from four to six or seven inches long, obliquely ovate, (the young ones much plaited and edged with red,) lobed at the margin and doubly serrated, the minute teeth red. Stipules ovate, membranaceous, acuminate, red. Peduncles a span and more long, rather stout, terete, deep and bright red, bearing a panicle of six large handsome flowers, which as well as the ovaries and pedicels and ovate bracts are rather pale red or deep cinnamon colour. The ultimate pedicels are ternate, drooping, of which the central flower is male, the lateral ones female.—*Botanical Magazine*, t. 4483.

The interior of the ovary not being described we are uncertain whether this is a true Begonia or not.

107. UROPEDIUM LINDENII. *Lindley.* An extraordinary herbaceous orchid, with all the habit of the long-tailed Lady's-slipper. Native of New Grenada. Introduced by Linden. Flowered in May with M. Pescatore.

This, which is the most remarkable of the terrestrial orchids yet known, is thus described in the *Orchidaceæ Lindenianæ:* This singular and magnificent plant grows on the ground in the little woods of the Savannah, in that elevated part of the Cordillera which overlooks the vast forests at the bottom of the Lake of Maracaybo, and situated on the territory of the Indians of Chiguará, at the height of 8500 feet. Sepals oval-lanceolate, pale yellow, streaked with orange. Petals purple, orange at the base. The flower may be from fifteen to twenty inches long in its greatest diameter. Leaves thick and fleshy ; June 1843. The habit of this curious plant is exactly that of Cypripedium insigne. The leaves are a foot long, blunt, unequally two-toothed at the point, shining, spotless, and longer than the downy scape. The bracts are two, of which the exterior is spathaceous, compressed, blunt, coriaceous, and much longer than the inner. The peduncle is six inches long, downy and one-flowered. The upper sepal is ovate-lanceolate, and four inches long ; the lower are united into one of the same form, but rather wider. The petals are linear-lanceolate, extended into a long, narrow tail, and are probably eight or nine inches long, but in my specimens they are broken. The lip is of exactly the same form, but broader, and like the sepals is shaggy at the base.

We learn from Mons. Pescatore that it has now produced two flowers with him, in his great collection at the Château of Celle St. Cloud, near Paris. The sepals are white streaked with green, and more than $3\frac{1}{4}$ inches long ; the petals and lip full 21 inches long, very velvety at the base, white streaked with green ; the tails have the colour of wine lees.

108. WARREA BIDENTATA. *Lindley; (aliàs* W. Lindeniana, *Henfrey*). A handsome terrestrial Orchid from New Grenada, with the habit of Warrea tricolor. Flowers pale cream-colour, with a purple lip. Introduced by Mr. Rucker before 1844. (Fig 47, the lip magnified.)

This well-marked species was originally described in the *Botanical Begister* for 1844, at p. 76 of the miscellaneous matter. It has lately been reproduced in the *Gardener's Magazine of Botany*, p. 177, under the new name of *W. Lindeniana.* It is not a native of Peru, as is stated in that work, but was found by Mr. Linden "on the ground in the thick forests at the foot of the peak of Tolima, at the height of 4000 feet," as is stated in the *Orchidaceæ Lindenianæ*, No. 96. It is said to have some pink in its flowers when wild, but that colour has not been yet observed in cultivation. The form of the lip, which is remarkable, is shown in the annexed cut.

109. WARREA WAILESIANA. *Lindley.* A one-flowered Orchid, with little beauty. Flower cream-coloured, with a violet lip. Native of Brazil. Introduced by George Wailes, Esq. (Fig. 48, the lip magnified.)

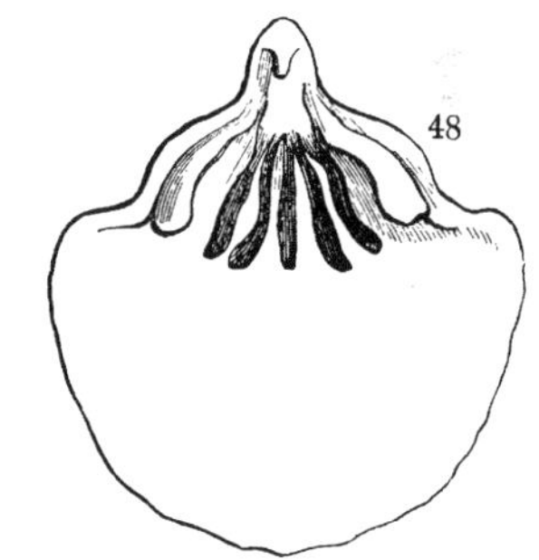

A fresh flower of this pretty species has been sent me from George Wailes, Esq., of Newcastle-on-Tyne, who received it from the late Mr. Gardner, it having been found by that lamented botanist in an excursion to the river Parahyba in search of *Huntleya Meleagris.* It appears, like that plant, to have a one-flowered scape, and is not a species of much beauty. The flowers, which smell of sweet peas, are cream-coloured, and about as large as those of *Warrea cyanea.* The sepals are all somewhat reflexed, the lateral not being straighter than the rest; the petals are also bent back, so that no arch can be formed over the column. The lip is tinted with delicate violet along the middle, is roundish, concave, wedge-shaped at the base, not at all lobed, but so turned upwards at the edges as to look as if it was furnished with basal auricles. Its appendage consists of five slender radiating violet fingers, which are perfectly free from the lip except at their origin; at the sides the edge of the lip is also furnished with a thin, linear, inflexed membrane. The column and pollen masses are those of *W. discolor.—Journ. Hort. Soc.*, vol. iv.

110. WARREA DISCOLOR. *Lindley.* A one-flowered Orchid from Costa Rica: sepals and petals pale lemon-colour, tinged with purple; lip dull purple. Introduced by Mr. Warcsiewitz. (Fig. 49, the lip magnified.)

A very distinct species, apparently one-flowered, the leaves, &c. of which I have not seen. [Mr. Bassett, the gardener to Mr. Holford, states that the habit is that of *Huntleya violacea*, the leaves, however, being only about 5 inches long and 1 inch wide.] The sepals, which are $1\frac{1}{4}$ inch long, are straw-coloured, the lower straight, concave and deflexed, the upper erect, rolled back at the point, pressed close to the petals, and with them forming an arch over the column and lip. The petals are straw-coloured at the base, dull purple at the upper part. The lip has a nearly circular outline, but is so concave as not to present that form until flattened; it is slightly 3-lobed, of a deep, dull, velvety purple colour, with, at the base, a roundish oblong yellow appendage, which adheres to the lip, and is divided at the edge into strong diverging teeth, five of which terminate so many distinct ribs. The column is yellow, shaggy in front, with an anther sloping forward, and a subulate rostel. The pollen masses are four, plano-convex, in pairs at the end of a broad, flat, thin caudicle, furnished on either side with a lateral tooth. (A singular monstrosity here occurred in the two posterior pollen masses, which had grown together into one by a narrow neck.) A remarkable species, the single flowers of which resemble a Lycaste, but their pollen-apparatus and lip-appendage are exactly those of Warrea. Upon this point it may be useful to explain that in Lycaste the caudicle is subulate, and the lip-appendage a truncate plate near the middle lobe of the lip, while in Warrea the caudicle is broad and flat, and the lip-appendage ribbed, fringed, and stationed at the very base of the lip.—*Journ. Hort. Soc.*, vol. iv.

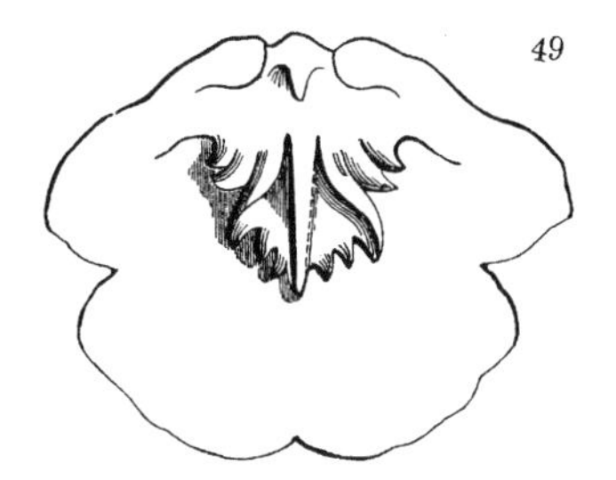

111. CEANOTHUS PAPILLOSUS. *Torrey and Gray*. A hardy Californian bush, with bright blue flowers, belonging to the order of Rhamnads. Flowers in June and July. (Fig. 50.)

An evergreen bush, covered with coarse hair and resinous tubercles, in a wild state forming a compact mass of branches, in cultivation growing longer and weaker. Leaves small, deep green, narrow-oblong, obtuse, with a single mid-rib, and numerous lateral veins, covered with down on the under side. Flowers in small roundish terminal stalked heads, bright blue as in C. azureus.—*Journ. Hort. Soc.*

This has now been ascertained to be capable of bearing our London winters without protection. But in places exposed to the sun it suffers from frost much more than under a north wall or at the back of rock-work. Very pretty.

112. CEANOTHUS RIGIDUS. *Nuttall*. A hardy evergreen purple-flowered Californian bush, belonging to the Natural Order of Rhamnads. Introduced by the Horticultural Society. (Fig. 51.)

A stiff branching dark green evergreen bush ; said to grow 4 feet high when wild. Young branches downy. Leaves small, truncate, spiny-toothed, subsessile, very shining and smooth on the upper side ; on the under pale and netted. This network is produced by numerous short branching veins, in the interspaces between which are deep pits, reaching half through the parenchym, and each closed up by a dense ring of white converging hairs. Such pits are placed pretty generally in a double row between each of the principal lateral veins. The flowers appear in small clusters or umbels at the end of very short spurs. They are deep purplish violet, not blue, and less showy than those of C. dentatus or C. papillosus. The species seems to be even more hardy than the two last-named sorts, for it has borne the winter uninjured and unprotected both in sunny and in northern aspects ; and, in fact, the specimens left unprotected are quite as healthy as those left under glass all the winter. The only blossoms that have yet appeared were in a greenhouse. It seems as if, in the open air, the shrub would prove an autumnal flowerer.—*Journ. Hort. Soc.*, vol. v.

113. DIPTERACANTHUS SPECTABILIS. *Hooker*. A very fine herbaceous Acanthad from Peru, with deep purple blue flowers of large size. It requires a warm green-house, or stove. Flowers in August. Introduced by Messrs. Veitch and Son. (Fig. 52.)

Sir W. Hooker states this to be unquestionably the largest flowered plant of the genus, if not of the order. It grows 2 feet or more high, much branched, and erect. Leaves nearly sessile, ovate, acuminate, ciliated, slightly pubescent on the surface, rather strongly veined and reticulated. Flowers sessile or very nearly so, two together from the axils of the upper leaves, large, very showy ; more than two inches across. Calyx quite without bracts, deeply cut into 5 erect, subulate lobes, much shorter than the funnel-shaped curved tube of the corolla. The limb of the latter very large, purple-blue, veined, the 5 lobes rounded, spreading, crenate, and somewhat waved at the margins. This is found to succeed in a temperature intermediate between that of the stove and greenhouse, and grows freely in any kind of light garden soil. Like many of the tropical *Acanthads*, after flowering, it becomes thin and naked. It propagates freely by cuttings. The young plants should be kept in small pots during winter, and receive very little water. In the spring they require to be shifted into a large pot, where they will soon make rapid progress, and produce a succession of large fine blue flowers.—*Botanical Magazine*, t. 4494.

114. THIEBAUDIA SCABRIUSCULA. *Humboldt and Bonpland.* A greenhouse evergreen bush, belonging to the order of Cranberries (*Vacciniaceæ*). Native of New Grenada. Flowers crimson, tipped with green. Flowered at Syon in April. (Fig. 53.)

A very pretty spreading evergreen shrub with slender downy branches, and broad oblong almost cordate triple or quintuple ribbed leaves, slightly downy on the under side. The flowers appear at the ends of the branches, in drooping

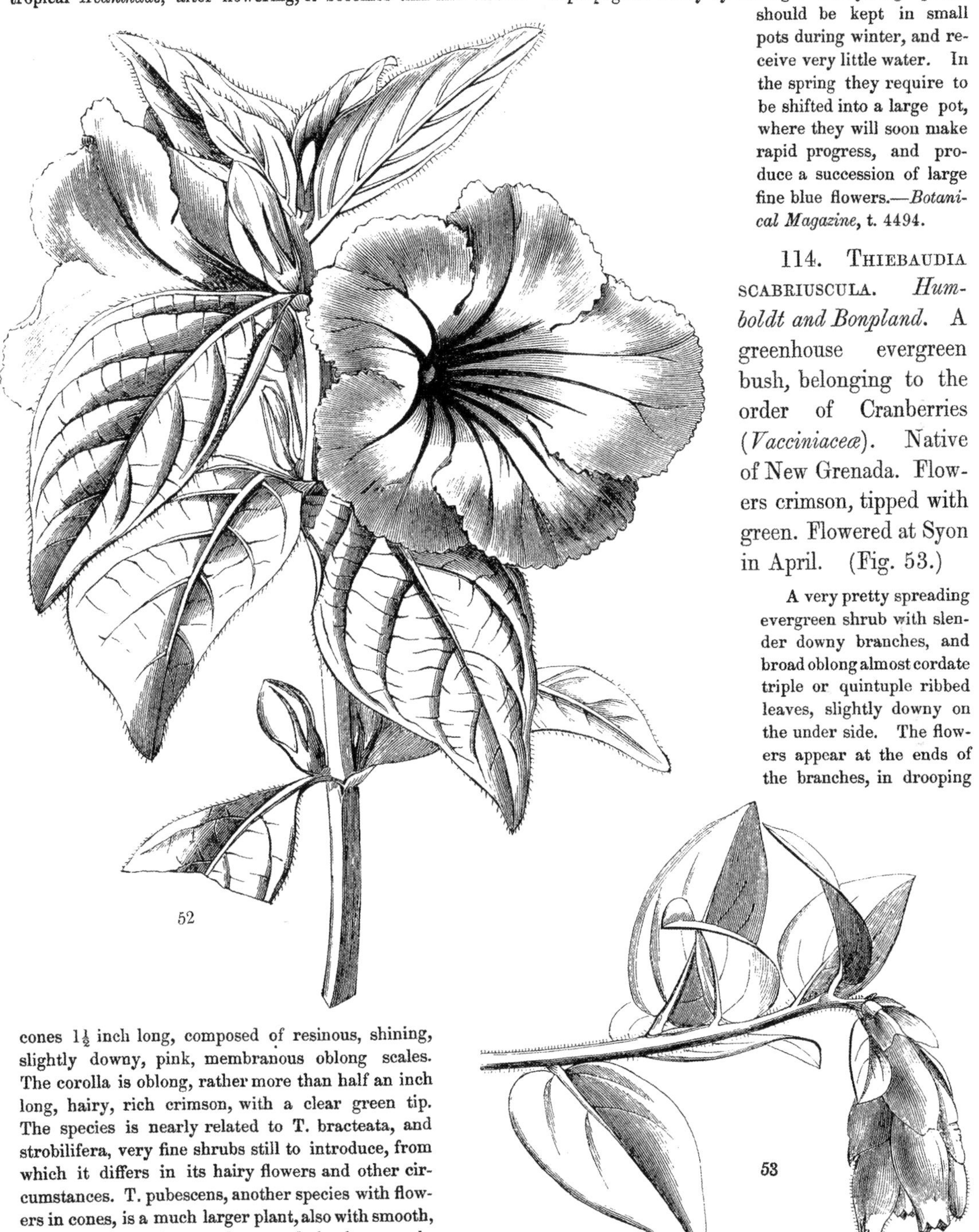

52

53

cones 1½ inch long, composed of resinous, shining, slightly downy, pink, membranous oblong scales. The corolla is oblong, rather more than half an inch long, hairy, rich crimson, with a clear green tip. The species is nearly related to T. bracteata, and strobilifera, very fine shrubs still to introduce, from which it differs in its hairy flowers and other circumstances. T. pubescens, another species with flowers in cones, is a much larger plant, also with smooth, not downy, corollas; at least such is the case in specimens now before us from Hartweg's Collections. This should form a very useful gay addition to spring shrubs of its class. It was raised at Syon from seeds received from Mr. Purdie.

115. Gynoxys fragrans. *Hooker.* A hothouse perennial plant, from Guatemala, with very fragrant yellow flowers, appearing in December. Stems trailing. Belongs to Composites. Introduced by Mr. Skinner. (Fig. 54.)

Stems long, climbing, perennial, with succulent branches, showing a disposition to root at their base. Leaves rather distant, on long petioles, ovate or approaching to lanceolate, acute, of a rather fleshy texture, dark green. The flower-heads are rather large, very fragrant, and form a terminal, and in the lower part leafy, corymbose raceme. A coarse soft-wooded scandent plant, having a large, thick, fleshy root, of the nature of a tuber. It grows freely in a mixture of light loam and peat or leaf-mould, and, by its rapid growth and clean habit, is well adapted for covering trellis-work in the hothouse, especially as it is not liable to be attacked by insects. It increases readily by cuttings; but these, on account of their soft, succulent, nature, must not be kept too close, or they will damp off before they produce roots.—*Botanical Magazine*, t. 4511.

116. Hoya coriacea. *Blume.* A Java climbing shrub, with the habit of Hoya carnosa, and umbels of yellowish flowers. A stove plant, flowering in August. Introduced by Messrs. Veitch and Co. (Fig. 55.)

Discovered by Dr. Blume in mountain woods on the western side of Java. Mr. Thomas Lobb detected it in the same island, on Mount Salak. Everywhere glabrous. Stem branched, twining, taper. Leaves on short thick petioles, which are glandular above at the setting on of the blade, which latter is almost exactly elliptical, or approaching to ovate, acute, between coriaceous and fleshy, acute or shortly acuminated, ribbed, with rather indistinct veins. Peduncles longer than the leaf, pendent, bearing a large umbel of numerous flowers, brown in the state of the bud, much paler when fully expanded. Pedicels very obscurely villous. Sepals subulate, much shorter than the corolla, which is glabrous and glossy externally, within pale tawny, and downy. The lobes triangular, acute. Coronet white, with a dark brown eye: leaflets ovate, gibbous at the base, obtuse, the apex a little curved down.—*Botanical Magazine*, t. 4518.

117. Hoya purpureo-fusca. *Hooker.* A remarkable twining stove plant, with small umbels of richly tinted purple and grey flowers. A native of Java. Flowers in September. Introduced by Messrs. Veitch and Son. (Fig. 56.)

Said to be common in the woods of Java. Sir W. Hooker compares it with the Cinnamon-leaved Hoya, and with the great-leaved (*H. macrophylla*) "but in the latter the leaf is reticulated between the nerves, the staminal crown (coronet) has the leaflets much more acuminated, and the colour of the flowers is quite different." It is a glabrous twining and branching shrub, everywhere (except the corolla) glabrous. Branches often throwing out short fibrous roots. Leaves on very thick brownish petioles, 4 to 5 inches long, exactly ovate, acute, or shortly acuminate, thick, fleshy, 5-nerved, the nerves all diverging from the base, and having a gland at the base where set on to the petiole. Peduncles axillary, shorter than the leaf, occasionally rooting, and bearing a dense many-flowered umbel. Corolla rotate, ashy-brown, downy and hirsute above, cut into 5 roundish and shortly acuminated lobes. Coronet of 5 ovate, fleshy, rich purple-brown, acute leaflets, nearly plane at the top, convex below.—*Botanical Magazine*, t. 4520.

118. Aotus cordifolius. *Bentham.* (*aliàs* Gastrolobium Hugelii *Henfrey.*) A pretty green-house leguminous shrub from Swan River, with glaucous heart-shaped leaves in threes, and large yellow axillary flowers. Introduced by Messrs. Knight and Perry.

This well-known plant, long ago published by Mr. Bentham under the name here quoted, is reproduced as a novelty in the *Gardeners' Magazine of Botany*. It is rather a nice plant, but its grey leaves are a disadvantage, and its yellow flowers are too much like those of a Genista. It must rank with Pultenæas and plants of that kind, and requires the same sort of management; that is to say, it wants to be potted in loose turfy soil, more loamy than peaty, to be grown in a brisk heat, with plenty of water applied with a syringe, in order to keep the air damp, and then when the growth is completed to be carefully hardened off. If they grow over fast the shoots will bear to be stopped; but not till the lengthening process is at an end.

119. Tropæolum Beuthii. *Klotzsch.* A tuberous climbing herbaceous plant from Bolivia. Flowers yellow. Introduced by Messrs. Low and Co.

Found by Bridges in Bolivia. Leaves deeply cut, peltate, roundish, bright green above, pale green beneath; leaflets 5-6 obovate; divisions of the calyx elliptical, apiculate, as long as the straight spur; petals obcordate, twice as long as the calyx. Near Tropæolum brachyceras.—*Allgem. Gart. Zeit.*, No. 21, 1850.

120. Centradenia floribunda. *Planchon.* (*aliàs* Donkelaaria floribunda of *Gardens.*) A dwarf half-shrubby plant from Guatemala, belonging to the Melastomads; with numerous lilac flowers. Introduced by Van Houtte and Co.

A very pretty species, much more worth growing than the Rosy Centradene, now common in gardens. The leaves are long, deep green, delicately tinged with violet on the underside, and full $2\frac{1}{2}$ inches long; having a pendent position they present both surfaces to the eye. The flowers are produced in much abundance, exhibit various tints of lilac, and produce a charming effect.—*Flore des Serres*, No. 453.

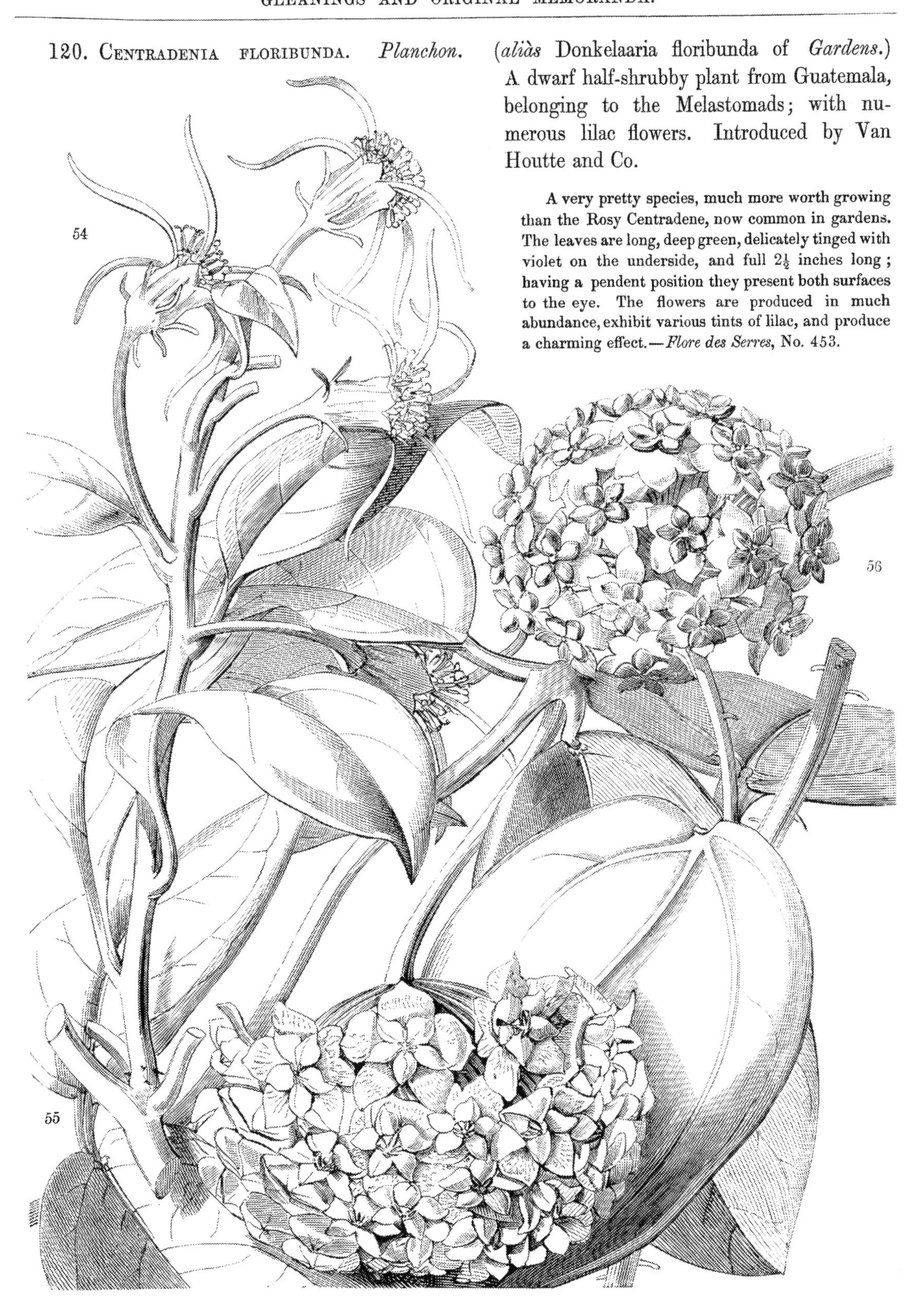

121. ACHIMENES GHIESBREGHTII *of the Gardens.* Origin unknown. A stove herbaceous plant with handsome scarlet flowers. Belongs to the Gesnerads. Introduced by Mr. A. Henderson.

Stems erect, deep purple brown, with a few scattered hairs. Leaves opposite, stalked, oblong-lanceolate, rugose, convex, coarsely serrated, not unlike those of the larger stinging-nettle. Flowers solitary, axillary, with a slender hairy peduncle, twice as long as the leafstalks. Calyx smooth, equally 5-parted. Corolla deflexed, nearly cylindrical, gibbous at the base on the upper side, 1½ inch long, bright scarlet, with an oblique regular limb, and a circular throat. Disk, a lobed fleshy ring. Stigma large, two-lobed, very hairy. This is a neat, distinct, and rather slender kind, requiring the same treatment as the old A. coccinea, and easily increased by the small scaly rhizomes. It grows about 8 or 10 inches in height, and flowers from June to August. It is very handsome.—*Journ. Hort. Soc.*, vol. v. *With a figure.*

122. ONCIDIUM NIGRATUM. An orchid from Guiana, with cream-coloured flowers spotted with blackish-brown, arranged in a branched panicle. Introduced by Mr. Loddiges.

O. nigratum (BASILATA) paniculâ ramosâ, sepalis linearilanceolatis undulatis acutis æqualibus, labello triangulari postice rotundato apice angustato acuto, cristâ multituberculatâ, columnæ alis angustis subdentatis basi productis.

A very curious and distinct species, received from Sir Robert Schomburgk many years since, and at last flowered by Mr. Loddiges. It is nearly allied to O. phymatochilum. The blossoms grow in branched panicles, and are about as large as those of O. incurvum. The colour of the sepals and petals is pale yellow or cream colour, with a few irregular brownish black blotches. The lip is brighter yellow, with a brown stain or two below the point.

123. ONCIDIUM PHYMATOCHILUM. A beautiful orchid, supposed to be derived from Mexico, with long green sepals and a white lip. Flowers in April.

O. *phymatochilum* (BASILATA) racemo subpaniculato, sepalis linearibus acuminatis apice recurvis lateralibus longissimis, labelli auriculis convexis dilatatis crenatis lobo intermedio unguiculato ovato acuminato basi multituberculato, columnæ alis semicordatis acuminatis.

Under this name is now not uncommon in gardens a charming orchid, supposed to have been obtained from Mexico, with erect, narrow, somewhat panicled racemes of greenish flowers having a snow-white lip. Three years since we received it from Messrs. Loddiges and the late Mr. Clowes. It has oblong, 2 edged, not furrowed, olive green pseudobulbs slightly tinged with purple, and surrounded by scales as long as themselves, which, when young, are olive green spotted with crimson. The leaves are of thin texture and vary in form from linear-lanceolate to oblong. The flowers are remarkable for the great extension of the lateral sepals, on which account, and because of their green colour spotted with chocolate brown, they have much the appearance of belonging to some Brassia. The lip is pure white, with yellow tubercles and a few stains of the same colour near the base.

124. CUPHEA IGNEA. *Alphonse De Candolle.* (*aliàs* C. platycentra *of Gardens.*) A Mexican perennial, with long scarlet flowers.

It is stated in the *Flore des Serres* that the true Broad-spurred Cuphea (*platycentra*, Bentham) is not the plant known under that name in Gardens; and consequently M. Alphonse De Candolle has given the latter the appropriate name of the Fiery Cuphea (*C. ignea*).

125. AUDIBERTIA POLYSTACHYA. *Bentham.* A half-hardy herbaceous plant from California, with white leaves, and racemes of white flowers. Belongs to the Labiate order. Introduced by the Horticultural Society.

A white, sage-like, herbaceous plant, growing about 2 feet high. Leaves on long stalks, oblong, blunt, crenate, having a strong and by no means agreeable odour, proceeding apparently from numerous point-like dark brown glittering glands with which they are covered, especially on the under side. Stem erect, producing a great number of white labiate flowers, on short, lateral, one-sided racemes. Stamens long and prominent. This seems to be unable to bear an English winter without protection; for it has perished among rockwork in that of 1849-50. The flowers have no beauty; but the snow-white leaves and stems produce an appearance sufficiently remarkable to give it a claim to cultivation where the climate agrees with it.—*Journ. Hort. Soc.*, vol. v.

126. FUCHSIA VENUSTA. *Humboldt.* A handsome greenhouse shrub, with lanceolate leaves in threes, and long solitary pendent salmon-coloured flowers tipped with pink. A native of Peru. Introduced by Mr. Linden. (Fig. 57.)

This is one of the best of the Peruvian Fuchsias, for the introduction of which we are indebted to Mr. Linden, from whom we received fresh flowers last autumn. Hartweg found it commonly near Santa Fé de Bogota, but, owing to the mismanagement of that collector, its seeds, like nearly all else that he brought home with him, perished in the hold of a sugar ship. It has long narrow lanceolate deep green leaves, quite toothless at the edge, usually growing in threes. The flowers appear singly in the axils, and are full 3 inches long, while the stalk measures $2\frac{1}{2}$ inches more. The tube of the calyx has the form of a lengthened cone, its lobes being tipped with light emerald green. The petals, which are also salmon-coloured, are lanceolate, wavy, not rolled up, but a little turned back at the point, and something longer than the calyx.

127. Berberis Wallichiana. *De Candolle.*

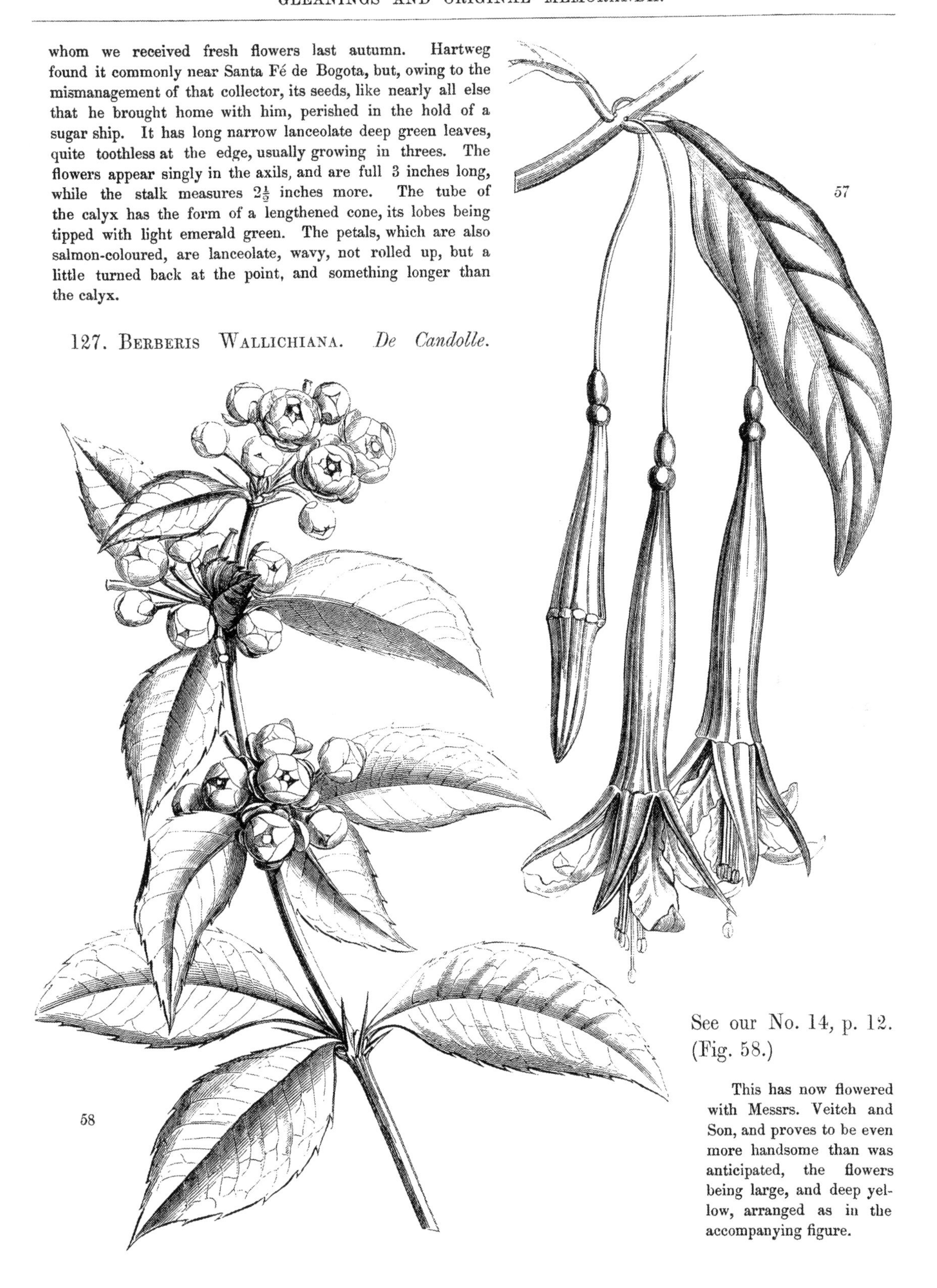

See our No. 14, p. 12. (Fig. 58.)

This has now flowered with Messrs. Veitch and Son, and proves to be even more handsome than was anticipated, the flowers being large, and deep yellow, arranged as in the accompanying figure.

128. Dodecatheon integrifolium. *Michaux.* A hardy herbaceous plant, belonging to the Order of Primworts. Flowers purple and yellow. Native of California. Introduced by the Horticultural Society.

A dwarf stemless plant, with a few long narrow, almost spathulate, undivided leaves, and a slender scape, bearing a single nodding flower, very like that of the common species, and of the same purple colour, with a yellow eye and dark purple anthers. Such was the plant in the Horticultural Garden. Upon looking, however, to the wild specimens, we find that it becomes much more vigorous when older, bearing as many as three flowers on a scape, or, according to Sir Wm. Hooker, eleven or twelve; in which case it becomes as interesting as the old and well-known species, so frequent in gardens. A damp, rich, shaded American border suits it best; and there it may be expected to grow without difficulty.—*Journ. Hort. Soc.*, vol. v. *With a figure.*

129. Ixora laxiflora. *Smith.* A graceful hothouse shrub from Sierra Leone, with panicles of long, slender, pink, sweet-scented flowers. From Lucombe & Co. Belongs to the order of Cinchonads.

Well worthy of general cultivation, for while small it has handsome foliage and flowers, which have a delicate and most agreeable fragrance. Leaves, the largest a span in length, oblong-lanceolate, acuminate, feather-veined, attenuated at the base into a very short petiole. Panicle terminal, large, and singularly trichotomous. Calyx deep red, the tube (or ovary) globose, red; the free portion or limb is very small and cleft into four erect, appressed teeth. Corolla white tinged with pink; the tube $1\frac{1}{2}$ inch long, slender; the limb cut to the base into four spreading obovate segments, hairy in the disk.—*Botanical Magazine*, t. 4402.

130. Espeletia argentea. *Humboldt and Bonpland.* A singular greenhouse herbaceous plant of the Composite order, with handsome silvery leaves and yellow heads of flowers. Blossomed at Kew and at Syon in the summer of 1848. Native of New Grenada.

The whole plant has a peculiar and somewhat terebinthine odour, and yields like the genus *Silphium* (to which *Espeletia* is allied in essential characters) a copious gum-resin, used in the preparation of ink, and for other purposes.

This is a beautiful plant, and a stately one when in flower, attaining then the height of five or six feet. Before flowering, however, the appearance is very different. A plant of three or four years old has a trunk six or eight inches high and as thick as one's wrist, rather bare below, but the rest forming a crown of dense spreading leaves a foot and more long, spreading all round like those of an Aloe. Leaves narrow-lanceolate, densely silky, and shaggy on both sides. At the flowering season the apex of the trunk lengthens out into an upright densely silky, nearly leafless corymboso-paniculate stem.—*Botanical Magazine*, t. 4480.

131. Arbutus xalapensis. *Humboldt.* A dwarf Mexican half-hardy shrub, with dull evergreen leaves, and close clusters of reddish flowers. Introduced by the Horticultural Society.

A low, dull brownish-green evergreen bush. Branches, petioles, and underside of leaves covered with a soft short down, without any trace of setæ. Leaves oblong, flat, long-stalked, rounded at the base, perfectly entire, or very slightly serrate, with a hard, firm, reddish edge, somewhat downy on the upper side. Flowers dirty reddish-white, in close downy terminal short pyramidal panicles. Peduncles glandular and woolly. Calyx nearly smooth. Corolla ovate, at the base, almost flat, and unequally gibbous, with a contraction below the middle, and a very small limb. Ovary with a granular surface. This little bush is by no means ornamental. It grows slowly, requires protection in winter, has dull spotted leaves, and remains in flower only for a week or two in April. Although a true Arbutus, it seems to have none of the beauty of its race, and must be consigned to the collectors of mere botanical curiosities.—*Journ. Hort. Soc.*, vol. v. *With a figure.*

PLATE 16

I. Constans. Pinx & Zinc

Printed by C.F. Cheffins, London.

[Plate 16.]

THE WHITE CUNNINGHAM RHODODENDRON.

(RHODODENDRON CINNAMOMEUM; VAR. CUNNINGHAMI.)

A hardy evergreen hybrid Shrub. R. cinnamomeum ♂, maximum ♀.

For the figure of this noble shrub we are indebted to Mr. George Cunningham, of the Nursery, Liverpool. It is probably the best hybrid Rhododendron yet raised, not possessing, indeed, the rich colours of the crimson mules, but quite as valuable to the cultivator on account of its large heads of pure white spotted blossoms. The history of the plant is thus given by Mr. Cunningham in his correspondence:—

"It was raised between Cinnamomeum and a late White Maximum, as you will at once see by the foliage. It is very remarkable for its strong ribbed leaf and brown under-surface. The white of the flower is very pure, and the dark purple spots contrast with it very beautifully. It is quite hardy; its maternal parent being the latest and hardiest of all our Rhododendrons, and Cinnamomeum, the father, will stand any severity of an English winter in January; but as it pushes early in the spring, it is liable to be cut by our late frosts.

"The object which I had in view in hybridising R. cinnamomeum with a pure White Maximum, was to improve the colour of each parent, keeping the purple spots of the former, and getting a later period of flowering from the latter. In this part of the kingdom the flowers from the hybrids with the Indian species and Ponticum, or Catawbiense, are in three seasons out of four destroyed by late frosts; the colour also of those between the true Scarlet Arboreum and the pink and purple species is diluted, and that between them and Cinnamomeum, or the White Arboreum, is often of a *muddy* pink, turning, as the flower gets old, into a dirty white. In the one I have sent you to figure, these objects have been obtained—the white colour has been preserved in all its purity, and a perfect hardiness also acquired. None of my plants of it have had any protection."

In form the leaves are exactly intermediate between the two parents. To the shape of the Cinnamon Tree Rhododendron they add the convexity of *R. maximum;* and the downy surface of the under-side is just half-way between the two. In both the mule and its ♀ parent, the hairiness consists of numerous much-entangled tubes, blunt, transparent, flat, thin-sided, and very often arranged in a starry manner. They are evidently the beginning of the raments (?) of Bejaria.

In one respect both leaves and stem are unlike either parent. The latter is of a rich crimson brown, and the former are covered with an abundant resinous secretion, which renders them sticky to the touch.

THE VERVAENE RHODODENDRON.

Although derived from a different source, and much less interesting than the preceding, the variety published by M. Van Houtte under the name of *Rhododendron ponticum*, var. *Vervaeneanum, flore pleno,* deserves mention in this place. It was no hybrid, but was an accidental seedling obtained by a M. Vervaene, " dont les heureuses tentatives de semis ont doté l'horticulture de cette riche acquisition," from *Rhododendron ponticum.* According to M. Van Houtte, it is no less remarkable for the elegance of its habit, than for the abundance of its flowers, the great breadth of its heads and of its corolla, and for its delicate tints. His very fine figure represents it as forming a head about as large as that shown in the annexed plate; the flowers measure full three inches in diameter, are semi-double, of a rich lilac colour, with the upper lip white, spotted with yellow. See *Flore des Serres,* tt. 492, 493.

I. Constans. Pinx & Zinc

Printed by C.F.Cheffins London.

[PLATE 17.]

THE CLOSE-HEADED BEJARIA.

(BEJARIA COARCTATA.)

A half-hardy evergreen Shrub, with crimson flowers, from the ANDES OF NEW GRENADA, *belonging to the Order of* HEATHWORTS.

Specific Character.

THE CLOSE-HEADED BEJARIA.—Branches shaggy with spreading hairs. Leaves oval, acute, on short stalks, closely imbricated, glaucous beneath ; the stalk and midrib shaggy, otherwise smooth. Flowers deep crimson, in very close corymbs ; stalks short, covered with rusty wool ; the calyx nearly smooth. Petals erect, nearly parallel, (not spreading). Style long, projecting.

BEJARIA *COARCTATA* ; ramis patentim villosis, foliis ovalibus acutis breviter petiolatis densè imbricatis margine revolutis subtus glaucis petiolo costâque villosis cæterum glabris, corymbis densissimis abbreviatis, pedunculis brevibus ferrugineo-tomentosis, calycibus glabriusculis, petalis rectis subparallelis, stylo longè exserto.

B. coarctata : *Humboldt and Bonpland, Plantæ æquinoctiales*, vol. ii. p. 125, t. 121.

THIS genus is little known in Europe. Mutis named it after his friend Professor Bejar, of Cadiz : but Linnæus, misreading j for f, published it under the erroneous name of Befaria. It should be written as above and sounded Beharia. It is nearly related to the Rhododendron, from which it differs in its petals being all distinct, overlapping each other, and not united into a tube. The species inhabit the Alps of Peru and Mexico, where their beauty becomes fully developed, and rivals that of the Azaleas and Rhododendrons of the United States and India.

The plant now figured seems to have found its way to Europe both through England and Belgium. To our own country it was sent by Mr. Purdie for His Grace the late Duke of Northumberland ; and it was at Syon that it flowered, for the first time in Europe, in May last under the care of Mr. Ivison ; we also believe that Mr. Linden's collectors, who found it near Pamplona, at the height of 8500 feet, also furnished a supply of fresh seeds. A third traveller from whom it has been derived was Messrs. Veitch's collector Lobb, who found it on the mountains of Peru. From one of his specimens a short account of it was given in the Gardeners' Chronicle for 1848, with a woodcut which we reproduce for the sake of showing the very inferior appearance of the plant in a wild state, and the nature of the hairiness, which is merely represented by colour in M. Constans' figure. It has hirsute branches, woolly flower-stalks, and a nearly smooth calyx, with seven or eight smoothish, blunt, ovate sepals, whose edges are a little woolly. The flowers are deep rich crimson, and very closely arranged. Each consists of seven or eight smooth petals. The leaves, when very young, are in the wild plant woolly on the under-side; when full grown are perfectly smooth, shining, rather convex, nearly sessile, and glaucous on the under-side. The nature of the longer hairiness is peculiar, and is more like

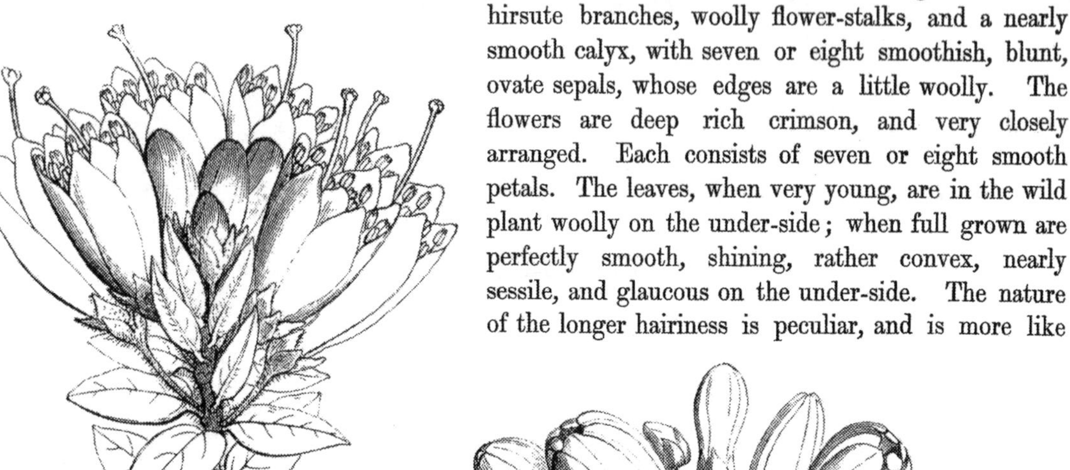

Bejaria coarctata, from a wild specimen.

what Botanists call raments than ordinary hairs, that is to say, it consists of long narrow thin plates tapering to a point, filled with a brown fluid, and composed of many rows of cells. Mixed up with them is a close wool or fur, much shorter, and composed of curved, or hooked, entangled, also brown, hairs.

B. Lindeniana.

We have little doubt that this is the plant represented by Humboldt and Bonpland under the name of *B. coarctata,* notwithstanding some small discrepancy in their description of the hairiness; for we know that such mountain plants vary much in the amount and nature of the wool that invests them at different seasons. The species is, however, totally different from what is published in the Botanical Magazine, t. 4433, under the same name, which Sir William Hooker does not appear to have recollected had been previously given to the subject of this plate. This error was pointed out by M. Hérincq, who, in reproducing the figure, called the species *B. Lindeniana.* This plant has also flowered in the great collection at Syon, and was exhibited by Mr. Ivison at one of the late exhibitions in the garden of the Horticultural Society, when the accompanying figure was made. It has much shorter hairs on the stem even when young, and they soon give way to a mere ferruginous down. The leaves are perfectly smooth, longer-stalked, flat, spreading, oblong, becoming blunt, although often sharp-pointed when young; and instead of the rich deep green of the close-headed Bejaria, they have a yellowish cast. The flowers, which are in loose corymbs, are pale pink, streaked with a darker rose-colour. In the Botanical Magazine their petals are represented as spreading as flat as those of a Mallow; but in the Syon plant they are closed, as in our cut. We suspect this to be very near Mr. Linden's *B. tricolor,* which is, however, said to be yellow at the base of the corolla.

Closely related to these, but perhaps finer than any, is a plant raised by Messrs. Veitch & Co. of Exeter. We presume it to be that which Mutis called *æstuans,* because, it would seem, it glows like a fire. Mr. William Lobb found it in the province of Chachapoyas, at the height of 8000 feet, and describes the flowers as rose-coloured. Messrs. Veitch of Exeter have raised it. The branches are covered with coarse hairs. The leaves are fringed with blackish bristles; when young they are covered beneath with a rusty secretion; when full grown they are very glaucous on the under-side, and dark green on the upper. The calyx and flower-stalks are shaggy with coarse hairs, and clammy with a sticky juice which oozes out from the surface.

Bejaria æstuans, Mutis.

Although we venture to attach to this species the name of *æstuans,* judging from the definition of it in books, yet it is quite possible that it may be another species. Indeed, if M. Hérincq is right in stating that the plant of Mutis has the habit of *Rhododendron ferrugineum,* it must be something quite different. No doubt it is distinct from Mr. Linden's *B. æstuans,* which Hérincq calls *myrtifolia,* and which is said to have long lanceolate leaves, very much narrowed towards the point.

It may be worth while to add to these memoranda a list of the Bejarias now or formerly in cultivation, with their supposed aliases :—

1. B. racemosa *Vent.*—Probably lost.
2. B. glauca *H. B.*—Formerly flowered at Ghent.
3. B. ledifolia *H. B.*—Fl. des Serres, t. 194.
4. B. Lindeniana *Hérincq* (*aliàs* B. coarctata *Hooker*).—Bot. Mag., t. 4433.
5. B. coarctata *H. B.*

6. B. myrtifolia *Hérincq* (*aliàs* B. æstuans *Linden*).
7. B. æstuans *Mutis.*
8. B. cinnamomea *Lindley.**
9. B. drymifolia *Linden.*
10. B. densa *Planchon* (*aliàs* B. microphylla).
11. B. tricolor *Linden.*
12. B. ——, an unknown species at Syon, with lanceolate leaves, and red branches covered with viscid stiff hairs.

The proper mode of managing these Bejarias is still uncertain. They are charming plants, and worth any amount of care and trouble. We believe that the treatment of Indian Azaleas will suit the strongest, and that of Rhododendron Chamæ-Cistus the weakest. A damp atmosphere, and free circulation of air in summer, are no doubt essential. Mr. Linden cuts the matter short, as will be seen by the following extract from his priced Catalogue :—

Bejaria (Befaria) æstuans	30 francs.	Bejaria (Befaria) glauca	10 francs.
" coarctata	15 "	" ledifolia	10–50 "
" densa (microphylla)	25 "	" tricolor	40 "
" drymifolia	40 "	" sp. nova.	" "

" Réputé à tort comme étant d'une culture difficile, ce magnifique genre réclame au contraire *peu de soins*. Planté en pleine terre, il fleurit abondamment et n'exige en hiver qu'une température très-basse et peu d'humidité."

* " Messrs. Veitch are also in possession of a third species of this genus, with purple flowers, found on the Andes of Caxamarca, at the height of 8000 feet. Its flowers are very much injured in the specimen before us, but appear to be smaller than in the species now figured (B. æstuans), and are arranged in a close panicle. The leaves are remarkable for being covered on the lower side with a bright brown wool, on which account it may be named The Cinnamon Bejaria (*Bejaria cinnamomea*).

" Sp. Char.—Branches downy and hispid. Leaves slightly downy above, covered beneath with thick ferruginous wool. Flowers in a close terminal panicle, with very woolly and hispid stalks and calyxes."—*Gardeners' Chronicle.*

PLATE 18

I. Constans. Pinx. & Zinc

Printed by C.F. Cheffins, London

[Plate 18.]

THE SPECKLED ODONTOGLOT.

(ODONTOGLOSSUM NÆVIUM.)

A stove Epiphyte, from the Andes of New Grenada, *belonging to the Order of* Orchids.

Specific Character.

THE SPECKLED ODONTOGLOT.—Pseudo-bulbs ribbed. Leaves thin, lanceolate, narrowed to the base. Panicles spreading. Sepals and petals narrow, ovate-lanceolate, acuminate, wavy. Lip of the same form, with a slight tendency to become hastate, with the 2 teeth of the crest large, downy, somewhat 3-lobed. Processes of the column subulate, spreading.

ODONTOGLOSSUM *NÆVIUM*.—Pseudo-bulbis costatis, foliis tenuibus lanceolatis basi angustatis, paniculâ diffusâ, sepalis petalisque angustis ovato-lanceolatis acuminatis undulatis, labello subconformi vix hastato: cristæ dentibus 2 grossis subtrilobis pubescentibus, columnæ cirrhis subulatis patulis.

In Central America there exists a herd of Odontoglots the distinctions between which can hardly be settled, in the first instance at least, by dried specimens. They have all a similar habit, branching panicles, and white-lipped flowers spotted with crimson, with long narrow wavy divisions. At present there are only two in cultivation, viz. that now published and the Sweet Odontoglot (*Odontoglossum odoratum*) good plants of which we see are offered for sale by Mr. Linden at the modest price of two guineas each. In that plant the sepals and petals are yellow, while the lip alone, which is distinctly halberd-shaped, is white. In this, on the contrary, there is no yellow, but all the ground is pure white.

The plant before us was sent to England several years since by Sir R. Schomburgk, and was exhibited by Mr. Loddiges at one of the Spring meetings of the Horticultural Society in the present year. What appears to be the same species is No. 721 of Mr. Linden's herbarium of 1846, found by his collectors, Funck and Schlim, at the height of 6000 feet, at St. Lazaro and la Peña, in the province of Truxillo, and said to have a *yellow* lip spotted with crimson; a circumstance possibly connected with the colour of the fading flowers. Another supposed variety of this same plant was flowered by Messrs. Rollisson in June 1847, with rather larger blossoms: and in that particular it would appear as if these Odontoglots were subject to considerable differences, just as we have large and small states of the Ample Oncid (*Onc. ampliatum*), the Sphacelated Oncid, and even the Wentworth Oncid, of which last Sir Philip Egerton has lately flowered a magnificent form.

Pseudo-bulbs ovate, compressed, rather strongly but bluntly ribbed. Leaves narrowly oblong, tapering to the base, single on the pseudo-bulbs, shorter than the panicle. Flowers pure white, speckled everywhere with rich crimson, arranged in the garden plant in a narrow racemose panicle; in what appears to be the same thing wild they form a loose branched panicle of considerable size. Bracts very short, scale-like. Sepals and petals from an ovate base linear-lanceolate, acuminate, spreading equally and very wavy. Lip of the same form and colour, but shorter, downy, very slightly halberd-shaped near the base which is yellow, with the edges of the claw clasping the column. Teeth of the crest yellow, rather small, distinct, with about 3 unequal blunt lobes to each; downy. Column downy, narrowed to the base, with a pair of awl-shaped ears near the summit, below the anther-bed.

The resemblance of this to the Long-tailed Oncid (*O. phymatochilum*) is so great as to raise a question as to the distinction between Oncids and Odontoglots. We have often opened this discussion, and endeavoured to show how the two genera could be certainly separated; but it must be owned that, after all, there is something vague and unsatisfactory in the characters usually assigned to the genera. Species, indeed, have been indifferently placed in one or the other, or species stationed in the Oncids by one botanist have been referred to the Odontoglots by another. It will therefore be useful to explain that, in addition to any other distinction, this may be taken as unexceptionable, namely, that the Oncids have a short column, tumid at the base in front, as in the annexed cut of *Oncidium phymatochilum*, while the Odontoglots have a lengthened column without any such tumour.

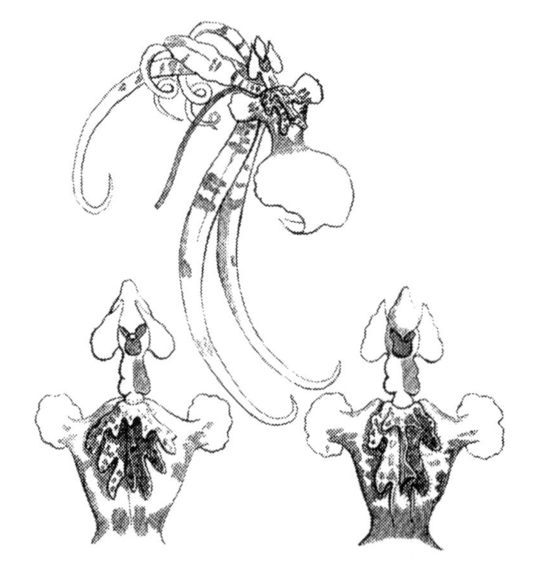

The management of this, and all such plants, is precisely what is required for the Spotted Oncid (*O. maculatum*).

GLEANINGS AND ORIGINAL MEMORANDA.

132. Passiflora Medusæa. *Lemaire.* A stove species of unknown origin, with red and yellow flowers. Introduced by M. Van Houtte. (Fig. 59.)

One of the slit-leaved species, with a slender habit, and pretty starry flowers, the rays of which are orange the first day, and lilac or rose the next. According to M. Lemaire, who named it on account of "quelque ressemblance avec la tête de la fille de Phorcus, après que ses cheveux eussent été changés en serpents par Minerve," these flowers have a strong penetrating odour in which there is nothing disagreeable. But M. Planchon, whose nose seems to be differently constituted, asserts that this smell, which becomes perceptible at the period of the change of colour, is most disgusting.—*Flore des Serres*, 528.

133. Cuphea cinnabarina. *Planchon.* A half-shrubby plant from Guatemala, belonging to the order of

Loosestrifes (Lythraceæ). Flowers crimson or deep purple. Introduced by M. Van Houtte. (Fig. 60.)

M. Planchon thinks this different from the *C. Llavea*, long since known in gardens, distinguishing it by its panicled flowers, the colour of the anthers, and some other circumstances. It seems to be a good bedding plant. Two varieties are figured, one with rich crimson, the other with purple flowers.—*Flore des Serres*, 527.

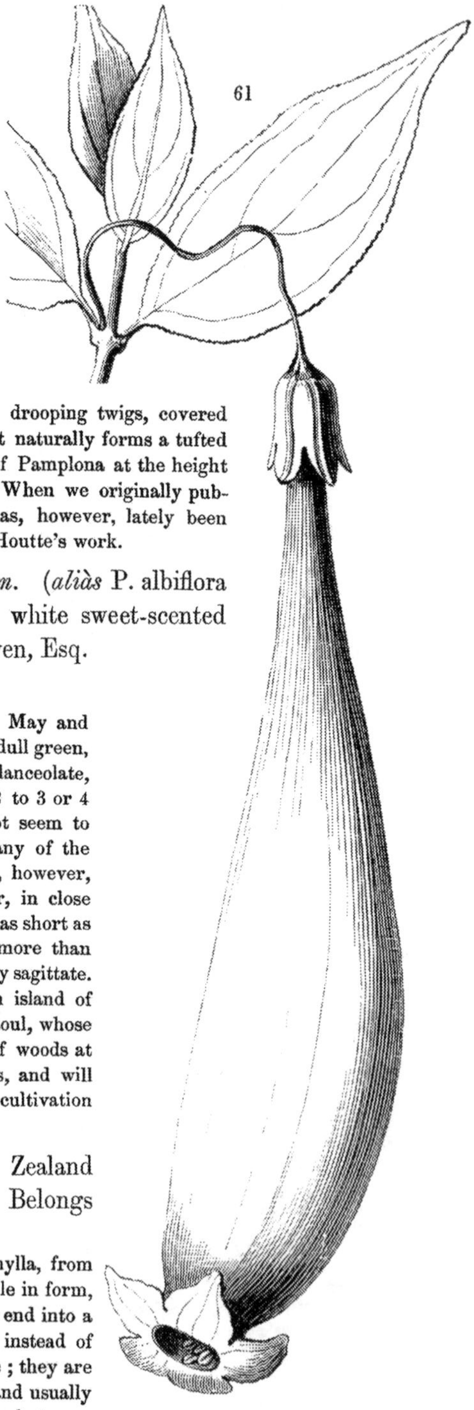

61

134. Lisianthus Princeps. *Lindley*. A greenhouse shrub, with very long scarlet, yellow and green flowers. A native of New Grenada. Belongs to the Gentianworts. Introduced by Mr. Linden. (Fig. 61.)

This must be one of the noblest plants in existence. Its long flowers, the size of the accompanying figure, are rich scarlet melting into yellow at either end, with an emerald green 5-lobed limb; they hang in clusters of four from the ends of the drooping twigs, covered with firm deep green opposite leaves. According to Mr. Linden, it naturally forms a tufted shrub 2 or 3 feet high, growing at the entrance of the table land of Pamplona at the height of 10,000—11,000 feet above the sea. *Flore des Serres*, t. 557. When we originally published this plant we knew it only from dried specimens. It has, however, lately been flowered by Mr. Linden, and is beautifully represented in M. Van Houtte's work.

135. Parsonsia heterophylla. *Allan Cunningham*. (*aliàs* P. albiflora *Raoul.*) A New Zealand twining evergreen shrub with white sweet-scented flowers. Belongs to the Dogbanes. Introduced by J. R. Gowen, Esq. (Fig. 62.)

A twining evergreen greenhouse plant, flowering abundantly in May and June. Stem covered with fine down, pale yellow; leaves leathery, dull green, slightly downy, wavy, very variable in form; linear-lanceolate, ovate-lanceolate, obovate, or even spathulate, often repand, varying in length from 2 to 3 or 4 inches. These singular diversities in the form of the leaves do not seem to be confined to any particular parts of the plant, but appear on any of the branches, and all intermingled; the short spathulate leaves are, however, most usual on short lateral shoots. Flowers pale cream-colour, in close one-sided naked panicles, rather sweet-scented. Calyx three times as short as the corolla. Corolla urceolate, with a revolute 5-cleft border, not more than a quarter as long as the tube. Anthers without any tails, but simply sagittate. According to Cunningham, this plant is common in the northern island of New Zealand, at Hokianga and Wangaroa, in shady woods. M. Raoul, whose P. albiflora can scarcely be different, found it on the outskirts of woods at Akaroa. It is rather a nice addition to our greenhouse climbers, and will probably prove hardy in the south of England. For purposes of cultivation it is much superior to P. variabilis.—*Journ. Hort. Soc.*, vol. v.

136. Parsonsia variabilis. *Lindley*. A New Zealand twining evergreen shrub, with white fragrant flowers. Belongs to Dogbanes. Introduced with the last.

A small twining greenhouse plant, very much like P. heterophylla, from which it differs in its leaves being shining and much more variable in form, the linear ones being far narrower, and often expanded at the very end into a circular blade. The flowers are not more than half the size, and instead of being contracted at the mouth or urceolate, are exactly campanulate; they are also far less hairy, by no means so numerous or densely arranged, and usually intermingled with long narrow leaves. It is a very curious thing, but possesses little claim to beauty. Its flowers are, however, much sweeter than in P. heterophylla.—*Journ. Hort. Soc.*, vol. v.

137. ACINETA DENSA. An epiphyte from Costa Rica, with a pendulous short close raceme of yellowish, somewhat fragrant flowers. Blossomed in July, in the nursery of Messrs. Lane and Son, of Berkhampstead. (Fig. 63.)

A. *densa;* racemo oblongo denso nigro-furfuraceo, bracteis ovario duplò brevioribus, labelli hypochilio concavo intus versus apicem dente obtuso tomentoso aucto, metachilii lobis lateralibus truncatis basi angustioribus appendice plano ovato subtridentato angulis posticis sinuatis, epichilio lineari-oblongo basi verrucoso, columnâ dorso tomentosâ.

This is one of M. Warczewitz's collection, imported in 1849 by Mr. Skinner. It was found at Turialba, in Costa Rica, and is very near the Barker Acinete, notwithstanding the appearance of dissimilarity caused by its dense, not long and narrow, raceme. The flowers are pale yellow, slightly spotted externally with crimson. The lip is yellow at the point, spotted with broad blotches on the lateral lobes, deep crimson in the space between the lobes occupied by the appendage. As regards structure, this Close-flowered Acinete differs from the Barker A. in having a more concave hypochil, the tooth of which is not notched at the point; an entire epichil remarkably warted at the base, and the lateral lobes of the metachil not at all rounded, with the posterior angles of the intermediate appendage sinuous, and not extended into a long subulate process.

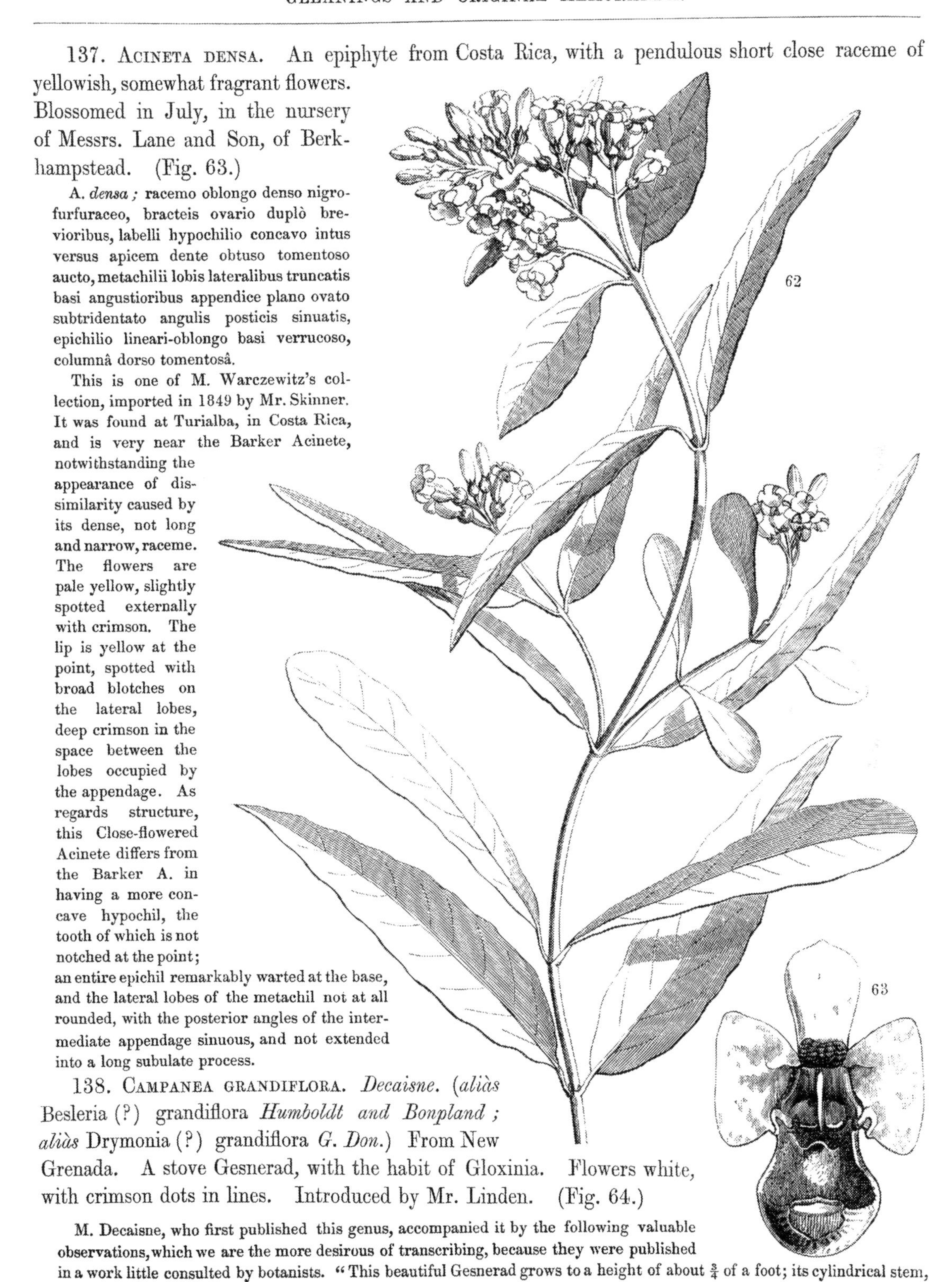

138. CAMPANEA GRANDIFLORA. *Decaisne.* (*aliàs* Besleria (?) grandiflora *Humboldt and Bonpland ; aliàs* Drymonia (?) grandiflora *G. Don.*) From New Grenada. A stove Gesnerad, with the habit of Gloxinia. Flowers white, with crimson dots in lines. Introduced by Mr. Linden. (Fig. 64.)

M. Decaisne, who first published this genus, accompanied it by the following valuable observations, which we are the more desirous of transcribing, because they were published in a work little consulted by botanists. "This beautiful Gesnerad grows to a height of about $\frac{3}{4}$ of a foot; its cylindrical stem, somewhat woody at the base, herbaceous above, is covered with long white hairs. The leaves are opposite, oval, more or less acuminate, sometimes oblique, soft, crenated at their edges, stalked and covered with long hairs like those which cover the branches. The flowers grow in a tuft at the end of a long axillary or terminal peduncle ; each flower being supported on a pedicel furnished with a lanceolate bract. The calyx is herbaceous ; its 5 divisions are oval, acuminate, nearly equal in size, and traversed by 3 nerves. The corolla, somewhat like that of *Ligeria*, is largely campanulate, and has 5 rounded lobes : the tube white, hairy outside, covered on the inside with red spots tolerably symmetrically arranged, is somewhat analogous to that of *Drymonia punctata*. The stamens, though included, appear at the mouth of the tube ; their 4 glabrous filaments carry heart-shaped anthers, which are firmly joined together, and form at the mouth of the tube a sort of pale yellow star, with which the stigma is in contact. The disk, formed of 5 fleshy, obtuse yellow bodies, surrounds a roundish hairy ovary.

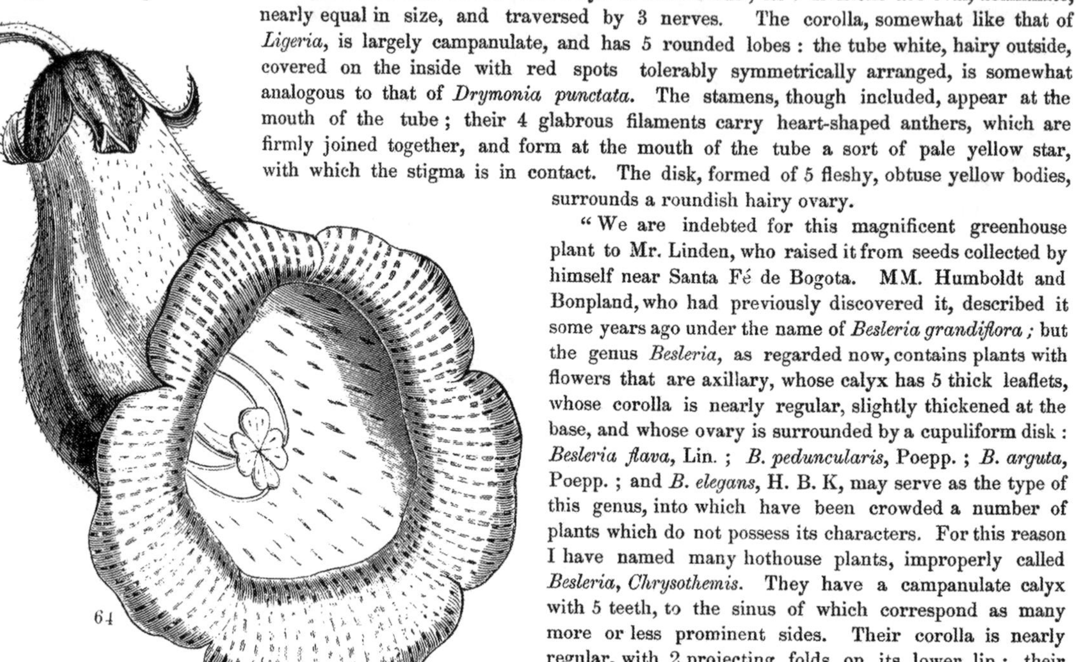
64

"We are indebted for this magnificent greenhouse plant to Mr. Linden, who raised it from seeds collected by himself near Santa Fé de Bogota. MM. Humboldt and Bonpland, who had previously discovered it, described it some years ago under the name of *Besleria grandiflora ;* but the genus *Besleria,* as regarded now, contains plants with flowers that are axillary, whose calyx has 5 thick leaflets, whose corolla is nearly regular, slightly thickened at the base, and whose ovary is surrounded by a cupuliform disk : *Besleria flava*, Lin. ; *B. peduncularis*, Poepp. ; *B. arguta*, Poepp. ; and *B. elegans*, H. B. K, may serve as the type of this genus, into which have been crowded a number of plants which do not possess its characters. For this reason I have named many hothouse plants, improperly called *Besleria*, *Chrysothemis*. They have a campanulate calyx with 5 teeth, to the sinus of which correspond as many more or less prominent sides. Their corolla is nearly regular, with 2 projecting folds on its lower lip : their stamens are included, their stigmas are 2-lobed, and their disk is a notched fleshy gland. This genus at present includes 2 species, viz., *C. pulchella* Dne.=*Besleria pulchella* Lodd., *Bot. Cab.*1028 ; *C.venosa* Dne.=*Besleria melissæfolia* Hortul. Each has an orange yellow corolla, streaked with carmine,and enclosed in a calyx with 5 wings,like that of *Sinningia*.

"The confusion which I have noticed is not confined to *Besleria*, but extends to the genus *Columnea*, which I reduce to those species the corolla of which reminds one of that of *Dircæa ;* its long tube is parted into 4 lobes : the upper broad and notched, the lateral ones oval and turned back, the lower like a small pendent tongue. An exact idea of this structure can be obtained from *Columnea Schiediana*. The other species belonging to this genus are, *C. Lindeniana* Brongniart. ; *C. flava* Mart. and Gall.; *C. crassifolia* Brongniart; *C. scandens* L.; *C. hirsuta* L.

"M. Lemaire has separated from this genus *Columnea*, for the purpose of making a new genus, *Collandra*, the species described and figured in the *Bot. Mag.*, 4294, under the name of Gesneria auro-nitens. I have adopted the separation correctly pointed out in the *Flore des Serres* (vol. 3, May, 1847, p. 225), although I have not been able to discover by analysis that the hypogynous disk is constantly formed of unequal glands; but the separation of this plant from *Columnea* was so natural, that it had been already made by Tussac and by M. Reichenbach ; unfortunately, the names substituted by these gentlemen could not be retained. I unite then to the genus *Collandra*, *Alloplectus sanguineus*, as well as other species, which are easily distinguished by the extreme inequality presented by each pair of leaves, one of which, constantly very much reduced, reminds one of the disposition of these same organs in *Ruellia anisophylla*.

"Lastly, many species of this group have at the end of the limb of their leaves a large blood-coloured spot, from which they are called in the colonies *Yerba de la Doncella*. Such is *Collandra phœnicea* Dne. = Dalbergaria phœnicea *Tuss. Fl. Antill.* i. p. 141, t. 19. The names of *Dalbergaria* Tussac, or of *Tussacia* Reichenbach, proposed for this plant, cannot be adopted, inasmuch as they already apply to other plants ; the name of *Collandra* ought therefore, I think, to be retained.

"The genus *Alloplectus* comprehends a great number of species, and many very different looking plants, which require to be grouped. The *Alloplectus*, properly so called, has a more or less bulging corolla, with a limb of five rounded nearly equal divisions ; a calyx with deep segments, coloured, entire or toothed, and a disk reduced to a great notched gland. In this group I place *Alloplectus speciosus* Linden, Cat. ; *A. pendulus* Endl. and Poepp., t. 205; *A. dichrous*, *Bot. Mag.* 4216 ; *A. Pinelianus* Hortul. ; *A. glaber* Dne. (Hypocyrta glabra, *Bot. Mag.* 4346); *A. strigulosus* Dne. (Hypocyrta strigulosa, Hort.); *A. splendens* Dne. (Hypocyrta splendens, Hort., et Columnea zebrina, Hort.); *A. congestus* Dne.; *A. bicolor* Dne. (Besleria bicolor, Hook.); *A. concolor*, *Bot. Mag.* 4371; *A. cristatus* Mart. (Besleria cristata, L.). The species with coriaceous, glabrous leaves, and in which the calyx is surrounded with large petaloid bracts, ought, I think, to constitute a natural group, for which I propose the name of *Macrochlamys*. This group comprehends *Macrochlamys Patrisii* Dne. (Alloplectus Patrisii, DC.); *M. involucratus* Dne. MSS.; *M. Miquelii* Dne. (Alloplectus Patrisii Miq. non DC.); *M. speciosus* Dne. MSS.; *M. guttatus* Dne MSS., Linden, No. 547.

"The other genera of the tribe of Gesnerads, with a free ovary and named by Brown Besleria, are *Hypocyrta*, with an hypogynous, cupule-shaped disk ; *Episcia*, the type of which may be taken to be *E. bicolor*.—*Bot. Mag.* 4590 ; *Drymonia*, of which one species, *D. punctata*, is cultivated ; *Nematanthus*, figured in the *Bot. Mag.* 4080 ; and in Paxton, under the wrong names of *Columnea splendens grandiflora*, vol. x.; *Tapeinotes;* and *Trichanthe*, of which we have no species in our gardens."

According to M. Van Houtte, who is the sole possessor of this fine species, it grows perfectly in a greenhouse in a mixture of leaf-mould and loam. M. Planchon mentions a second Campanea, with sea-green flowers, speckled with purple, and current under the name of *Gloxinia tigridia*, concerning which we have no further information.

139. Abutilon insigne. *Planchon.* A greenhouse shrub, with large round heart-shaped leaves,

and pendulous flowers with broad rich crimson veins, almost covering a white ground. A Mallowwort from New Grenada. Introduced by Mr. Linden. (Fig. 65.)

A very fine species, with the habit of the other kinds now so common in gardens, but with large bell-shaped flowers remarkable for the very deep rich crimson of the veins, which scarcely leave any white perceptible between them or on the edges. It is said to be a native of the cold regions of the Andes of N. Grenada, and to succeed perfectly in the open air in summer.—*Flore des Serres*, t. 551.

140. ACROPERA ARMENIACA. An epiphyte from Nicaragua, with rich apricot-coloured flowers in pendent racemes. Belongs to the Orchids. Flowers in July. Introduced by M. Warczewitz. (Fig. 66.)

A. armeniaca; racemo laxo multifloro, sepalis apiculatis lateralibus obliquis apice rotundatis, petalis liberis columnâ duplò brevioribus, labello calceato carnoso apice libero ovato plano acuminato intus pone basin cristâ tuberculatâ aucto.

For this curious and really pretty species we are indebted to Sir Philip de Malpas Grey Egerton, Bart. It differs from the *A. Batemanni* in its petals being much shorter than the column, and perfectly distinct from it, and in the point of the lip being undivided, free and flat. There is nothing peculiar in its habit; but its large flowers, coloured like the sunny side of a ripe apricot, are very remarkable.

66

141. CAMPYLOBOTRYS DISCOLOR. *Lemaire.* A handsome dwarf half-shrubby plant, belonging to Cinchonads. Flowers rich red in axillary clusters. Introduced by M. Van Houtte, of Ghent. (Fig. 67.)

This appears to be a dwarf soft-wooded plant, flowering freely when only a few inches high, and not exceeding a foot in stature. The stems are crimson-purple; the leaves have deep rose-coloured stalks, a satiny shining surface raised between the lateral veins, and a rich tint of purple on the under side. The flowers are said to form a short nodding spike, placed on an axillary crimson stalk 2 or 3 inches long. The corolla is a deep rich red.—*Flore des Serres*, t. 427.

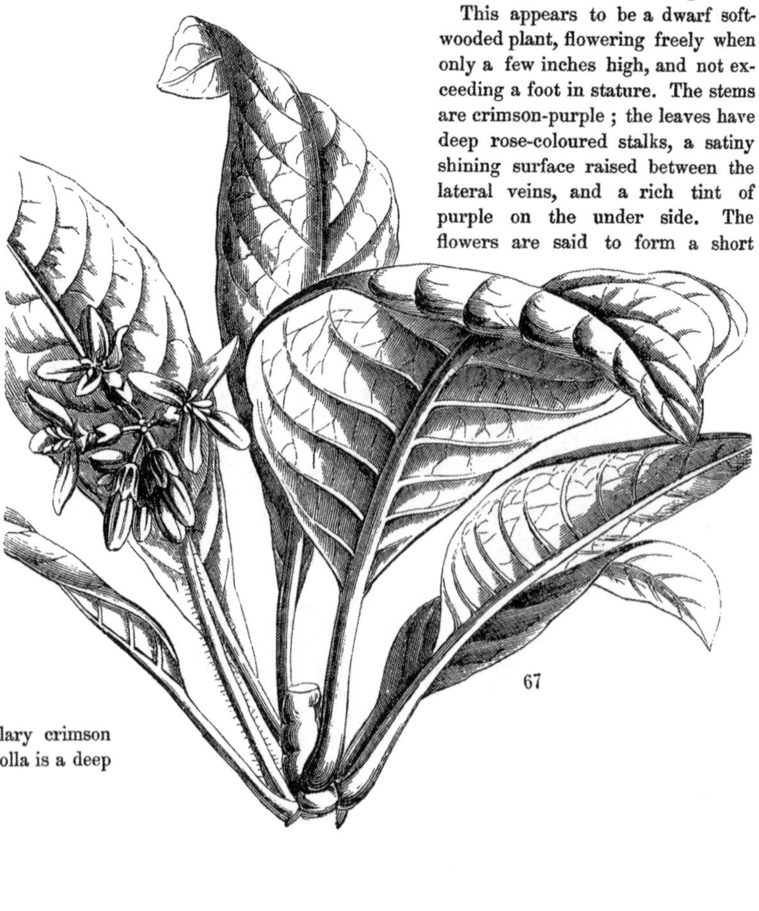

67

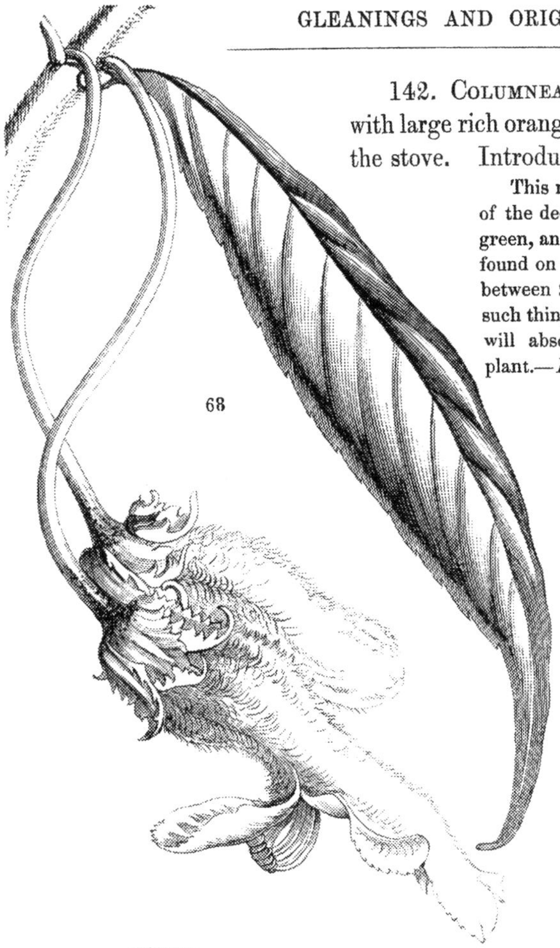

142. Columnea aurantiaca. *Decaisne*. A climbing Gesnerad, with large rich orange-coloured flowers, from New Grenada. Requires the stove. Introduced by Mr. Linden. (Fig. 68.)

This must be one of the handsomest of its race, the flowers being of the deepest and richest orange colour; the calyx pale yellowish green, and the stalk richly spotted with purple at the point. It was found on the Andes of Merida, in a temperate region, forming a zone between 9000 and 10,000 feet of elevation above the sea. Like all such things, it grows well upon a lump of nearly rotten wood, which will absorb water like a sponge, and give it back gradually to the plant.—*Flore des Serres*, t. 552.

68

143. Arctocalyx Endlicherianus. *Planchon*. A stove Gesnerad, with a shaggy brownish-black stem, and long yellow sessile flowers. From Mexico. Introduced by M. Abel, of Vienna. (Fig. 69.)

A remarkable plant said to have been found by the traveller Carl Heller, in the forests near Mirador, in the province of Vera Cruz, at the height of 2000 feet above the sea. It has the habit of an Alloplect. The leaves are fleshy, oval, unequal at the base, doubly serrated, and shaggy with long hairs on the veins of the under-side. The flowers are represented as springing from various parts of the surface of the stem, and not from the axils of the leaves exclusively. The shaggy calyx is nearly smooth at its upper end and glaucous. The corolla is golden yellow, with a regularly lacerated 5-lobed limb, streaked inside with lines of large crimson spots.—*Flore des Serres*, 546.

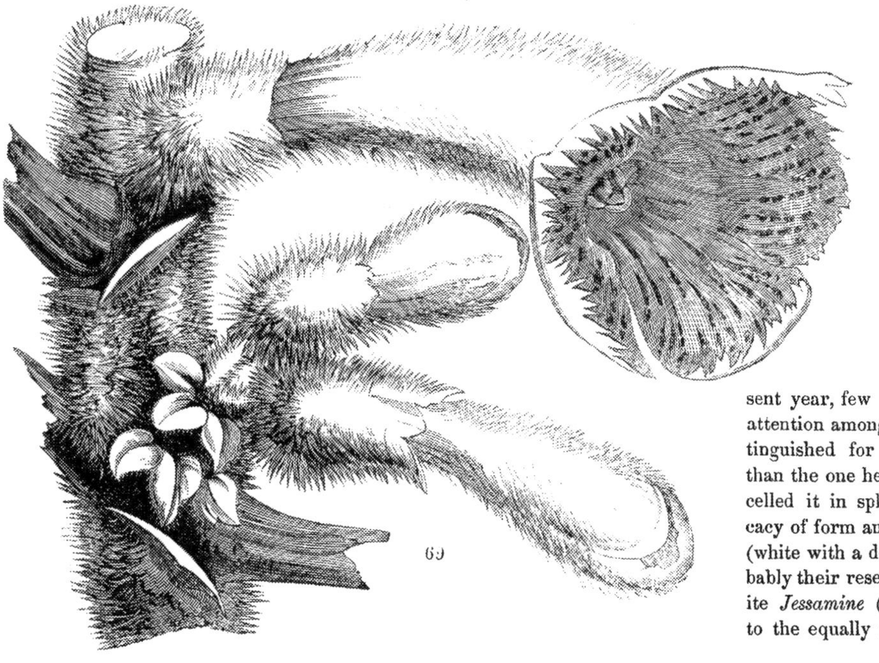

144. Rhododendron jasminiflorum. *Hooker*. A greenhouse shrub, with fragrant white flowers. Native of Malacca. Introduced by Messrs. Veitch & Co. (Fig. 70.)

69

"At the first, and truly splendid, Exhibition of flowers at the Chiswick Gardens of the present year, few plants excited greater attention among the visitors most distinguished for taste and judgment, than the one here figured. Many excelled it in splendour; but the delicacy of form and colour of the flowers (white with a deep pink eye), and probably their resemblance to the favourite *Jessamine* (some compared them to the equally favourite *Stephanotis*), attracted general notice. So unlike, indeed, are they to the ordinary form of Rhododendron blossoms, that the 'Gardeners' Chronicle,' in recounting the prizes of the day, seemed to imply that this was probably no Rhododendron at all!" It is a native of Mount Ophir, Malacca; elev. 5000 feet, and seems a ready flowerer. Branches bare of leaves below, and knotted where they had been inserted. Leaves crowded towards the upper part of the branches, lowermost ones subverticillate, on short petioles, obovate-oblong, rather acute, glabrous, nearly coriaceous. Umbel terminal, many-flowered. Peduncles 1-flowered, short, with small reddish bracteas at the base, and, as well as the very small, shallow, obscurely 5-lobed calyx, lepidote. Corolla salver-shaped, white, slightly tinged with rose below the limb; the tube two inches long, straight, scarcely gibbous at the base: the limb spreading, of five obovate wavy lobes, almost exactly equal. Stamens 10. Filaments filiform, downy, as long as the tube. Anthers red (forming a red eye, as seen at the mouth of the white corolla). Ovary oblong-cylindrical, lepidote, 5-celled, glandular at the base. Style rather shorter than the stamens, filiform, downy. Stigma dilated, obtuse, green.—*Botanical Magazine*, t. 4524.

We do not think that the "Gardeners' Chronicle" expressed an opinion adverse to this plant being a Rhododendron. And we can answer for this, that any observation which was made had no relation to the mere form of the corolla. There are points connected with the alpine Indian Rhododendrons which have attracted no attention, and yet deserve serious examination. What, for instance, is the meaning of the *continuation* of the style and ovary, instead of the usual *articulation?* And what is the equivalent among true Rhododendrons of the epidermoidal glands, capped with scurfs, which lie everywhere among the stomates of this and some other Rhododendrons? These matters are of higher interest than the comparative length of the tube of a corolla.

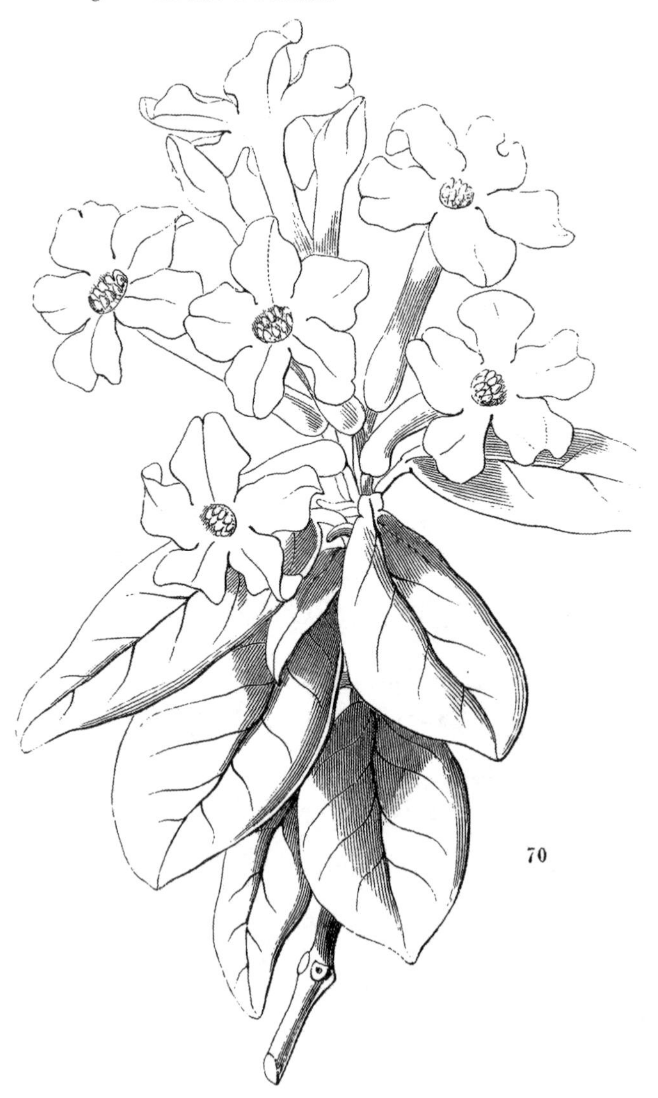

70

PLATE 19.

L. Constans, Pinx. & Zinc.

Printed by C.F. Cheffins, London.

[PLATE 19.]

THE UPRIGHT BRYANTH.

(BRYANTHUS ERECTUS.)

A hardy evergreen dwarf Shrub, of UNCERTAIN ORIGIN, *belonging to the Order of* HEATHWORTS.

Specific Character.

THE UPRIGHT BRYANTH.—Stem erect much branched. Leaves linear, obtuse, obscurely serrated. Flower-stalks hairy. Flowers solitary, corymbose. Sepals acute, smooth. Corolla campanulate, tubeless, acutely five-lobed. Style projecting.

BRYANTHUS *ERECTUS.*—Caule ramoso erecto, foliis linearibus obtusis obsolete serratis, pedunculis pilosis, floribus solitariis corymbosis, sepalis acutissimis glabris, corollâ campanulatâ acutè 5-lobâ tubo omninò nullo, stylo longè exserto.

Bryanthus erectus : *of the gardens.*

THIS charming little bush is said to be a hybrid, obtained by Mr. Cunningham of Comely bank, Edinburgh, between the blue Phyllodoce (*P. taxifolia,* alias *Menziesia cærulea*) and the Cistus Rhododendron (*Rhodothamnus Chamæcistus.*) Whatever its origin, it is certainly one of the most lovely plants that our gardens know. The specimen from which the accompanying figure was taken formed a round compact bush as large as a man's head, covered for a long time with the most delicate rose-coloured flowers, resembling miniature Kalmias. It was in perfection in April in the Garden of the Horticultural Society, where it was grown with the Cistus Rhododendron itself.

Such plants, although capable of bearing any degree of cold, are found difficult to cultivate on account of their impatience of dry air. Hence it is impossible to keep them in health in the open ground in ordinary places in London. The north side of walls, where the sun never shines, and low, but thoroughly drained places are where they succeed best. Better still are damp cold shaded pits in which the air always remains damp ; it is in such places that Mr. Gordon grows them in the Garden of the Horticultural Society.

And now for the question is this really a hybrid, or such a one as is pretended. A correspondent well acquainted with the practical results obtained by muling says that—" If Rhododendron Chamæcistus were to breed with Menziesia cærulea, the mule would differ from Bryanthus erectus, as will be evident from comparing the three." We have taken some pains to institute a fair comparison between them, and the result is that we believe the plant to be a mule, probably deriving its parentage in part from the Cistus Rhododendron, and in part from some sort of Phyllodoce. It is not, however, to the Blue Phyllodoce that we should refer it, but rather to the Crowberry Phyllodoce, published by Dr. Graham in the Botanical Magazine under the name of *Menziesia empetrifolia*, afterwards altered by Sir W. Hooker to *M. Grahamii*. At first sight, indeed, one would say that the Upright Bryanth was the same plant—leaves, manner of flowering, manner of growth being almost identical. But the flowers of this plant are twice too large; their sepals are very sharp-pointed instead of being blunt; and, above all, the corolla has no tube whatever, but expands regularly from the base upwards into its peculiar bell-shaped form. In this respect it *seems* to answer to the character of a Bryanth, to the lawful species of which we do not possess any access; and is at variance with all the Phyllodoces, which, the Crowberry Phyllodoce included, have a distinct separation, by means of a contraction, between the tube and the limb.

We therefore conclude that this Bryanth may be a cross between the Cistus Rhododendron and the Blue Phyllodoce; owing its larger flowers, with the more delicate colour, to the influence of the former.

L. Constans, Pinx & Zinc

Printed by C.F. Cheffins, London.

[Plate 20.]

THE SALMON-COLOURED MOUTAN.

(MOUTAN OFFICINALIS; SALMONEA.)

A hardy under-shrub from China, *belonging to the Natural Order of* Crowfoots.

Pæonia Moutan, Salmonea. *Journal of the Horticultural Society*, vol. iii., p. 236.

When Mr. Fortune first visited China, in the service of the Horticultural Society, the acquisition of new Moutans was one of the first objects to which he attended. In his "Wanderings" he mentions the beauty of the varieties seen by him at Shanghae, how he heard of yellow, and purple, and blue sorts, and at one time saw lilacs and purples, some nearly black, at another, dark purples, lilacs and deep reds. Afterwards, having discovered that these things came from a place only six or eight miles from Shanghae, Mr. Fortune tells us that he proceeded there daily during the time the different plants were coming into bloom, and secured some most striking and beautiful kinds for the Horticultural Society.

One of these, received by the Society in April 1846, is now figured. About its beauty and distinctness there can be only one opinion. With all the largeness and doubleness of varieties of the common Officinal Pæony, it combines that delicacy of texture and fineness of colour which exist among the Moutans alone. "The outer petals when fully blown are a pale salmon-colour; the inner have a deep rich tint of the same." The accompanying figure is in no respects an exaggeration of the beauty of this variety.

The name Moutan seems to be an alteration of the word *Botan*, the usual name of these plants in Japan, as we are told by Kæmpfer, who adds that it is also called *Fkamigusa* and *Hatskangusa*. As the Japanese name the common Pæony *Saku jaku* and *Kawu Junkusa*, they seem to think the Moutan and the Pæony distinct genera, in which we quite agree with them, for reasons that will be given on another occasion, when we figure a still finer variety than this. It is to be suspected also that more species than one is comprehended under the common name of Tree Pæony: even although, as is probable, the Poppy Moutan (*P. papaveracea*) should be a mere variety of the common kind; for some of the Japanese kinds are said to form rapidly a woody stem eight or ten feet high; a stature which the common Moutans would only gain after many years, in even favourable climates.

No English cold seems to affect these plants: and yet their beauty is usually impaired when in flower, by the coldness of our nights. An obvious remedy for this is to protect their blossoms with glass screens: but the same result may be had if they are grown under *north* walls, so as to retard their flowering and to lower their excitability. It will also be found that the gradual thaw which takes place when the vernal sun has no access is a powerful safeguard against the consequences of being frozen; while, on the contrary, the rapid elevation of temperature which occurs in a sunny border is invariably productive of bad consequences.

The Chinese and Japanese are said to reckon their varieties of Moutans by hundreds, as we do our Roses. It is not improbable, now that the single and very slightly double kinds are beginning to establish themselves in Europe, that we too shall have the same dominion over them as over Camellias and Chrysanthems. The double varieties sometimes seed; there is nothing whatever to prevent the single kinds from doing so; and it is only necessary for the imported plants to become common to secure abundance of seed, out of which a new European race is sure to arise. The largest collection of these plants yet brought to Europe is that of Dr. V. Siebold, who imported them from Japan in 1844. They are said to have been obtained from the Imperial Gardens of Jedo and Mijako, and include all the finest sorts known in that empire. They are distinguished by the form and colour of the petals, and of the disk, styles, and stamens. None of them are completely double; most are single; some only semi-double; and hence very likely to have seeds. The blossoms are described as being very large and in some cases very sweet-scented.

The following list of these Japanese Tree Pæonies has been circulated by Dr. V. Siebold, who cultivates them all, as well as others, in his Nursery at Leyden.

Reine Victoria. Petals white. Disk purple.
Reine des Belges. Petals white, greenish on the outside, with a pale rose-coloured spot at the base. Disk white.
Flora. Petals white, with a straw-coloured tinge, and a pale lilac spot at the base. Disk whitish green.
Duchesse d'Orleans. Petals white, with a straw-coloured tint, the outer streaked with green. Disk white.
Nymphæa. Petals pure white. Disk white.
Madame De Cock. Petals white (before expansion greenish straw-colour) dotted with dark lilac at their base. Disk yellowish.
Ida. Petals pale rose (streaked with straw-colour and tinged with green before expansion). Disk pink.
Helena. Petals pink (clear rose-colour before expansion). Disk purple.
Reinwardt. Petals dark rose, streaked with purple and carmine. Disk dark purple.
De Vriese. Petals dark rose, streaked with purple and carmine. Disk white.
Princesse Charlotte. Petals pale rose with darker streaks. Disk white.
Von Siebold. Flowers semi-double. Petals carmine red streaked with purple. Disk deep red.
Comte de Flandre. Flowers semi-double. Petals carmine streaked with purple. Disk crimson.
Van Hulthem. Petals purple red. Disk purple.
Duc de Devonshire. Petals carmine red. Disk dark purple.
Duc de Brabant. Petals pink with a lilac tint. Disk white.
Roi des Belges. Petals dark crimson with a purple tinge. Disk carmine.
Alexandre Verschaffelt. Petals purple red, variegated, dotted with white and lilac. Disk purple.
Prince Albert. Petals dark brown red, the outer ones sometimes variegated with white and green. Disk purple.
The Wild Tree Pæony. On this are worked the varieties obtained by cultivation. It deserves attention as well for the colour and sweet scent of its flowers as in a horticultural point of view, for its easy propagation by the division of its root and its hardiness, it having borne several winters in the open air without any shelter. Its colour is bright scarlet; each petal has a black spot at its base, and the stamens are surmounted with golden yellow anthers.

PLATE 21.

L. Constans. Pinx & Zinc.

Printed by C.F.Cheffins, London.

[Plate 21.]

THE SESSILE ONCID.

(ONCIDIUM SESSILE.

A stove Epiphyte, from Peru, *belonging to the Natural Order of* Orchids.

Specific Character.

THE SESSILE ONCID.—Pseudo-bulbs 2-leaved, oblong, compressed, ribbed. Leaves strap-shaped, papery, blunt, shorter than the scape, which bears a panicled raceme. Sepals distinct and petals equal in size and form, all sessile. Lip-eared, dilated at the end and retuse; its re-entering angles slightly lobed; the crest hollowed out, smooth, 3-lobed, with two small edges in front. Wings of the calyx short and truncated.

ONCIDIUM *SESSILE.*—(Pentapetala macropetala); pseudobulbis diphyllis oblongis compressis costatis, foliis loratis pergameneis obtusiusculis scapo racemoso-paniculato brevioribus, sepalis distinctis petalisque oblongis obtusis planis æqualibus conformibus omninò sessilibus, labello auriculato apice dilatato retuso sinu sublobato, cristâ excavatâ lævi 3-lobâ anticè bilamellatâ, columnæ alis brevibus truncatis.

A native of the country at the back of Santa Martha, whence it was sent to His Grace the Duke of Northumberland by Mr. Purdie. It flowered at Syon in this last spring.

It is nearly related to the little known Excavated Oncid (*O. excavatum*), a Peruvian plant formerly in the possession of Messrs. Loddiges; but it is much handsomer and may be regarded as one of the best of the little group to which it belongs. The Excavated Oncid differs essentially in the following circumstances; the flowers form a loose, and not a close or racemose panicle; the sepals are narrower than the petals, not of the same breadth, they are distinctly stalked (unguiculate) not perfectly sessile, and they are acute not blunt like the petals; the hollow at the base of the lip is much more considerable, and covered with little frosty specks, but here it is quite smooth; there are a few scattered tubercles on each side of the hollow, but here there are none; and the wings of the column are much larger, rounded and not truncate.

The habit of this species is that of the Lofty Oncid (*O. altissimum*) on a small scale; the leaves have the same firm thin texture; and the flowers are in a narrow panicle. The sepals and petals are remarkable, in this genus, for their total want of the stalk or unguis so generally characteristic of Oncids; instead of which they sit close round the column, and give the flower something of the roundness and flatness obtained by art in what are called Florist's flowers; they are clear yellow with a few pale cinnamon-brown spots near the base. The lip has one curved stain of the same colour on each side near the base.

This, the Excavated Oncid, with several others, forms a group in the genus readily known by the sepals being perfectly distinct from each other, and smaller, or at least not larger, than the petals,—or if broader considerably shorter—contrary to what is usual among the neighbouring species. Hence, in the table published in our VIth plate they stood as a sixth section under the name of Pentapetala macropetala. Of that section we will avail ourselves of the opportunity to give an enumeration.

VI. PENTAPETALA MACROPETALA.

1. O. convolvulaceum.

O. rhizomate volubili filiformi, pseudobulbis secus rhizoma distantibus compressis subrotundis monophyllis, folio plano sessili ovato-oblongo obtuso mucronulato, pedunculis basi squamatis unifloris folio subæqualibus, sepalis liberis petalisque latioribus oblongis acutis patentissimis, labello maximo bilobo baseos auriculis linearibus apice dilatatis rotundatis, cristâ elevatâ truncatâ utrinque lobatâ verrucis a 2 fronte, columnæ alis acutè truncatis.

Native country, *Venezuela*. Herb. Linden, No. 1444, from the voyage of Funck and Schlim.

This most curious plant has the habit of a Bolbophyl rather than of an Oncid, agreeing in that respect with the very different *O. serpens*. On a hard twining rhizome appear at the distance of 3 or 4 inches, one-leaved pseudobulbs usually springing from the axil of a small leaf; these pseudobulbs are thin, nearly round, scarcely an inch long, and each bear a solitary flat leaf about 2 inches long. The flowers, which are nearly 2 inches in diameter, grow singly on peduncles scarcely longer than the leaves; they appear to be spotless, but their colour is unknown. *Not in cultivation.*

2. O. excavatum *Lindl. in Sert. orch. sub t.* 25. *B. Reg.* 1839., *misc.* 150.

O. pseudobulbis . . . ., foliis oblongo-ligulatis . . . ., scapo paniculato, bracteis squamiformibus membranaceis acutis, sepalis lateralibus obovatis liberis supremo concavo acuto, petalis membranaceis oblongis basi angustatis, labello sessili pandurato apice rotundato emarginato sellæformi basi cordato convexo fornicatim excavato, cristâ tuberculatâ, columnæ alis oblongis rotundatis.

Native country, *Peru.*

This has yellow flowers, spotted with brown, and is easily known by the base of the labellum being very convex, a little hollowed out in front, and excavated with a deep pit on the under side.

3. O sessile *of this Plate.*

4. O. sarcodes *Lindl. in Journ. Hort. Soc.* iv. *p.* 260; *aliàs* O. Rigbyanum, *Paxton Mag., Oct.* 1849.

O. paniculâ racemosâ angustâ, sepalis liberis obovatis planiusculis, petalis majoribus unguiculatis obovato-spathulatis repandis, labelli lobis lateralibus nanis serrulatis intermedio maximo undulato repando emarginato, cristâ lineari apice bilobâ tuberculosâ pubescente, columnæ pubescentis clinandrio angustè marginato alis carnosis truncatis glabris.

Native country, *Brazil.*

The habit of this species is entirely that of *O. pubes* and *O. amictum*. The flowers are large, bright yellow, blotched with brown-red; the column white, with blood-red fleshy truncated wings. In structure it approaches nearly to *O. ampliatum* and *excavatum*, from which its downy column, serrated side lobes of the lip, and peculiar two-lobed hairy crest abundantly distinguish it. Like so many others it varies much in the size and colour of its flowers; the best variety we have seen was sent by Mrs. Lawrence.

5. O. ampliatum *Lindl. gen. et sp. orch., p.* 202. *B. Reg. t.* 1699.

O. pseudobulbis subrotundis ancipitibus rugosis maculatis diphyllis, foliis oblongis coriaceis planis subundulatis scapo paniculato brevioribus, sepalis omnibus liberis, labello bilobo subrotundo transverso: laciniis lateralibus brevissimis, callo baseos 5-lobo: lobis lateralibus patentissimis planis truncatis intermediis teretibus centrali compresso, alis columnæ cuneatis dentatis reflexis.

Native country, *Panama and Guatemala.*

A noble species, of which there are two varieties, one much larger than the other. According to Mr. Skinner it comes from Costa Rica, on the sea-shore in the Gulf of Nicaya; and is also found throughout the coasts of Nicaragua, and in the Escuintla, 15 leagues from Guatemala; growing in a climate the temperature of which does not rise above 80° or 85°; flowering in February.

6. O. onustum *Lindl. gen. et sp. orch., p.* 203.

O. foliis linearibus complicatis falcatis, scapo simplici, racemis cernuis secundis multifloris, sepalis omnibus liberis, labello bilobo transverso: lobis lateralibus linearibus apice subdilatatis, callo baseos oblongo cochleato anticè appendiculâ tuberculiformi instructo, alis columnæ 2 integerrimis.

Native country, *Panama and Colombia.*

Flowers (apparently whole-coloured) in a simple curved raceme 3 or 4 inches long. *Not yet in cultivation.*

7. O. stramineum *Lindl. B. Reg.* 1838, *misc.* 63. 1840, *t.* 14.

O. ebulbe, foliis crassis carnosis ovato-lanceolatis acutis dorso rotundatis scapo paniculato rigido erecto brevioribus, sepalis subrotundis unguiculatis concavis liberis integerrimis, petalis duplò majoribus oblongis obtusis emarginatis margine crispis, labelli lobis lateralibus oblongis carnosis acutis margine revolutis basi columnæ proximâ nectariferis intermedio reniformi plano emarginato longioribus, tuberculis disci 4 geminatis, columnæ alis carnosis linearibus obtusis elongatis genuflexis decurvis.

Native country, *Mexico.*

Leaves short, fleshy, stiff. Flowers in a dense panicle, pale straw-colour, with a few dark dots on the lip.

8. O. pyramidale *Lindl. in. Ann. nat. hist.* xv.

O. pseudobulbis ovatis ancipitibus 2—3-phyllis, foliis oblongis tenuibus basi angustatis scapo erecto rigido paniculato pyramidali multo brevioribus, sepalis obtusis liberis dorsali ovali lateralibus linearibus, petalis duplo latioribus ovatis obtusis, labelli lobis lateralibus amplexicaulibus intermedio bilobo latioribus, crista anticè excavata processubus 7 (?) linearibus anticis longioribus, columna nana alis verticalibus lineari-cuneatis sublobatis, rostello subulato.

Native country, *Peru,* near Pasto in the woods of Menesco, on trees.

Allied to *O. excavatum,* but with the rostellum of *O. ornithorhynchum.* Scape six inches high. Panicle not more than 4 inches across, and fully a foot long. *Not in cultivation.*

9. O. lancifolium *Lindl. in Plant. Hartweg.,* p. 151.

O. pseudobulbis oblongis compressis, foliis lanceolatis acutis scapo stricto apice paniculato brevioribus, ramis valde flexuosis, sepalis linearibus obtusis, petalis obovatis subundulatis duplo latioribus, labelli basi obcuneati lobo intermedio transverso reniformi bilobo, cristæ tuberculis plurimis carnosis ramentaceis, columnæ recurvantis basi biauris alis maximis acinaciformibus subserratis undulatis.

Native country, *Peru,* on the Cordillera near Loxa.

Leaves not more than 6 inches long. Scape about 6 inches high. Panicle oblong, close, not quite so long. Flowers small, apparently yellow speckled with purple in the middle. *Not in cultivation.*

10. O. Jamiesoni.

O. pseudobulbis . . . . , folio carinato complicato, paniculâ effusâ ramulis divaricatis, floribus heteromorphis pluribus abortientibus, sepalis linearibus obtusis rectis, petalis duplo latioribus oblongis obtusis subundulatis, labello auriculato apice semicirculari bilobo, cristæ tuberculis 5 parvis duabus lateralibus patentissimis cæteris subparallelis intermedio productiore, columnæ alis oblongis erectis rotundatis.

Native country, *Peru,* near Quito.

A handsome species, with flat yellow flowers, having broken bands of brown at the base of the petals and nowhere else. The wings of the column are not unlike a bat's ears. Many of the flowers are abortive in this and some other Peruvian species, and form little irregular starry bodies among the rest. Received from Dr. Jamieson of Quito. *Not in cultivation.*

---

11. O. Papilio *L.,* p. 203. *B. Reg.,* t. 910. *B. M.,* t. 2795. *B. Cab.,* t. 1086.

O. pseudobulbis subrotundis compressis rugosis monophyllis, foliis oblongis coriaceis obtusis maculatis, scapo perennante debili ancipiti articulato apice paucifloro, sepalo supremo petalisque linearibus longissimis basi angustatis, sepalis lateralibus latis revolutis undulatis labello longioribus, labelli laciniâ intermediâ oblongâ emarginatâ subrotundâ crispâ basi valde angustatâ lateralibus rotundatis, cristæ glandulis formam ranæ cubantis referentibus, columnæ alis serratis.

Native country, *Trinidad.*

It must be confessed that this well-known species has no resemblance to the others here associated with it. It probably should form a section (or genus?) by itself.

GLEANINGS AND ORIGINAL MEMORANDA.

145. Trichosacme lanata. *Zuccarini.* A woolly climbing Asclepiad from Mexico. Flowers small, dark purple, with long tails. Introduced by Messrs. Knight and Perry. (Fig. 71.)

This singular plant is so buried in wool that no part of its surface, except the face of the corolla, can be seen. The leaves are white, like a lamb's fleece. The stem is in the same state. The minute flowers grow in pendulous umbels at the end of a woolly reflexed flower-stalk. The singularity of the flower resides in the production of long, weak, feathery, purple tails from each lobe of the corolla; not, however, from the apex, as Zuccarini supposed. On the contrary, each lobe of the corolla is cut into two equal triangular teeth, and it is from the right hand tooth of each lobe that the tails proceed. They spring forth abruptly, wave in the wind in the most curious manner, and do not separate from the corolla without the application of some force. No doubt they are analogous to the tails of Strophanths; but what can they be for? Messrs. Knight and Perry received it from the Imperial Botanic Garden, St. Petersburg.

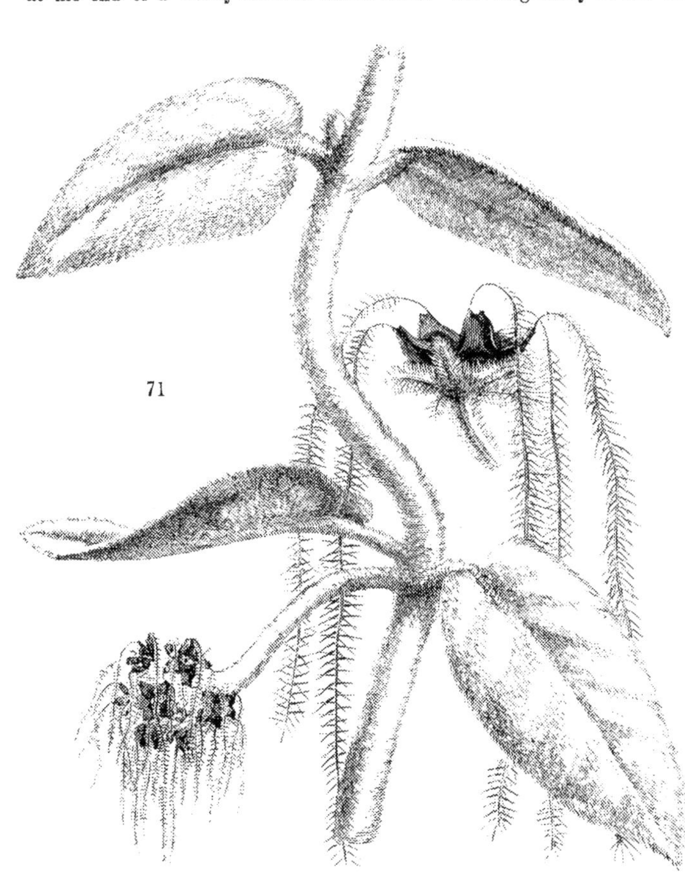
71

146. Calceolaria pavonii. *Bentham.* A herbaceous species, hardy in summer, but requiring protection in winter. Flowers yellow. Leaves large and coarse. Introduced by Lucombe and Co.

A rare and remarkably large species, originally detected at Chincao and Muña in the Andes of Peru, and afterwards discovered in the province of Chachapoyas. Messrs. Lucombe and Co. say, that when bedded out in the summer it makes a very striking appearance, with its noble and rather deep yellow flowers and ample foliage. Root perennial. Stem one and a half to two feet and more high, a good deal branched, herbaceous, succulent, taper, or but slightly angled, hairy, green, sometimes tinged with purple, and slightly viscid. Leaves often more than a span long, opposite and perfoliate; the stalk very broad, and winged at the base; the blade ovate, acute, or acuminate, often truncate or cordate at the base; the surface wrinkled, the margin doubly toothed, downy above, pale, almost white, and somewhat woolly beneath. Panicle ample. Flowers very large; lower lip almost orbicular, folded against the upper lip, but not so much as to exclude from view the deep blood-coloured spots in the inside. "We have hardly yet had the opportunity of testing its merits as a bedding plant, but we fear its tall and rude growth may be somewhat against it for that purpose. Its handsome flowers make it well worthy of being grown as a show-plant for the greenhouse."—*Botanical Magazine*, t. 4525.

147. Calanthe vestita. *Wallich.* A very handsome terrestrial Orchid, from Burma. Flowers white, with a deep stain of bright crimson in the middle of the lip. Flowers in November. Introduced by Messrs. Veitch. (Fig. 72, *a* & *b*.)

This is scarcely less beautiful than C. sylvatica, our No. 33 of the present volume: and must be classed among the

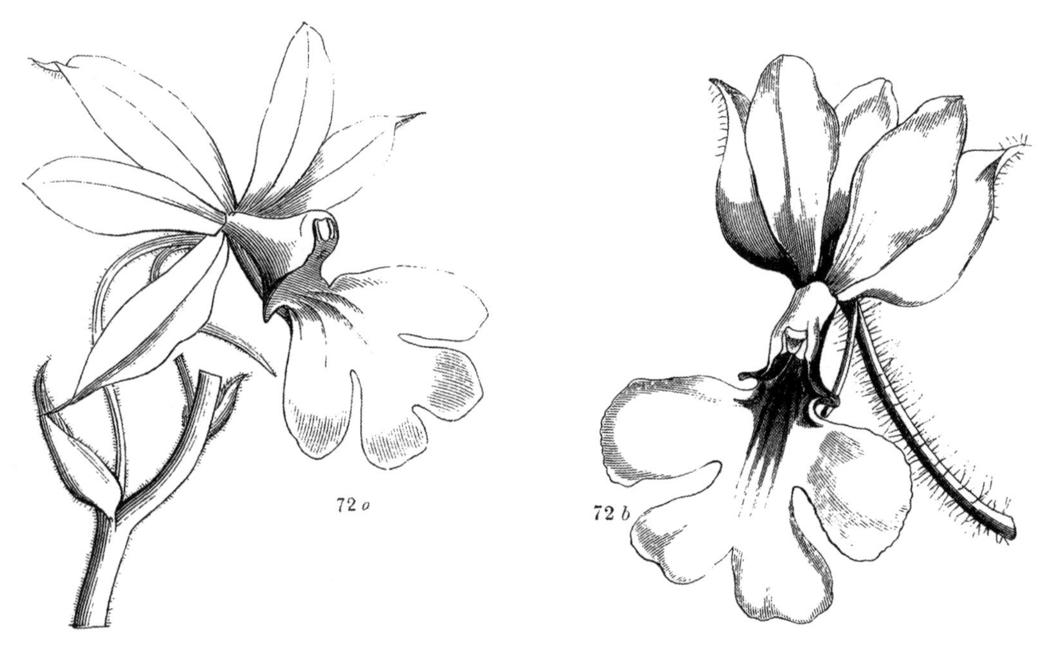

finest of the terrestrial Orchids. The stems are fully two feet high, and like all the other parts are clothed with long soft hairs—very slender, long-jointed, of unequal thickness, and blunt, containing in their interior a brown fluid. The flowers are in loose, zig-zag racemes, with conspicuous ovate acuminate bracts. The sepals and petals are finally turned back so as to be nearly parallel with each other; they are snow-white, with a few hairs on the back of the first. The lip is bluntly 4-lobed, with a narrow short ear on each side at the base. The spur is very slender, and abruptly bent upwards, so that its point touches the lip. A large silver medal, the highest ever given in Regent Street, was awarded to this plant by the Horticultural Society on the 7th of Nov., 1848, when it was exhibited by Messrs. Veitch for the first time.

148. Oncidium varicosum. *Lindley.* A fine stove Orchid from Brazil, with tall scapes covered with a glaucous bloom and bright yellow flowers. Introduced by M. de Jonghe, of Brussels. Flowered at Chiswick, in October, 1849.

A glaucous strong-growing species, of considerable beauty. The leaves are firm and ligulate-lanceolate. The scapes which are strong, very glaucous, and about 3 feet long, have a great branching panicle, loaded with from 80 to 90 large showy flowers. The sepals and petals are pale dull green banded with dull brown. The lip is large, very bright yellow, with two ovate lateral ears, somewhat crenate in front, and a 4-lobed central portion. The crest consists of two triple teeth, one standing before the other, and of a little ring of varicose veins placed on each side of it. The wings of the column are oblong, whole coloured, and finely notched.—*Journ. Hort. Soc.*, vol. v. p. 143.

73

149. STERIPHOMA PARADOXUM. *Endlicher; (aliàs* Capparis paradoxa *Jacquin; aliàs* Stephania cleomoides *Willdenow*). A small stove shrub of great beauty belonging to the Capparids, with bright yellow flowers. Native of Venezuela. (Fig. 73.)

A plant of ancient introduction, figured many years since by Jacquin in his account of the plants of the Imperial Garden at Schœnbrunn. Re-introduced by M. Karsten, it has found its way into modern gardens. It grows naturally to the height of a yard or two. The long-stalked, simple, ovate-lanceolate leaves are deep green. The flowers grow in a close raceme. The calyx is downy with star-shaped hairs, 2-lobed, and deep golden yellow. The petals, which extend a little beyond it, are 4, and much paler yellow. The stamens, 6 in number, are curved downwards and fully 3 inches long. The fruit appears to be cylindrical, and about 5 inches long, succulent like a berry. It requires a damp stove, plenty of pot room, and a good rest in the autumn. It strikes easily from cuttings.—*Flore des Serres*, 564.

150. CAMPANULA NOBILIS. *Lindley*. With white flowers.

This, which has been figured by Mr. Van Houtte, t. 563, is hardly so handsome as the original Chinese plant with purple flowers.

151. ACER VILLOSUM. *Wallich*. A hardy tree from the Himalayas, with broad downy deciduous leaves. Introduced by Messrs. Osborne and Co.

This Himalayan tree, the villous Sycamore, is said to be hardy, in the Nursery of Messrs. Osborne and Co., of Fulham. It has broad heart-shaped angular leaves, 5 inches long, with the 2 lower lobes *shorter* than the 3 upper. Their stalks are as long as themselves. Young wood, leafstalks and leaves on the under side, are clothed with a short hairiness which makes those parts quite soft to the touch. It has not flowered in this country; when it does it will produce close shaggy panicles of small green flowers. The Keys (*samaræ*) are rather more than 1½ inch long, hairy at the base, where they are also much wrinkled; but nearly smooth on the winged part.

152. MANDRAGORA AUTUMNALIS. *Bertoloni*. A hardy stemless perennial, with deep mazarine blue flowers, belonging to the Nightshades (*Solanaceæ.*) Native of the South of Italy and Levant.

The common Mandrake produces its pale lilac flowers in midwinter and early spring, and is a plant of no horticultural interest. This, on the contrary, which was probably the real Mandrake of Scripture, is a very handsome autumn flowering plant, with large dark-green sinuous leaves and flowers of the most intense blue. There is no English cold that it is incapable of supporting, provided it is kept dry ; but the great fleshy roots rot whenever a low temperature is accompanied by water in contact with them. We have found the common kind live for many years in sand among stones raised a a foot or so above the ground, in a south aspect and covered with a hand-glass, which is never removed till dangerous frosts are gone. It is, however, very subject to the attack of slugs. According to M. Van Houtte this species produces ripe fruit and seed abundantly at Ghent; we never saw the common Mandrake show any tendency that way. See *Flore des Serres*, t. 457.

153. Anigozanthus tyrianthinus. *Hooker.* A fine showy, herbaceous plant, from Swan River, with densely packed, deep, but dull, purple flowers, pale yellow inside. Belongs to the order of Bloodroots (Hæmodoraceæ). Not in cultivation.

One of the many fine things discovered by Mr. Drummond, during his excursions in the interior to the southwest of the Swan River settlement. He could not fail to be struck with the magnificence of this plant, three or four or more feet high, growing in masses, and bearing paniculated branches, and copious flowers clothed with dense tomentum of the richest Tyrian purple. Seeds have not yet germinated, but the dried specimens retain their form and colour almost equally with the living plant, and we are hence able to present an accurate figure to our readers. Its nearest affinity is perhaps with the *A. fuliginosa*, (Bot. Mag. t. 429,) but the flowers are very different in shape as well as in colour.—*Botanical Magazine*, t. 4507.

154. Eugenia Brasiliensis. *Lamarck.* (*aliàs* Myrtus Dombeyi *Sprengel.*) A stove evergreen tree with terminal tufts of large white flowers. Belongs to the Myrtle-blooms (Myrtaceæ.) Native of Brazil.

A small tree, found in the province of Rio de Janeiro, where, we are informed, it is also cultivated and the fruit brought to market, and sold under the name of *Grumichama.* It is handsome in its foliage and in its copious snowy flowers, which latter are remarkable for having their origin upon the lower portions of young terminal branches, or, in other words, upon partially developed leaf-buds, springing from the axils of opposite scales below the leafy portion. In this state the young leaves are deep purple-brown, contrasting prettily with the dark green of the old foliage and the pure white of the blossoms which are produced in April. Fruit, according to St. Hilaire, as large as a cherry, white or red, or black violet-coloured, esculent.

This is an old inhabitant of the Royal Gardens. Having been kept for many years in a small pot it never produced flowers ; but on being removed into the Palm-house, and shifted into a large pot, it grew vigorously, and in the spring of this year produced a profusion of flowers. It is now a handsome Laurel-like bush, six feet high. Light loam, mixed with a small quantity of leaf-mould, suits it ; and, as it is what may be termed a thirsty plant, it requires to be well supplied with water during the spring and summer months.—*Bot. Mag.*, t. 4526.

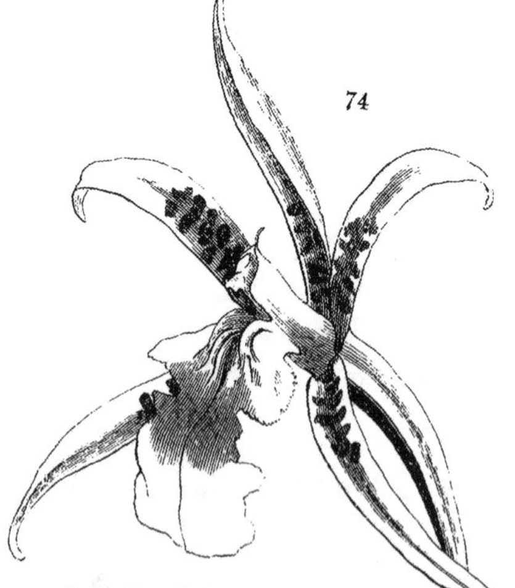

155. Aspasia lunata. *Lindley.* A stove epiphyte from Brazil, with pale-green speckled fragrant flowers. Lately blossomed with J. J. Blandy, Esq. (Fig. 74.)

This little-known species naturally bears a curved, somewhat crescent-shaped violet spot in the middle of a whitish lip. The sepals are green, spotted near the base with brown, like a Brassia. In drawings made in Brazil the crescent-shaped spot on the lip is represented as being much more distinct than it proves to be in cultivation.

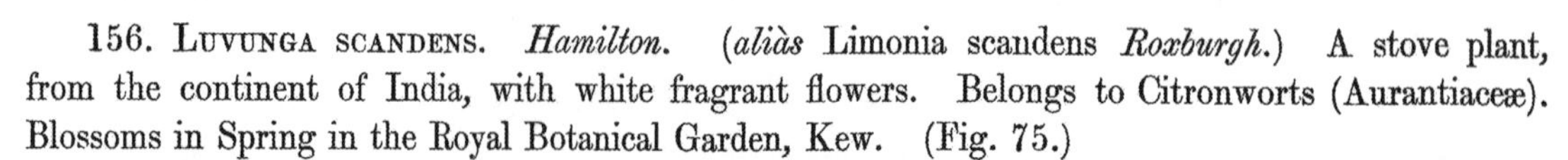

156. Luvunga scandens. *Hamilton.* (*aliàs* Limonia scandens *Roxburgh.*) A stove plant, from the continent of India, with white fragrant flowers. Belongs to Citronworts (Aurantiaceæ). Blossoms in Spring in the Royal Botanical Garden, Kew. (Fig. 75.)

A delicately fragrant plant from Silhet and Chittagong. Dr. Hamilton called it *Luvunga* (from its Sanscrit name, "*Luvungaluta*"). In cultivation, though attaining a height of nearly twenty feet, it hardly deserves to be called scandent. Leaves alternate, remote, each with three leaflets. Stalks two to three inches long. Leaflets five to six inches long, lanceolate, acuminate, entire, feather-veined, with clear transparent dots. Flowers axillary, in a dense short raceme, much resembling those of the Orange, and not less fragrant. Although this plant was introduced into the Royal Gardens in 1823, it never produced flowers till the present year ; which may be accounted for by its now being allowed greater freedom of growth in the Palm-house. The kind of soil is not important: any light loam suits it, so that it be not retentive of water.—*Botanical Magazine*, t. 4522.

157. ARNEBIA ECHIOIDES. *Alphonse De Candolle; (àlias* Lycopsis echioides *Linn.; alìàs* Anchusa echioides, *Bieberstein ; alìàs* Lithospermum erectum, *Fischer.*) A hardy herbaceous plant, with showy yellow flowers, belonging to Borageworts. Native of Armenia and the Caucasus. Flowers in June. (Fig. 76.)

Sir W. Hooker follows M. Alphonse De Candolle, who in a late volume of the Prodromus has referred this plant to *Arnebia*, although it does not accord in character. The species is a native of the Caucasian Alps, and of Armenia: is quite hardy, flowering in the open border, or in a pot, in June and July, where it makes a very pretty appearance

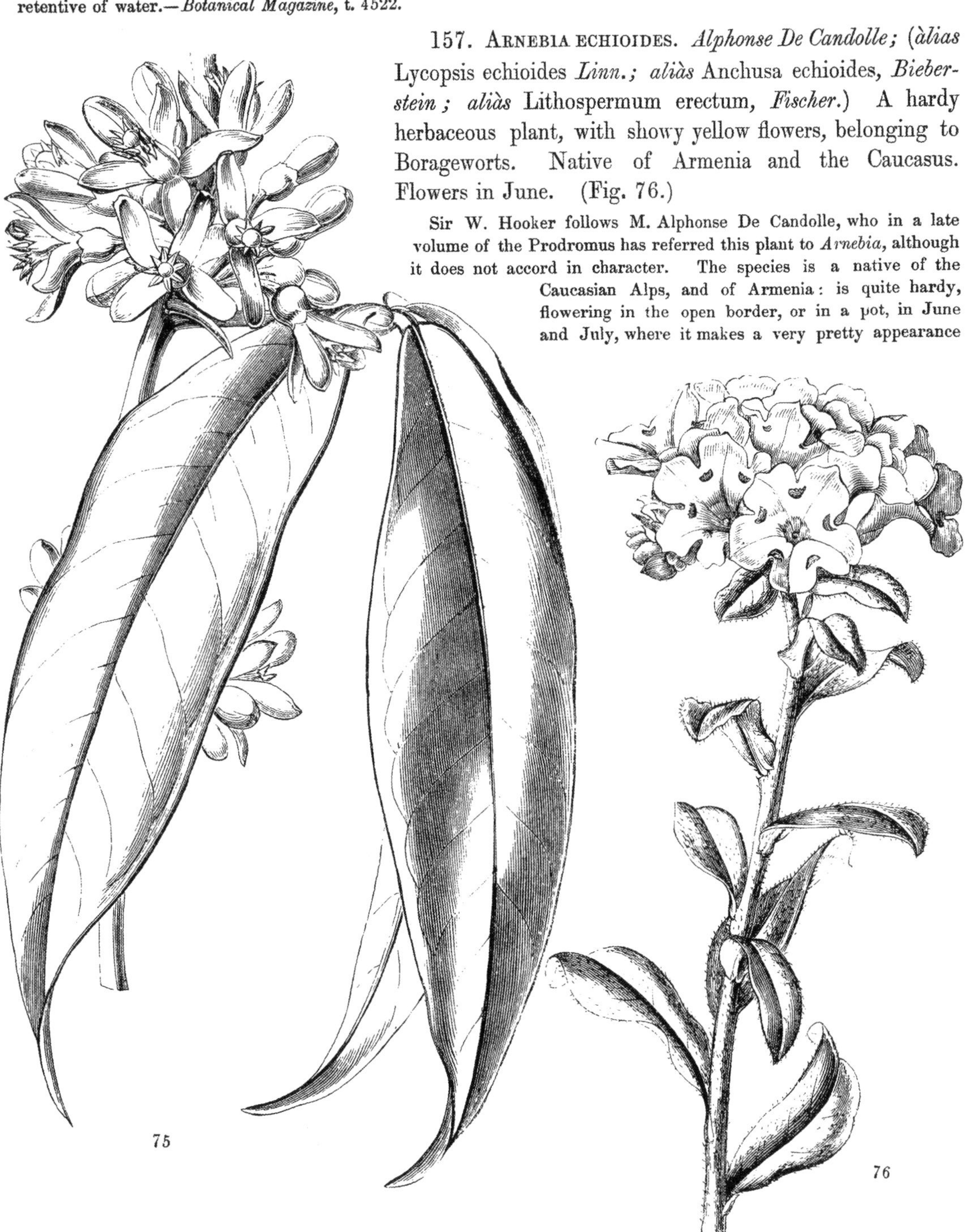

75

76

with its scorpioid spikes of large yellow flowers, with five deep purple, well-defined spots at the throat. These spots, however, in the cultivated plant, are sometimes obsolete—plants were raised in the Kew Gardens from seeds sent by Dr. Fischer, of St. Petersburg. Root fusiform, woody, throwing up two or more erect, leafy, herbaceous stems, about a span or more high, downy, with short hair. Leaves spreading, somewhat hoary, soft, all sessile; those from the roots, large, obovate, oblong; from the stem, obovate-lanceolate, all rather obtuse and becoming smaller upwards. The stems terminate in a branched, scorpioid, leafy spike of large yellow flowers. Calyx, cylindrical, hairy, cut almost to the base into five, erect, linear, obtuse segments. Corolla between funnel and salver-shaped; the mouth spreading; the tube nearly twice as long as the calyx, hairy within; the limb cut into five nearly equal, rounded lobes, having a dark orbicular purple spot at the re-entering angle of each pair of lobes. Style shorter than the tube. —*Bot. Mag.*, t. 4409.

Notwithstanding the number of aliàses under which this plant is already known, it is still unsatisfactorily named. It cannot with any propriety be placed in the same genus with *Arnebia cornuta*, whose style divides into 4 arms at the point, and which has 5 converging scales on the tube of the corolla near the base. It would rather seem to be an *Alkanna*, near *A. Græca*. At least it is identical in genus with *Alk. hirsutissima* from the Euphrates. We forbear, however, from disturbing the name, not possessing materials or leisure for investigating the different oriental species assembled by M. Alphonse De Candolle under the names of *Arnebia* and *Alkanna*. It was found in Persia by Major Willock, and we have recently remarked it among dried plants from the neighbourhood of Trebizond. The specimens from the former gentleman are nearly 18 inches high and loaded with flowers. M. Planchon, who has republished this plant in the "Flore des Serres," doubts its being an Arnebia, but throws no light upon its true genus.

158. ECHINOPSIS CRISTATA. *Salm Dyck.* (*aliàs* Echinocactus obrepandus *Salm Dyck.*) A beautiful white, or purple-flowered plant, belonging to the order of Indian Figs (Cactaceæ). Native of Bolivia.

No less remarkable for the large size of its flowers than for the deeply-lobed ribs of the stem; purchased of Mr. Bridges on his return from Bolivia, where he had gathered them and other fine species of *Cactaceæ* then first known in our gardens, in 1844. In 1846, the individual which blossomed, produced purple flowers; that which bloomed the following year (1847) bore white ones. This showy *Echinopsis* is a native of Chili, and, like its Mexican allies, thrives if potted in light loam with a little leaf-mould and a few nodules of lime-rubbish. The latter are for the purpose of keeping the soil open; it is also necessary that the pot should be well drained. In winter, water must be given very sparingly and the atmosphere of the house should be dry: the temperature need not exceed 50° during the night, and in very cold weather it may be allowed to fall 10° lower, provided a higher temperature be maintained during the day. As the season advances, the plants should receive the full influence of the increasing warmth of the sun; and during hot weather they will be benefited by frequent syringing over head, which should be done in the evening: it is, however, necessary to guard against the soil becoming saturated, for the soft fibrous roots suffer if they continue in a wet state for any length of time.—*Botanical Magazine*, t. 4521.

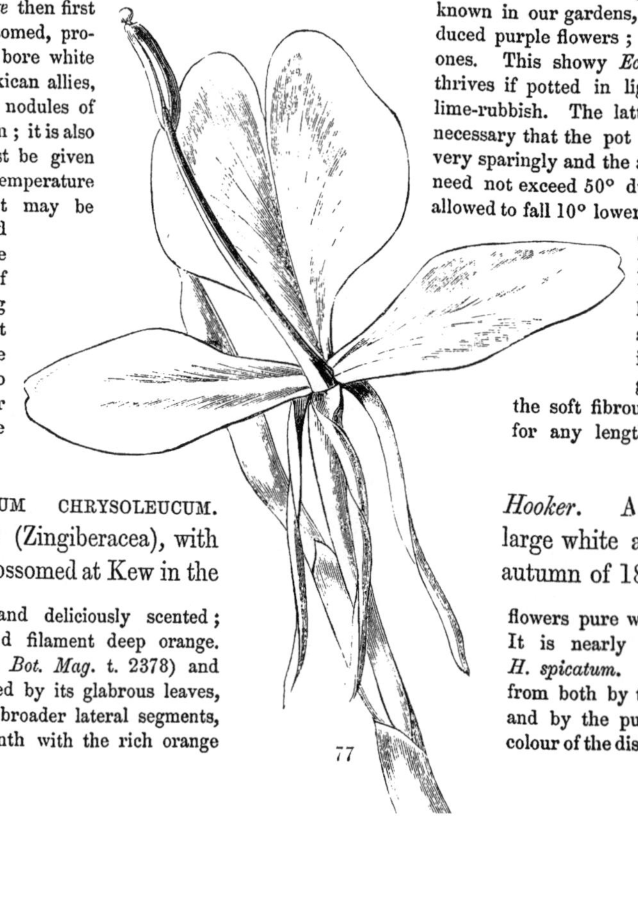

159. HEDYCHIUM CHRYSOLEUCUM. *Hooker.* A showy stove herbaceous Gingerwort (Zingiberacea), with large white and yellow flowers. Native of India. Blossomed at Kew in the autumn of 1849. (Fig. 77.)

Very handsome, and deliciously scented; flowers pure white, bright orange in the disk; anther and filament deep orange. It is nearly allied to *Hedychium flavescens* (*H. flavum, Bot. Mag.* t. 2378) and *H. spicatum*. From the former it is at once distinguished by its glabrous leaves, from both by the larger flowers and the much larger and broader lateral segments, and by the pure white of the inner segments of the perianth with the rich orange colour of the disk or centre.—*Botanical Magazine*, t. 4516.

77

160. SIPHOCAMPYLUS ORBIGNYANUS. *Alph. De Candolle.* A Bolivian (?) greenhouse plant, with broad dark-green leaves, and crimson and green flowers. Belongs to Lobeliads. Introduced by M. Van Houtte. (Fig. 78.)

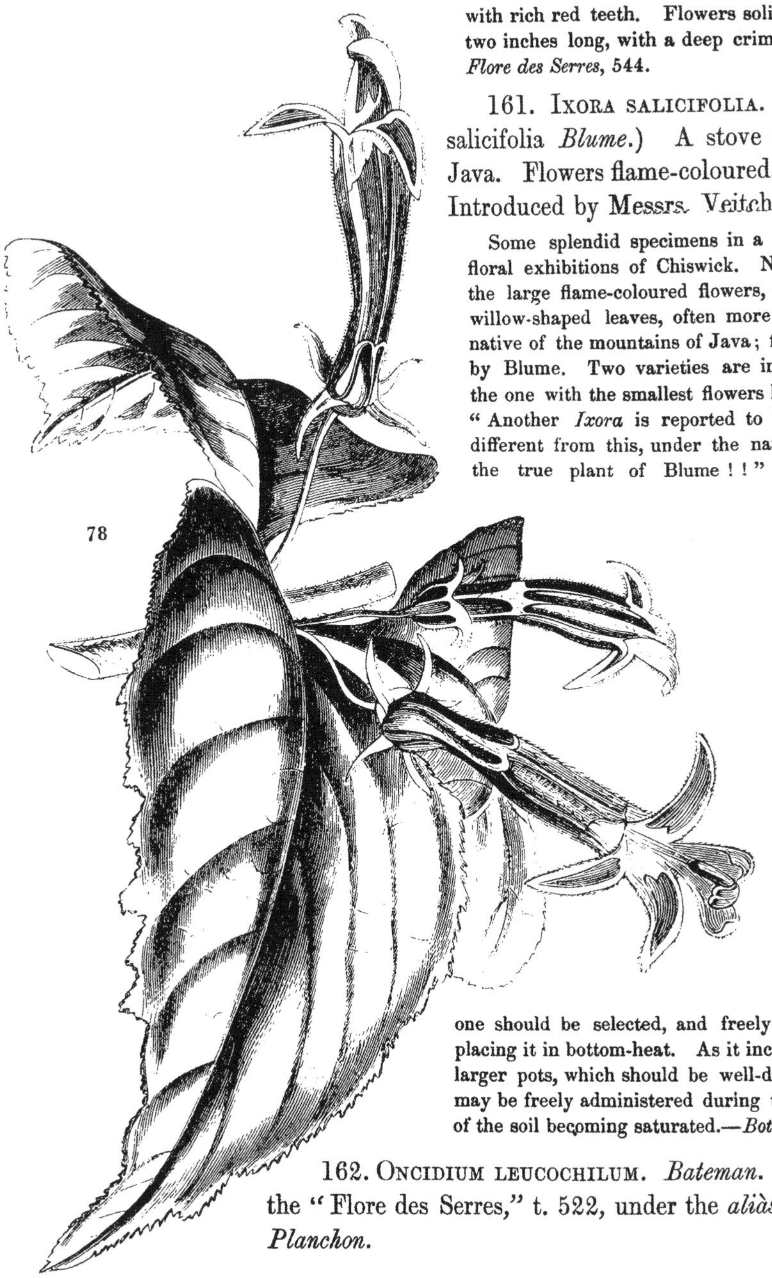

Branches and leaves covered with fine down. Leaves in threes, with rich red teeth. Flowers solitary in the axils, long-stalked, about two inches long, with a deep crimson tube, and a green-edged limb.—*Flore des Serres,* 544.

161. IXORA SALICIFOLIA. *De Candolle.* (*aliàs* Pavetta salicifolia *Blume.*) A stove shrub, of great beauty, from Java. Flowers flame-coloured. Belongs to the Cinchonads. Introduced by Messrs. Veitch and Co.

Some splendid specimens in a living state were exhibited at the floral exhibitions of Chiswick. Nothing can be more beautiful than the large flame-coloured flowers, or more graceful than the copious willow-shaped leaves, often more than a span in length. It is a native of the mountains of Java; first noticed there and characterised by Blume. Two varieties are in cultivation with Messrs. Veitch: the one with the smallest flowers has them the most deeply coloured. "Another *Ixora* is reported to be on sale in this country, quite different from this, under the name of *I. salicifolia* which may be the true plant of Blume!!" An erect shrub, 2-3 feet high, with rather closely-placed opposite leaves, borne on extremely short stalks, almost sessile, narrow-lanceolate, very much acuminated, often a span long, entire, smooth, dark shining green above, paler beneath. Corymb large,—when the flowers are fully expanded, forming a hemispherical head of deeply-coloured, orange-coloured flowers, or almost crimson. Style scarcely exserted. Stigma three-lobed. This showy *Ixora*, an abundant flowerer even when only six inches high, requires a warm and moist stove, and a soil composed of about half loam and half peat, with a portion of sharp sand. In order to form a handsome plant, a young healthy one should be selected, and freely encouraged into quick growth by placing it in bottom-heat. As it increases in size it must be shifted into larger pots, which should be well-drained, so that water and syringing may be freely administered during the summer-season without the risk of the soil becoming saturated.—*Botanical Magazine,* t. 4523.

78

162. ONCIDIUM LEUCOCHILUM. *Bateman.* This has been lately figured in the "Flore des Serres," t. 522, under the *aliàs* of Cyrtochilum leucochilum *Planchon.*

163. AMARYLLIS LATERITIA. *Dietrich.* A stove Amaryllid from Guinea, with red flowers. Introduced by Mr. Decker of Berlin.

It is uncertain to which of Herbert's genera this plant belongs; it seems intermediate between Vallota and Amaryllis.

The leaves appear later than the flowers, and are between lanceolate and strap-shaped. The scape is two feet high, taper, glaucous, and 2-flowered. The segments of the flower are spreading, but combined into a curved funnel-shaped tube, whose throat is destitute of appendages. The outer divisions of the flower are broadest. The stigma is very small and 3-cornered. The flower stalks are a full inch long; the flowers themselves about 3 inches.—*Allg. gartenzeit.* 1850. No. 9.

164. Hippeastrum (Amaryllis) robustum. *Dietrich.* A stove Amaryllid from Brazil, with deep red flowers. Introduced by Mr. Decker of Berlin.

Nearly related to *H. aulicum.* Leaves long, 2½ inches wide, strap-shaped, not glaucous, longer than the glaucous scape, which is nearly 3 feet high. Flowers in pairs, erect, deep carmine red, a little inclining to carmine, in form between bell-shaped and funnel-shaped, 5 inches long; the divisions separated quite to the base, flat, those on the outside lanceolate with a callous hooded point, on the inside oblong, acute. The coronet very short and cup-shaped, scarcely ¼ inch deep, and quite green.—*Allg. gartenzeit.* 1850. No. 6.

165. Gaultheria Lindeniana. *Planchon.* An evergreen greenhouse shrub, belonging to the order of Heathworts. Flowers small, pure white. Native of the mountains of Caraccas. Introduced by Mr. Linden. (Fig. 79.)

Found on the Silla de Caraccas, at an elevation of between 6000 and 7000 feet. Leaves said to resemble those of the Camellia in form, and of the Arbutus in texture. Flowers, although small, very conspicuous because of the pure whiteness of their calyx and corolla.—*Flore des Serres,* 501 d.

79

J. Constans, Pinx & Zinc

Printed by C.F.Cheffins, London

[PLATE 22.]

THE KAMTCHATKA RHODOTHAM.

(RHODOTHAMNUS KAMTCHATICUS)

A hardy evergreen dwarf Shrub, native of EASTERN SIBERIA, *belonging to the Order of* HEATHWORTS.

Specific Character.

THE KAMTCHATKA RHODOTHAM—Leaves oblong and obovate, fringed with coarse hairs, thin, blunt, tipped with a conspicuous gland. Sepals obovate, blunt. Corolla purple, with rounded lobes.

RHODOTHAMNUS *KAMTCHATICUS;* foliis oblongis obovatisque fimbriatis papyraceis obtusis apice glandulâ conspicuâ auctis, sepalis obovatis obtusis, corollæ purpureæ lobis rotundatis.

Rhododendron Kamtchaticum: *Pallas, Fl. Ross.*, I., p. 48, t. 33.

FOR this great rarity and exquisitely beautiful shrub we are indebted to Mr. Loddiges, whose predecessors raised it from seed about twenty years ago. It appears to be of slow growth, as the plant is now only about ten inches high, forming a compact bush. Mr. Loddiges finds it perfectly hardy, but it is best kept under a north wall. It is admirably adapted for rock-work in a shady situation.

According to Pallas this charming plant grows abundantly near the sea of Ochotsk, in the peninsula of Kamtchatka, and in Bhering's Island in muddy mountainous places. There it begins to blossom from the end of July, grows vigorously to the end of August, and ripens its seeds about the end of September. The root, he says, is woody, dry, as thick as a quill, and forms creeping runners. From this arise a great many leafy stems, which every here and there break into flower. The leaves are close together, alternate, sessile, somewhat ovate, tapering downwards, somewhat 5-nerved, rather sharp-pointed, perfectly entire, and fringed with very perceptible hairs. The peduncles are two or three inches long, closely surrounded by small leaves, besides which there are generally about two ovate sessile leaves; they are 2-flowered, or occasionally 1-3-flowered, and very hairy. The flowers are nodding, and deep-purple. The sepals leafy, 3-nerved, two being nearer to each other than to the others. The corolla is irregular, rotate, with a very short funnel-shaped tube, and a deeply

5-lobed limb; the segments lanceolate, downy at the throat, unequal, the three uppermost rather the smallest, and less deeply divided, spotted with crimson at the base, standing up like a hood, the two lower very much spreading and spotless. The stamens, which arise from the bottom of the flower, are ten, curved downwards, the upper shortest, the lower twice as long as the others, not so long as the corolla, with ovate, double, deep-purple anthers. *Fl. Rossica*, vol. i., p. 40.

To the locality given by Gmelin and Pallas, Ledebour adds the following: Mount Marekan, according to Turczaninoff; the country of the Tschuktskes in the Bay of St. Lawrence; Kamtchatka and Unalashka. Sir W. Hooker gives Banks's Island and Port Edgcombe, on the north-west coast of N. America. It is, therefore, clear that it belongs to climates far more rigorous than our own, and with much worse summers. And this is the key to its cultivation. Like the *R. Chamæcistus*, it is unable to endure the drier air and brighter summer sky of England; but shrinks from our heats, and withers beneath such evaporation as leaves undergo in this climate. Hence the wisdom of the treatment which consists in keeping such plants in a cold pit closed up all day, and uncovered all night. Mr. Loddiges's cultivators made nothing of it till they put it under a north wall where Liverworts and such soft flabby plants delight to dwell.

We do not believe that any botanist would have thought of calling this a Rhododendron, had not Linnæus set the example by including the *Chamæcistus* in that genus. Its great leafy calyx, flat corolla divided almost to the base, and nearly equally spreading although very unequal stamens, are quite at variance with Rhododendron. Neither has it the scurfs or stellate hairs observable, we believe, in all the genuine species in which hairs are ever found. On the contrary, the hairs are always simple, in which respect it agrees with the Chinese Azaleas, to which it is more nearly related than to Rhododendrons, but from which its corolla, almost divided into separate petals, sufficiently divides it. To this may be added, the singular gland at the end of the leaves, a nearer approach to which is to be found in the scaly Azalea (*A. squamata*) than in any Rhododendron we have examined.

In the accompanying figure, 1, represents an anther previous to its bursting by two pores at the end; and 2, the underside of a leaf with the terminal gland.

I. Constans. Pinx & Zinc

Printed by C.F Cheffins, London.

[Plate 23.]

THE OVAL AND THE PALLID HOYAS.

(HOYA OVALIFOLIA AND PALLIDA.)

Stove climbers from Tropical India, *belonging to the Natural Order of* Asclepiads.

Specific Characters.

I. *THE OVAL* HOYA.—Leaves fleshy, narrow, oval, 3-nerved, rolled back at the edge. Peduncle rather shorter than the leaf, and smooth. Corolla fleshy, with ovate acute segments. Coronet-lobes acute, revolute at edge. *Left-hand figure.*

I. HOYA *OVALIFOLIA*.—Foliis carnosis angustis ovalibus trinerviis margine revolutis, pedunculo folio paulò breviore glabro, corollâ carnosâ glabrâ laciniis ovatis acutis, coronæ foliolis acutis margine revolutis. *Fig. sinistr.*

Hoya ovalifolia : *Wight and Arnott, contributions to the Flora of India.* p. 37 ?

II. *THE PALLID* HOYA.—Leaves fleshy, ovate, feather-veined, turned back at the edge. Peduncle rather shorter than the leaf. Corolla fleshy, smooth, with ovate acute segments. Coronet-lobes acute, revolute at edge. *Right-hand figure.*

II. HOYA *PALLIDA*.—Foliis carnosis ovatis penniveniis margine revolutis, pedunculo folio paulò breviore glabro, corollà carnosâ glabrâ laciniis ovatis acutis, coronæ foliolis acutis margine revolutis. *Fig. dextr.*

Hoya pallida : *Lindley in Botanical Register*, t. 951.

For the knowledge of the first of these species we are indebted to the Chatsworth collection, where it flowered in June last, from among Mr. Gibson's Indian collection. Along with it is represented on the right hand the Pallid Hoya, which blossomed at Chatsworth at the same time. A comparison of the two figures will show their differences better than mere description.

The Pallid Hoya was originally observed at Syon, whence, in 1825, materials were supplied for a figure in the Botanical Register. Its origin was then unknown; but the Chatsworth plant now proves it to be a native of India, and we possess specimens from the Burmese Empire collected by the late Mr. Griffith. It is distinguished from the Fleshy Hoya (*H. carnosa*) not only by a yellowish tint which replaces the dark heavy green of that species and by its sweeter smell, but also by the form of its leaves, which are acute and exactly ovate; that is to say, similar in figure to an egg divided longitudinally, while in the Fleshy Hoya they are as nearly as possible truly elliptical. The umbels of flowers also are smaller. In the Botanical Register the artist has made the stalk of the umbel appear far too short in an unsuccessful attempt at foreshortening.

The Oval-leaved Hoya has much the appearance of the last; but differs in its flowers being distinctly yellow instead of straw-coloured; and in the form and construction of the foliage. The leaves are about 6 inches long, in the form of a narrow ellipse, differing very little in width near either end. Instead of the veins diverging regularly from the midrib in the same way as in the Pallid Hoya, there are three principal veins which proceed together from a little above the base, giving the leaf a triple-nerved venation. So that in fact these two species belong to two different types of structure, and stand in two different sections of M. Decaisne's classification of the genus.

These charming species each require the same treatment as the Fleshy Hoya, and trained with it along the rafters of a house, grow in perfect harmony, and produce an extremely agreeable variety without occupying more room than one of them would require.

PLATE 24.

L. Constans Pinx. & Zinc

Printed by C.F. Cheffins, London.

[PLATE 24.]

VARIETIES OF THE RUBY-LIPPED CATTLEYA.

(CATTLEYA LABIATA.)

Stove Epiphytes, natives of the CARACCAS, *belonging to the Order of* ORCHIDS.

Specific Character.

THE *RUBY-LIPPED* CATTLEYA.—Stems between club-shaped and spindle-shaped, furrowed. Leaves solitary, oblong. Spathe as long as the peduncle. Sepals linear-lanceolate, acute, coloured. Petals membranous, oblong-lanceolate, wavy, much broader. Lip obovate, crisp and wavy, emarginate, smooth on the disk.

CATTLEYA *LABIATA*.—Caulibus clavato-fusiformibus sulcatis, foliis solitariis oblongis, spathâ pedunculi longitudine, sepalis lineari-lanceolatis acutis coloratis, petalis membranaceis oblongo-lanceolatis undulatis multò latioribus, labello obovato crispo-undulato emarginato disco lævi.

C. labiata *Lindley Collectanea Botanica* t. 33 ; *aliàs* C. Mossiæ *Hooker in Bot. Mag.* t. 3669.

THESE magnificent varieties of the Ruby-lipped Cattleya are quite new and at present among the rarities of Horticulture. For the white one we are indebted to the noble collection at Syon; for the blotched sort to J. J. Blandy, Esq., of Reading.

The Ruby-lipped Cattleya is that on which the genus was founded. It was first sent to Europe by Mr. Swainson, who discovered it in Brazil and used its stems as a kind of "dunnage" to set fast certain chip boxes of lichens &c., which he transmitted to Sir William (then Mr.) Hooker. Where he gathered it we are not informed, but we learn something precise on the subject from Mr. Gardner. This lamented Botanist found it on the edge of a precipice on the eastern side of the Pedro Bonita Mountain, about fifteen miles from Rio Janeiro, where it grew along with Vellozias, the Mackay Zygopetalum and Dipladenes (*Journ. of Hort. Soc.*, vol. i. p. 196); and also on the Gavea, or Topsail Mountain, so called from its square shape, and well known to English sailors by the name of Lord Hood's nose. *Travels in Brazil*, p. 28. This plant has a pale lilac tint with a very broad rich stain of ruby red over-spreading all the front half of the lip except the very edge.

Since that time large importations have been made from the Caraccas and New Grenada of a Cattleya with pinker flowers, of much larger size, the veins of whose lip alone were crimson, while the spaces between were yellowish or white or both; some of them had crimson veins run together. Upon these specimens Sir William Hooker proposed to establish a new species, to which he gave the name of *Mossiæ;* and it must be owned that the peculiarities of the Caraccas plants seemed sufficient to justify that conclusion. We are however obliged to say, after a most careful comparison of large numbers of this *Cattleya Mossiæ,* that we can find no distinctive characters in it except size and colour.

It would be useless to attempt an enumeration of the varieties that exist of this plant, unless for the purposes of a Florist. We therefore merely present those now figured with the names of the White Ruby-lipped Cattleya (*C. labiata candida*) and the Blotched (*C. l. picta*).

The following account of the climate in which *Cattleya labiata* grows, furnishes cultivators with hints which they will readily apply to practice. "At this elevation (2000 feet) the climate is very much cooler than it is at Rio. In the months of May and June the thermometer has been known to be as low as 32° just before day-break: the lowest at which I observed it myself was one morning at the end of May, when, at 8 o'clock a.m., it indicated 39°. The highest to which it rose during the six months I resided there, was in the end of February, when, one day, it indicated 84° at noon. The hot season is also the season of rains, and it is then that the mass of the Orchids, and almost every other tribe of plants, come into flower. From these facts cultivators ought to take a lesson in the cultivation of the productions of this and of similar regions. If the difference of temperature between the season of wet and that of flowering be so great in the state of nature, it must be obvious that to grow them well, artificially, a somewhat similar state of things ought to be observed. The greater part of the Orchids which are sent to England from the Organ Mountains, grow in the region of the above temperature, the elevation being from 3000 to 3500 feet above the level of the sea. In the account which I shall presently give of my visit to the summit of those mountains, which is more than double that elevation, I shall have occasion to mention several species which may be cultivated in a much cooler temperature. Another reason why no general rule can be laid down for the cultivation of these plants, is, the great variety of soil and situation which they affect in their native country; some, like *Zygopetalum Mackaii,* are terrestrial, and grow in open exposed places; others, like *Warrea tricolor,* are also terrestrial, but grow in the deep virgin forests; some, like *Zygopetalum maxillare,* are only found to inhabit a particular tree, while others are found indiscriminately on all kinds of trees, on rocks, and even on the ground; some, like *Laëlia cinnabarina,* grow in moist places on exposed rocks; while others, like *Cyrtopera Woodfordii,* grow in a similar soil, but in shaded places; some, like *Maxillari apicta,* grow on the most dry and exposed rocks; while others, like *Grobya Amherstiæ,* grow also on dry rocks, but generally in the shade."—*Gardner in Journal of Hort. Soc.,* i. 277.

GLEANINGS AND ORIGINAL MEMORANDA.

166. Dianthus cruentus. *Fischer*. A hardy herbaceous plant, with deep rose-coloured flowers. Introduced by Dr. Fischer. Flowered with Mr. Van Houtte. (Fig. 80.)

This charming Pink has been received by Mr. Van Houtte from the Botanic Garden, St. Petersburgh, under the name of D. cruentus. It is supposed to come from the Caucasus, or from Siberia (rather distant stations it must be confessed). Perhaps less brilliant than some varieties of Sweet William (*D. barbatus*), but quite as ornamental. The leaves form tufts of light green, from which rise simple stems terminated by a nearly globular flower-head, which produces from the midst of a curious mixture of scarious, rusty, long-pointed bracts, numerous blossoms with a violet calyx, and wedge-shaped petals elegantly toothletted, resplendent with vivid carmine, relieved by certain violet hairs which adorn the base of each limb. Allied botanically to *D. carthusianorum* and *barbatus*.—*Flore des Serres*, t. 488.

80

167. Calamintha mimuloides. *Bentham*. A hardy Californian perennial belonging to the Labiates, with dull red flowers. Introduced by the Horticultural Society. Blossoms in August and September.

A hairy, half-shrubby, herbaceous plant, covered all over with viscid glands. Stems erect, regularly and simply branched, about 1½ foot high. Leaves stalked, ovate, acute, coarsely crenate-serrate except at the base, which is entire. From the axils of the upper leaves rise solitary stalked labiate flowers, about 2 inches long, with a somewhat cylindrical striated, 5-toothed, hairy, and glandular calyx, and a yellow corolla deeply stained with orange at the upper part. One of the best of the hardy herbaceous plants obtained from Hartweg's expedition to California, but too leafy for a bedding out species. It seems best adapted to cultivation apart from other plants, when it forms a deep green summer bush of some beauty. Although the flowers are described above as growing singly in the axils of the leaves, yet it is to be observed that each flower is succeeded by five or six others, so that there is a long succession of bloom.—*Journ. Hort. Soc.*

168. ECHEANDIA TERNIFLORA. *Ortega; (aliàs* Conanthera Echeandia *Persoon; aliàs* Anthericum reflexum *Cavanilles; aliàs* Phalangium reflexum *Poiret*). A half-hardy Mexican Lilywort, with fugitive yellow flowers. Blossoms in August. (Fig. 81.)

It seems worth while to reproduce this plant, which, although long known in gardens, is rare, and has never been figured in English works. It was sent us on the 6th of August by Edward Leeds, Esq., of St. Ann's, Manchester, with the following note :—

"I send you specimen of a plant raised from Mexican seeds : it was marked 'Asphodelus sp.,' but is more like an Anthericum. The roots are thick and fleshy, and I think it will make a fine border plant, treated the same as Commelyna cœlestis ; keeping the roots in dry sand, and out of frost in winter. The flowers last only one day, but come out in succession for a long time ; and when the plants become strong, it will be as ornamental as some species of Ixia ; hitherto I have kept it in a pot. The seeds were round and black ; and were given to me by the lady of Dr. Robinson, of this place."

The filaments are club-shaped bodies covered near the upper end with rings of blunt projections hooked back, which may be regarded as an incomplete state of the hairs on such plants as Bulbine, no doubt nearly related to Echeandia. Examined with the microscope, these projections are found to be caused by the free ends of long loose club-shaped cells hooked back and placed in a whorled manner around a central cord of spiral vessels. They are filled with a yellow grumous fluid.

169. LILIUM WALLICHIANUM. *Schultes.* A very fine hardy bulbous plant, with white flowers, from the N. of India. Introduced by Major Madden. Blossoms in August. (Fig. 82.)

Asia has furnished us with four distinct kinds of tube-flowered white lilies ; namely *candidum*, the common white, *japonicum*, *longiflorum* with its dwarf 1-flowered variety, and *Wallichianum*. The first has a short tube and flowers in racemes. The others have them varying in number from 1 to 3, with a very long tube. Of these *Japonicum* has broad leaves, and leathery flowers stained outside with olive brown ; the two others have the flowers perfectly white, with a much thinner texture. Between themselves *L. longiflorum* and *Wallichianum* differ in the latter having very long narrow leaves, of which the uppermost are extended into a linear point, and flowers as much as 8 inches long ; while *longiflorum* has leaves twice as broad, and flowers generally much smaller. These are, we believe, the only real distinctions between the two, and seem hardly sufficient to justify the creation of two species ; the distinctions are however permanent, and affect considerably the general appearance of the plants.

According to Schultes, *L. Wallichianum* must have been long in cultivation, for he refers to it two figures in the Botanical Register and Botanical Cabinet, both of which well represent the dwarf 1-flowered form of *L. longiflorum*, sometimes called *eximium* in Gardens, and bear no resemblance to the Indian Lily.

The accompanying figure of *L. Wallichianum* is certainly the first that has been given from a European specimen. We received it last August from Mr. D. Moore, the skilful curator of the Botanic

82

Garden, Glasnevin, to which establishment it had been sent, with many other Himalayan varieties, by Major Madden, from Almorah. The bulbs reached Mr. Moore in April last, and on the 10th of August the plants were 4 feet high. Each stem bore but one flower; but our wild specimens from Mount Sheopore, for which we are indebted to Dr. Wallich, are 2-flowered; and he speaks of as many as three flowers on a stem, 9 inches long, and sweet-scented. The following is Dr. Wallich's account of the Lily, extracted from his Tentamen, where it stands under the name of *L. longiflorum.*

"This is a very distinct and noble species, with a tall and slender stem, two-thirds of which are thickly furnished with long and linear leaves. The flowers are white, fragrant, extremely large, with a very long and narrow tube, which is gradually widening into an ample spreading limb; there are generally two or three at the apex of the stem; sometimes only one. In size they exceed those of Lilium giganteum. The claws of the three exterior sepals are closely united to those within, in consequence of their sharp margins being confined within the deep furrow, which is formed on each side of the dorsal rib of the latter. The base of the stem I have repeatedly found horizontal, creeping and scaly like that of a fern, without any remainder of a bulb, but marked with a number of vestiges of old stems. This lily is also found towards Sirinuggur; I have received plenty of fine specimens collected by Mr. Robert Blinkworth."

170. Erythrina Erythrostachya. *Morren.* A stove shrub of unknown origin, belonging to the leguminous order. Flowers scarlet, very handsome. Introduced by the Belgians.

The genus *Erythrina* of Linnæus is composed of shrubs or shrub-like plants, occasionally having a subterraneous stem with annual sub-herbaceous branches. They are indigenous to the tropical and sub-tropical regions of the whole globe Their stem and leaves are often furnished with prickles; their leaflets are trifoliate and pinnated, the terminal leaflet being at some distance from the other two; instead of stipules there are stalked glands, small stipules distinct from the petioles; the spikes of the flowers are long; the pedicels are often in threes; the flowers are generally red and scarlet, and most beautiful; the seeds are often black, or variegated with black, and brilliant. This splendid species is not like any hitherto described, and enumerated in the repertorium of M. Walpers. It approaches *Erythrina reticulata* Presl., but the leaves are glabrous, not wrinkled or downy. Besides, the thick tuberculiform tooth of the calyx separates it from all the other species of the genus. The spike is more than 6 inches long; the flowers arranged in threes, are very numerous, and an inch and a half or two inches in length. Their colour is very brilliant, and it is no doubt one of the prettiest plants that can be cultivated. It was found in the collection of M. Verleuwen of Ghent, from whom it was bought by M. Cachet of Angers, under the erroneous name of *Erythrina speciosa.* This was in 1832. We have given it one which recalls the beauty of its spike. The cultivation does not differ from what is required for the *Erythrina Corallodendron.* The trunk, when well cut in, is placed in a large pot in a temperate house, where it begins to grow after February, if, that is to say, it is not wished to force it. In fine weather it may be planted out, and in summer it forms a great ornament in our gardens.—*Annales de Gand*, t. 291.

171. Malesherbia thyrsiflora. *Ruiz and Pavon.* A greenhouse herbaceous or half-shrubby plant, with long spikes of dull red and yellow flowers. Belongs to the Crownworts (*Malesherbiaceæ*). Flowers in August. Introduced by Messrs. Veitch and Sons.

An erect plant, covered rather thickly with long yellow hairs. The leaves are linear-oblong, somewhat crenate, rugose, shorter than the axillary sessile flowers. The calyx forms a rusty reddish yellow tube about 1½ inch long, with 10 strong veins; its five divisions are narrowly triangular, a little longer than the petals which have the same form and surface. The coronet or "crown" is a narrow membrane at the orifice of the calyx with five lobes usually 3-notched with a small tooth between each lobe. Stamens prominent. No doubt there are several perfectly distinct species included under the name of *thyrsiflora*, but as we have no means of settling to which the name most properly belongs, we leave the question as we find it. A mere botanical curiosity.

172. Conoclinium Ianthinum. *Morren.* A stove herbaceous plant from Brazil, belonging to the Composites. Flowers in broad violet flat-headed panicles. Introduced by M. Alex. Verschaffelt.

According to Professor Morren, this is a plant of great beauty. It forms a low soft-wooded shrub, covered with short brown down. The leaves are heart-shaped, acute, stalked, serrated, strongly marked with pale veins. The flat heads of violet flowers are full six inches across, and appear to consist of numerous entangled many-pointed stars. They have a mild honey-like fragrance, with a peculiar aroma. In Ghent it has been regarded with great favour; but it will hardly meet with much notice in this country. The colour of the flowers is too dull, and the] habit too weedy for English taste. *Annales de Gand*, t. 253. If this is to be brought into a state of considerable beauty, it will require a damp stove, and to be kept carefully from red spider. It is very like a Cœlestine.

173. HYPOCYRTA GRACILIS. *Martius.* A pretty creeping stove Gesnerad with cream-coloured flowers, from Brazil. Introduced by Messrs. Backhouse of York. (Fig. 83.)

Plant minutely pubescent, creeping, sometimes bearing ascending shoots. Stem purplish-brown, rooting from below the insertion of the leaves. Leaves on short petioles, opposite, thick, fleshy, ovate, subacute, dark green and slightly concave above, pale and often blotched with red and convex beneath. Flowers on short red peduncles, solitary or in pairs, single-flowered. Calyx of five, deep, linear-lanceolate segments, red at the base. Corolla moderately large, cream-white, spotted with orange on the underside of the tube within, between bell-shaped and funnel-shaped: tube decurved, and again curved upward at the mouth; limb of five, nearly equal, rounded segments. Ovary ovate, with a large gland at the base of the back.

A soft-wooded suffruticose plant, of a trailing scandent habit, emitting roots from below the axils of the leaves, and growing as an epiphyte on trees in the moist forests of Tropical America. It should be kept in such an atmosphere as that appropriate for the cultivation of tropical Orchids, and if there is sufficient accommodation, it may be allowed to grow in a natural manner over any elevated surface, covered with turfy sods, kept moist; or may be planted in a pot or basket filled with loose turfy soil and suspended from the root.—*Bot. Mag.*, t. 4531.

This is not a Hypocyrta, as Decaisne limits the genus, but would rather belong to what he understands by Alloplectus.

174. CYCNOCHES PESCATOREI (*aliàs* Acineta glauca *Linden.*) A stove Orchid from New Grenada. Flowers yellow and brown. Introduced by M. Linden in 1848. Blossomed with M. Pescatore.

C. Pescatorei, foliis coriaceis subtùs glaucis, racemo multifloro pendulo, ovario tomentoso, sepalis oblongis acutis, petalis minoribus lanceolatis basi angustatis, labello plano trilobo medio tomentoso lobo intermedio carnosiore acuto.

This noble plant is only known to us from the inspection of two dried flowers sent from M. Pescatore's rich collection by M. Luddeman, who describes it thus. " A much stronger plant than Acineta Humboldti, with a pseudo-bulb 0.16 of a yard long and 0.09 of a yard broad. The leaves are leathery, lanceolate, glaucous beneath, 0.60 to 0.80 of a yard long on the young pseudo-bulbs, which are not more than half the size of the imported ones. The flower stem hangs down perpendicularly, a yard long, with ninety-six flowers. These last about a fortnight, but for several months the long string of buds excited the curiosity of visitors. The sepals are dull yellow, a little brown inside; the petals and lip are bright yellow." The specimens sent us measured $1\frac{3}{4}$ inch in diameter. The species seems to be closely allied to the bearded Cycnoches (*C. barbatum.*)

175. Catasetum fimbriatum (*aliàs* Myanthus fimbriatus *Morren in Ann. de Gand*, t. 231). A terrestrial Orchid of unknown origin, with dirty white and pink flowers. Introduced by the Belgians. (Fig. 84.)

C. fimbriatum ; racemo cernuo multifloro, sepalis petalisque linearibus acuminatis lateralibus longioribus, labello plano cordato membranaceo dentato vel fimbriato basi saccato conico, dente prominente in discum.

All that is known to us regarding this plant is what we find in Professor Morren's account, published in the work above quoted. It appears to be a species of no great beauty, with the habit of *C. cernuum*, but with pink sepals and petals speckled with red, and a broad heart-shaped dirty white lip strongly cut at the edge. It is said to have obtained an extra gold medal at the National Horticultural and Agricultural Exhibition at Brussels in 1848, when we are told "Pendant trois jours plus de cent mille yeux se fixèrent sur cette étrange et admirable gynandre dont le parfum embaumait la salle." In this country people would have hardly remarked it. Two varieties are mentioned ; one green and white, the other rose and yellow. It is not improbable that they are identical, their supposed differences being due merely to the mode of cultivation.

84

176. Medinilla Sieboldiana. *Planchon* (*aliàs* M. eximia *Siebold.*) A handsome stove-plant from Java, belonging to the order of Melastomads. Introduced by M. Van Houtte. Flowers white and rose-colour.

The habit of this plant, and the manner in which it is to be cultivated are the same as those of our *Medinilla magnifica* (Plate 12 of the present volume). The branches are perfectly taper, or very slightly four-cornered when quite young. The leaves are deep green, triple-nerved, brownish underneath, oblong, tapering into a short footstalk. The flowers are white, of the texture of wax, in short naked divaricating panicles, with a yellowish brown calyx and deep rose-coloured stamens. It appears to be a handsome species even although it wants the brilliant bracts of the Magnificent Medinil. *M. eximia* of Blume is a different species. *Flore des Serres*, t. 482.

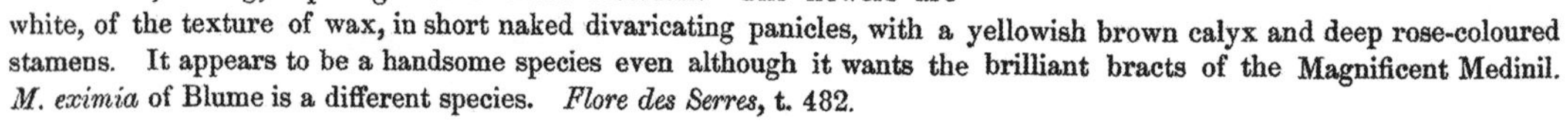

177. Puya maidifolia. *Decaisne.* A very handsome stove herbaceous plant belonging to the Bromeliads, spikes crimson and green. Native of the Caraccas. Introduced by M. Linden.

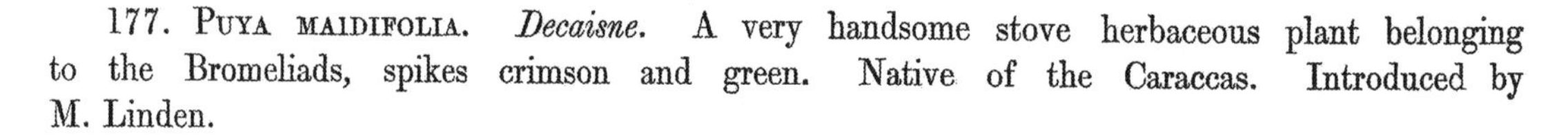

Leaves broad, thin, ribbed, resembling those of Indian corn, but apparently rather glaucous. Spike long, cone-shaped, consisting of brilliant crimson bracts tipped with green. Flowers pale cream-colour about 2 inches long. *Annales de Gand.* t. 289. This takes rank by the side of the Vriesias and Gusmannia, and seems well worth the having.

178. Bessera miniata. *Lemaire.* A beautiful bulbous plant from Mexico, with scarlet and white flowers in umbels. Belongs to the Lilyworts. Introduced by M. Van Houtte.

According to M. Lemaire, this differs from *B. elegans* in having a toothed coronet, and one-celled anthers. To us, it appears to be identical with that species. According to M. Van Houtte, these beautiful bulbs, hardly known in English gardens, require no other protection than a cold frame, the sash of which is removed in summer. While growing they are watered rather abundantly, but they are kept perfectly dry in winter. *Flore des Serres*, t. 424.

179. Hakea cucullata. *R. Brown.* A Swan River Protead with great coriaceous leaves and pink axillary flowers, produced in April. Requires a greenhouse. (Fig. 85.)

Discovered by the late Mr. Baxter at King George's Sound. Mr. Drummond has also found flowering individuals at the Swan River Settlement, and has sent seeds, from flowering plants of which our figure was taken at the Royal Gardens, in April, 1850.

An erect shrub, 4 to 5 feet high, the branches pale brown, shaggy. Leaves leathery, cordate, sessile, concave, waved and rather minutely toothed at the edge, glaucous green, distinctly reticulated both above and below. From the axils of the upper leaves the flowers appear in copious clusters: at first surrounded by imbricated deciduous bracts. Sepals red, unequal linear, smooth. Style twice as long as the longest sepals.—*Bot. Mag.*, t. 4528.

Upon the cultivation of this and other Proteads, Mr. Smith has the following useful observations :—

"Before the introduction and high state of cultivation of the splendid flowering plants now annually exhibited in the

85

vicinity of London, it was customary to estimate the value of public and private collections by the number and rarity of the species, without regard to the circumstance of their producing fine flowers. Perhaps no plants were in higher repute than those of the family to which this belongs, as is amply shown by the early volumes of the Botanical Magazine. Within the last twenty or thirty years, however, the cultivation of *Proteaceæ* has declined ; the species have gradually disappeared from most of the private collections around London ; and but few nurserymen now take interest in them. This change may be partly owing to the supposed difficulty of preserving them, for under certain circumstances the plants suddenly die, even when in vigorous health. In the Royal Gardens *Proteaceæ* have maintained their place, more especially those that are natives of Australia ; and as there are some at this time between forty and fifty years of age, and others of a large size half that age, it may be inferred that *Proteaceæ* are not so short-lived in a state of cultivation as they are generally supposed to be. Within our recollection it was the common practice to grow them in some kind of light soil, usually peat. The hygrometric condition of such soil is easily affected by changes of the surrounding atmosphere ; becoming quickly dry during hot weather, and apt to become sodden with moisture in winter, and the spongioles or rootlets of *Proteaceæ* are very sensitive to either extreme ; the use of light soil, therefore, in our opinion, accounts for the frequent sudden death of plants of this kind. We use good yellow loam, to which, for small plants, we add a little sharp sand. In shifting or repotting a plant we make it a rule to keep the ball of roots a little elevated above the surface of the new mould, to prevent any superabundance of water from lodging round the base of the stem. In the winter, care must be taken to give no more water than is required to keep the soil moderately moist; but in summer, water may be given freely in the evening or early in the morning. It is important that the plants should be so placed that the sun's rays do not strike the sides of the pot. The species here figured, being a native of the Swan River Colony, requires to be treated as a greenhouse plant. It does not readily propagate by cuttings, but may be increased by grafting on any of the more common free-growing species. Imported seeds germinate freely."

86

180. VERONICA FORMOSA. *Bentham.* (*aliàs* V. diosmæfolia *Knowles and Westcott.*) A little half-hardy evergreen bush from Van Diemen's Land. Flowers bright blue. Belongs to the Linariads (*Scrophulariaceæ*). Very pretty. A native of Mount Wellington; and found to stand the winter at Kew, planted against an east wall. (Fig. 86.)

It forms a shrub about 2 feet high, erect, much branched, with two obscure lines of hairs between the leaves. Leaves rather crowded, arranged somewhat in four rows, oblong, lanceolate, spreading, scarcely stalked, single-nerved. Flowers in terminal racemes, not many of which open at one time, though there is a succession of them. Corolla bright and deep purplish blue, somewhat 2-lipped; upper lip of one broad oval lobe, lower of three narrower segments, the middle one the smallest. This with a few others belongs to a section of Veronica characterised as evergreen shrubs, having small closely-set decussate leaves, and forming myrtle-like bushes. The old and well-known Veronica decussata may be viewed as the type of the group. They are natives of high southern latitudes; being found in Van Diemen's Land, New Zealand, Falkland Islands, and Lord Auckland's and Campbell's Islands, in latitude 53°. The two species known to us in a living state prove sufficiently hardy to bear the winter of this climate, when planted in sheltered situations and protected during severe frosts. That now figured is worthy of being kept in the greenhouse, where it produces its pretty racemes of light blue flowers in the spring. It grows readily in light loam and leaf mould, and is easily propagated by cuttings, treated in the usual way; it also freely produces seeds.—*Botanical Magazine*, t. 4512.

181. LYCASTE CHRYSOPTERA. *Morren.* A stove epiphyte from Mexico, with deep orange-yellow flowers. Belongs to the Orchids. Introduced by the Belgian Government.

We only know this plant by the figure in the *Annales de Gand.* t. 232. It seems very like *L. cruenta*, but according to Professor Morren, its flowers are much larger, the colours more brilliant, and the details of the lip essentially different, the appendix being 3-lobed, and the middle division of the lip lanceolate, acuminate, and toothletted. The yellow flowered Lycastes related to *cruenta* approach each other so nearly that without knowing exactly on what their differences depend, the one may be easily confounded with the other. We trust that the following memorandum will assist in clearing up the difficulty surrounding them. *Lyc. cruenta* is taken for the standard of comparison.

1. *L. cruenta* Lindley; (aliàs *L. balsamea* A. Richard). LIP roundish, spotted with crimson at the base, the lateral lobes short, the central oblong and rounded; appendix minute, emarginate. COLUMN hairy all over. PETALS naked—Guatemala.

2. *L. chrysoptera.* Morren. LIP roundish, spotted, the lateral lobes short, the central lanceolate acute toothletted; appendix 3-lobed. COLUMN hairy. PETALS naked. Mexico.

3. *L. macrobulbon* (aliàs *Maxillaria macrobulbon*, Hooker in Bot. Mag. t. 4228.) LIP much longer than broad, spotted with crimson on inside; the lateral lobes short; the central ovate-oblong, rolled back, crisp, broader than the laterals; appendix acute entire. COLUMN ? PETALS naked ? (Description and figure imperfect.)—Native of Santa Martha. Said to have very large pseudo-bulbs.

4. *L. cochleata.* LIP nearly circular, not spotted; the lateral lobes long, rather acute; the central flat, circular, emarginate, slightly crisp; appendix entire, as large as the lateral lobes. COLUMN long, hairy. PETALS hairy.—Native country unknown. Flowers whole-coloured, deep orange; the sepals and petals ovate, the latter obtuse and not much smaller than the former.

5. *L. crinita*. Lindley. LIP narrowly oblong, slightly speckled ; the lateral lobes linear, blunt, nearly as long as the equally narrow hairy oval central one ; appendix inconspicuous, terminating a narrow shaggy elevation. COLUMN long, slightly hairy. PETALS very hairy—Mexico. Petals yellow, very acute, much smaller than the greenish sepals.

6. *L. aromatica*. Lindley (*aliàs* Maxillaria aromatica, *Hooker*). LIP oblong, narrowed to the base, spotless, hairy inside ; the lateral lobes ovate, slightly curved, obtuse ; the central unguiculate, dilated at the end ; appendix very large 2-lobed, concave. COLUMN long, narrow, hairy. PETALS naked—Mexico ? Peru.

87

182. OCHNA ATRO-PURPUREA. *De Candolle*. (*aliàs* Diporidium atro-purpureum *Wendl.* ; *aliàs* Ochna arborea *Burchell* ; *aliàs* O. serrulata *Hochstetter* ; *aliàs* O. Natalitia *Meisner* ; *aliàs* O. Delagoensis *Ecklon*.) A greenhouse shrub, of some beauty, from Southern Africa. Belongs to the Ochnads. Lately produced its handsome yellow flowers in the Royal Garden, Kew. Said, however, to have been introduced in 1823. (Fig. 87.)

A native of South Africa, east of the Cape, as far as Delagoa Bay, varying in size, in the solitary or racemose flowers and in the size and notches of the leaves, which are sometimes sharply serrated, sometimes nearly entire. It derives its name from the dried state of the plant, when the large persistent calyxes become purple-brown, especially when in fruit. In the living plant, the bright yellow flowers with pale yellow-green calyx enliven the greenhouse in the month of March.

The history of its having at last flowered, after refusing to do so for twenty-seven years, is thus given by Mr. Smith:—" Thinking it would be benefited by a greater warmth during winter, and having accommodation in the Palm-house, it was placed there last Autumn. The result was, that in April we were agreeably surprised to see it profusely covered with its pretty, sweet-scented flowers. Several other plants have flowered similarly for the first time on being placed in a greater degree of heat, which shows that with our long-continued low temperature in winter and spring, and deficiency of bright sunshine in summer (as compared with the Cape), our usual greenhouse climate is not adapted for the perfect development of this and other slow-growing Cape and New Holland plants."—*Botanical Magazine*, t. 4519.

183. MOUSSONIA ELEGANS. *Decaisne*. A hothouse Gesnerad, with orange and yellow flowers, from Guatemala. Introduced by M. Van Houtte. (Fig. 88.)

Stems and leaves covered with soft hairs. Leaves ovate, oblong, acuminate, crenel-toothed. Umbels three or four-flowered. Corolla scarlet with a yellow limb, spotted in lines with purple. Being a native of the mountains of Guatemala, it will flower in the open ground (in Belgium) in summer.

" The genus Moussonia was established in 1848 by M. Regel upon the *Gesnera elongata* of Graham, a plant evidently allied to, although quite distinct from the species here described, as well as from the Peruvian species described by Kunth, under the name of *Gesneria sylvatica* in Humboldt and Bonpland's *Nova gen. et sp. Amer.* One of us (M. J. Decaisne) having carefully studied the whole family of *Gesneraceæ*, the results of which examination have been partially made public in the *Revue Horticole* for 1848, has been able to confirm the creation of the genus in question, and to include in it three species. He thinks he can also settle two synonyms which arise from a second article on *Gesneraceæ* published by M. Regel in the *Flora*, March 28, 1849, No. 12. First, the genus *Giesleria* Reg., established on the *Achimenes picta* of our hothouses is nothing but the *Tydæa* Dne, previously created ; second, in proposing the name *Salicia* for the genus *Gloxinia* as founded by l'Héritier, M. Regel departs from the rule of nomenclature which invariably attaches a generic name to the species which first served as a type ; to conform to this rule the name of *Gloxinia* should be reserved for *Gloxinia maculata* l'Hérit. and to its true analogies, whilst *Gloxinia speciosa*, *caulescens*, and the species and varieties analogous to them should be designated by the name of *Ligeria* Dne."—*Flore des Serres*, t. 489.

184. Metrosideros buxifolia. *Allan Cunningham.* (*aliàs* M. scandens *Forster.*) An evergreen greenhouse bush from New Zealand, with box-like leaves, and heads of pale yellowish-white flowers. Belongs to Myrtleblooms (Myrtaceæ). Flowered at Kew in August. (Fig. 89.)

Rather a pretty plant, said to be a climber, but not evincing any tendency that way, in cultivation. It would seem that this and other plants in the damp woods of New Zealand produce, like ivy, roots from the branches, by which they scramble up the trunks of forest trees. The native name is said to be *Aki*—that of the English settlers *Lignum vitæ.* Young branches hoary. Leaves close set, spreading in four rows, ½-inch long, almost sessile, elliptical or ovato-rotundate, very blunt, leathery, glossy, rolled back at the edge, dark green above, somewhat hoary with minute hairs beneath, where they are also dotted. Principal veins about five, the lateral ones from near the base. Peduncles very short, 3-flowered from the axils of the upper leaves, and thence forming a sort of capitate leafy corymb. Calyx turbinate, slightly hairy, with five obtuse lobes. Petals elliptical, small, white. Filaments white, four times as long as the erect petals. Anther yellow. *Botanical Magazine*, t. 4515.

PLATE 25.

I. Constans. Pinx & Zinc

Printed by C.F. Cheffins, London.

[PLATE 25.]

THE ACUMINATE ONION.

(ALLIUM ACUMINATUM.)

A Hardy Bulb, from CALIFORNIA, *belonging to the Order of* LILYWORTS.

Specific Character.

THE ACUMINATE ONION.—Stem leafy at the base. Leaves subulate, as long as the scape. Umbels lax; the pedicels much longer than the spathe; not bulbiferous. Sepals and petals acuminate, erect, recurved at the point, the latter much smaller than the former. Filaments shorter, entire, free. Ovary and capsule obovate, without appendages.

ALLIUM *ACUMINATUM.*—Caule basi folioso, foliis subulatis scapo æqualibus, umbellâ laxâ, pedicellis spathâ multò longioribus haud bulbiferis, sepalis petalisque acuminatis erectis apice recurvis his multo minoribus, filamentis brevioribus integris liberis, ovario capsulâque obovatis inappendiculatis.

Allium acuminatum: *Hooker, Flora Boreali-Americana.* Vol. II. 184, t. 196.

A few bulbs of this charming plant were sent from California to the Horticultural Society by Mr. Hartweg, and flowered last spring in the Chiswick Garden, in a greenhouse. It is, however, in all probability hardy, if kept in a place dry in winter.

The name Onion conveys to an English ear ideas of anything but beauty, for many common species are as ugly as plants well can be, and the handsome kinds are almost unknown in gardens. Nevertheless, in a genus consisting of nearly a couple of hundred species, many may be found which ought to take rank with Hyacinths and Jonquils; of these Moly and the Magical Onion are well-known examples, though now-a-days confined to curious collections; and the rare species here figured is another, much handsomer than either, and probably the Queen of the family. Its gay flowers, almost transparent when colourless, and stained with the richest rose-colour near the points, can scarcely be regarded as inferior in beauty to the Guernsey Lily itself, and they are far less fugitive.

The plant grows about a foot high, with narrow taper rushy leaves, about as long as the scape.

The flowers are arranged in loose umbels, or stalks, very much longer than the spathe. The sepals are much larger than the petals, and rather broader; otherwise they are both of the same form and colour—sharp-pointed and richly stained with crimson at the point, while the lower half is colourless and semi-transparent; they all cohere near the base. The stamens are inserted a little below the middle of the petals, and just above the base of the sepals; but they are in both cases easily detached; at the base they are united in the smallest possible degree; the filaments are flat, in no degree lobed, awl-shaped from a broad base: those opposite the petals, the longest. The ovary is obovate, depressed at the apex, and terminated by a sunken awl-shaped style, 3-celled, with two erect ovules in each cell; the stigma is nearly simple. The capsule is papery, and opens through the back of the cells. Seeds thin, black, with a soft skin; the greater part abortive.

Were it permitted to suppose that a plant so similar to Onions in most respects could form a separate genus, one would be tempted to place this apart, for it wants their smell, and is most remarkable for its petals being considerably smaller than the sepals. But no other difference being perceptible we must believe it to belong to the group of which Allium roseum forms one.

At first sight it would seem to differ from the Acuminate Allium described by Sir W. Hooker in his Flora Boreali-Americana, in the absence of toothings from the petals, in the smallness of those parts, and in stature: being a much larger and more handsome plant than Sir W. Hooker's figure represents. We have, however, ascertained, from the examination of authentic specimens, that there is no real distinction. In our wild plant from Douglas the petals are smaller than the sepals, as in this, and we are unable to detect the toothings above referred to.

I. Constans. Pinx & Zinc.

Printed by C.F.Cheffins London.

[Plate 26.]

THE GAUNTLETTED TACSONIA.

(TACSONIA MANICATA.)

A Greenhouse Creeper from Peru, *belonging to the Order of* Passionworts.

Specific Character.

THE GAUNTLETTED TACSONIA. — Bracts entire, united at the base, downy, longer than the tube of the calyx. Leaves downy on the under side, smooth on the upper, divided below the middle into 3 serrated lobes. Leafstalks with several glands. Stipules roundish, toothed in a crested manner.

TACSONIA *MANICATA*.—Bracteis integris basi connatis tomentosis calycis tubo longioribus, foliis subtus tomentosis suprà glabris ultra medium 3-fidis; lobis ovali-oblongis serratis, petiolis pluriglandulosis, stipulis subrotundis cristato-dentatis.

Tacsonia manicata: *Jussieu in Annales du Muséum.* Vol. VI., t. 59, f. 2.

We believe this species to be unrivalled among climbers, for the brilliant scarlet of its gorgeous blossoms. Placed by their side, the red coat of an English soldier becomes dull and pale. It is a native of Peru, and probably common there, for many botanical travellers have observed it. Humboldt and Bonpland brought some varieties from the city of Loxa; Hartweg says that it is found in hedges near that place; and it forms No. 1294 of Linden's Herbarium, gathered by his collectors Funck and Schlim, in the province of Merida, at the elevation of 7000 feet above the sea.

It forms a rambling climber, with grey 3-lobed leaves and large scarlet flowers, whose tube is almost concealed by 3 downy bracts, from which circumstance we presume that Jussieu gave it the name of the gauntletted (or manicate); it must be owned that the tube of the flower may not unaptly be compared to an arm thrust into a large loose glove. The coronet consists of two principal rows of short violet teeth planted on the green tube of the calyx-lining.

Upon what precise ground the Tacsonias are separated from the Passionflowers is by no means clear. De Candolle relies upon the former having a very long tube to the calyx and a scaly coronet;

but in this plant the tube is as short as in any Passionflower, and there is nothing peculiar in the coronet. Meisner's analysis brings out no more; and it is impossible to gather any distinction after comparing Endlicher's prolix descriptions. Nevertheless, there is something very peculiar in the appearance of Tacsonias, and we trust that in time a real distinctive character will be discovered.

The species was introduced by the Horticultural Society. Its flowers have been produced abundantly in the conservatory of A. F. Slade, Esq., of Chiselhurst, from whom we received specimens on the day of the June Exhibition at Chiswick. Upon comparing them with the finest colours there, nothing could be found to equal them in brilliancy. Others have been less fortunate; and it is understood that the plant is a bad bloomer. We understand, however, from Mr. Ansell, the gardener at Chiselhurst, that it only requires plenty of room, when it soon becomes loaded with flowers. No doubt it refuses to produce anything more than leaves when pruned much, as it must be if restricted in space. In this respect it behaves exactly like other climbers—Bougainvillæa for instance.

PLATE 27.

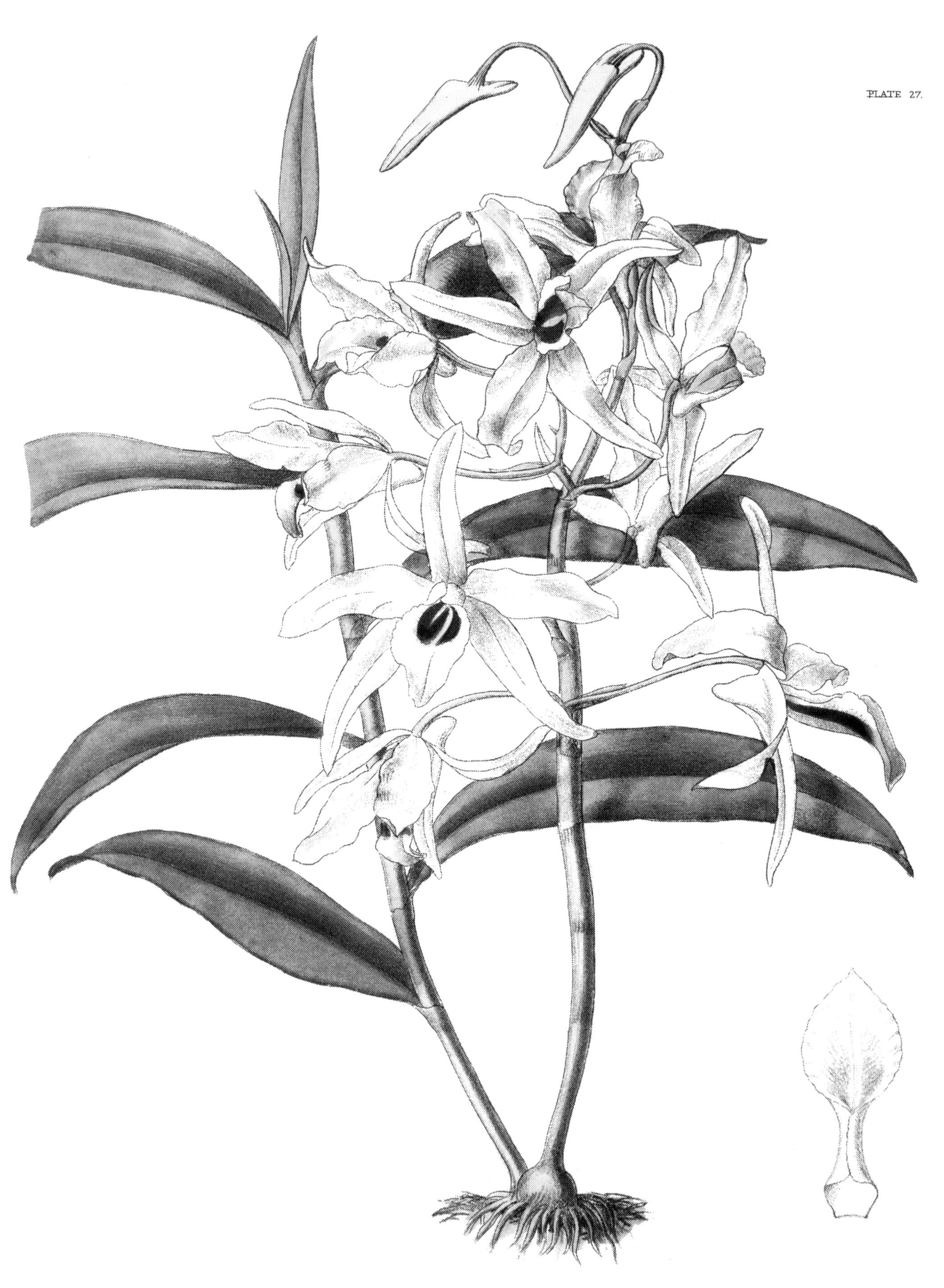

I. Constans. Pinx & Zinc.

Printed by C.F. Cheffins, London.

[PLATE 27.]

THE TRANSPARENT DENDROBE.

(DENDROBIUM TRANSPARENS.)

A Stove Epiphyte from NORTHERN HINDOSTAN, *belonging to the Natural Order of* ORCHIDS.

Specific Characters.

THE TRANSPARENT DENDROBE.—(TRUE DENDROBES). Stems erect, tapering, smooth. Leaves ovate-lanceolate, acuminate, oblique a the point. Flowers in pairs or threes. Sepals linear-oblong; petals broader, blunt. Lip acute, oblong, downy, with the sides erect and rolled inwards.

DENDROBIUM *TRANSPARENS*; (EUDENDROBIUM.) Caulibus erectis teretibus glabris, foliis ovato-lanceolatis acuminatis apice obliquis, floribus 2-3nis, sepalis lineari-oblongis, petalis latioribus obtusis, labello acuto oblongo pubescente lateribus versus basin erectis involutis.

Dendrobium transparens: *Wallich, Catalogue*, No. 2008: *Lindley, Genera & Species Orchid.*, p. 79.

ONE of the most delicate and beautiful of a delicate and beautiful genus. It was first made known by Dr. Wallich, whose collectors found it in Nepal; and from very imperfect specimens it was incorrectly described in the Genera and Species of Orchidaceous plants as a pendulous species, with the habit of the Pierard Dendrobe. Recently it has been introduced by Messrs. Veitch and Co., for whom it was collected by Mr. Thomas Lobb at a place called Myrong, on the Garrow Hills, at an elevation of 5300 feet. This Myrong, or Myrung, seems to be a wood abounding in plants; for in Griffith's "Itinerary Notes," thirty-four species are named as having been uncommon enough to be gathered by him, and among them are eleven orchids, of which this Dendrobe was probably his No. 1013, growing on rocks and trees; at least we find it among his Khasija plants. From the nature of the vegetation associated with it we may infer that it is by no means a tender kind.

It is readily known among its race by its short erect stems, obliquely emarginate leaves, and wide spreading pinkish flowers stained with crimson in the middle of the lip, and as transparent as anything vegetable well can be. It flowers most abundantly, and must be regarded as a great acquisition.

In this genus there is found to be so large a number of species, having such widely different habits, that Botanists, at an early period of their acquaintance with them, were led to create many supposed genera, the distinctions among which are now found to be unreal or unimportant. They, however, in some cases, form good sectional divisions, of which a vacant page enables us to present the following sketch, along with an enumeration of all the species known to us, and their more important synonyms :—

SECTIONS OF THE GENUS DENDROBIUM.

Folia equitantia	§ 1. *Aporum* Bl. (Macrostomium *Bl.*)
Folia teretia , ...	§ 2. *Strongyle.*
Folia plana, v. O.	
Labellum plumosum, aut pectinatum	§ 3. *Desmotrichum* Bl.
——— nec plumosum nec pectinatum	
a. Caules elongati undique foliosi	
Flores fasciculati	§ 4. *Eudendrobium.* (Grastidium *Bl.*)
Flores racemosi	
Petala nana	
Labellum elongatum, angustum, intùs nudum, .	§ 5. *Pedilonum* Bl.
——— brevius, dilatatum	§ 6. *Stachyobium.*
Petala antennæformia	§ 7. *Ceratobium.*
b. Caules clavati apice tantum foliosi	§ 8. *Dendrocoryne.*
c. Pseudobulbi tantùm aut caules brevissimi	§ 9. *Bolbodium.*
d. Rhizomata tantùm	§ 10. *Rhizobium.*

§ 1. APORUM.

This consists of species with erect or prostrate stems; succulent equitant leaves, and inconspicuous flowers. It includes the genera *Macrostomium* and *Sarcostoma* of Blume, and *Schismoceras* of Presl., which seems to be *Aporum Leonis*. The following are the species :—

1. A. micranthum *Griffith.* 2. A. anceps *Lindley.* 3. A. Leonis *Id.* 4. Dendrobium Sarcostoma *Id.* 5. Macrostomium aloefolium *Blume.* 6. A. sinuatum *Lindley.* 7. A. cuspidatum *Wallich.* 8. A. indivisum *Blume.* 9. A. lobatum *Id.* 10. A. incrassatum *Id.* 11. A. Serra *Lindley.* 12. A. subteres *Griffith.*

§ 2. STRONGYLE.

Here are found all the Dendrobes with tapering or awl-shaped leaves. The section is quite analogous to the Cebolletes, among Oncids, as the last was to the equitant division of that genus. Several of Blume's *Onychiums* must be referred to it. They are generally plants of no beauty.

13. D. gracile *Lindley.* 14. D. tenellum *Id.* 15. D. subulatum *Id.* 16. D. teretifolium *R. Br.* 17. D. acerosum *Lindley.* 18. D. schœninum *Id.* 19. D. teres *Id.* 20. D. crispatum *Swartz.* 21. D. aciculare *Lindley.* 22. D. junceum *Id.* 23. D. calamiforme *Loddiges.*

§ 3. DESMOTRICHUM.

With this section we enter upon the mass of the genus, with flat leaves, and more conspicuous blossoms. They have erect stems, often more or less distended into pseudobulbs, and are remarkable for the end of the lip being broken up into long tufted fringes, or in *D. planibulbe*, marginal threads.

24. D. Scopa *Lindley.* 25. D. criniferum *Id.* 26. D. comatum *Id.* 27. D. angulatum *Id.* (There is another species with this name in Eudendrobium.) 28. D. Blumei *Id.* 29. D. planibulbe *Id.*

§ 4. EUDENDROBIUM.

The centre of the genus, rich in species, among which are several of considerable beauty, although not of the greatest. They have long leafy stems, erect or pendulous, and flowers in lateral pairs or rarely in threes, with no trace of the feathery or tufted lip of the last section. Two divisions are conveniently made by attending to the form of the lip.

A. *Lip undivided.*

30. D. macrophyllum *Lindley.* 31. D. anosmum *Id.* 32. D. moniliforme *Swartz.* 33. D. cærulescens *Lindley;* (aliàs *D. Wallichii*). 34. D. nobile *Id.* 35. D. tortile *Id.* 36. D. pulchellum *Roxb.* 37. D. Devonianum *Paxton.* 38. D. Pierardi *Roxb.* 39. D. cretaceum *Lindley.* 40. D. cucullatum *R. Br.* 41. D. Egertoniæ *Lindl.* 42. D. mesochlorum *Id.* 43. D. crepidatum *Id.* 44. D. transparens *Wallich.* 45. D. amœnum *Id.* 46. D. macrostachyum *Lindley.* 47. D. gemellum *Id.* 48. D. foliosum *A. Brongniart*; (is this a Stachyobium? or a new genus? or an Appendicula?) 49. D. rugosum *Lindl.* 50. D. salaccense *Id.* 51. D. chrysanthum *Wallich.* 52. D. Paxtoni *Lindl.* 53. D. ochreatum *Id.* (aliàs D. Cambridgeanum *Paxton.*) 54. D. aureum *Id.;* (aliàs D. heterocarpum *Wallich*). 55. D. candidum *Wallich.* 56. D. nutans *Lindley.* 57. D. stuposum *Id.* 58. D. connatum *Id.*

B. *Lip three-lobed.*

59. D. longicornu *Lindley.* 60. D. Ruckeri *Id.* 61. D. sanguinolentum *Id.* 62. D. aqueum *Id.* 63. D. revolutum *Id.* 64. D. excisum *Id.* 65. D. bilobum *Id.* 66. D. calcaratum *Id.* 67. D. crumenatum *Swartz.* 68. D. angulatum *Wallich*; (see Desmotrichum No. 27). 69. D. biflorum *Swartz.* 70. D. acuminatissimum *Lindley.* 71. D. Cunninghamii *Id.* 72. D. Luzonense *Id.* 73. D. tridentiferum *Id.* 74. D. tetraedre *Lindl.*

§ 5. PEDILONUM.

The habit of Eudendrobium, together with flowers in racemes, diminutive petals, and a long narrow naked lip, distinguishes this small group, among which the beauty of *D. secundum* typifies that of the remainder.

75. D. secundum *Wallich.* 76. D. erosum *Lindley.* 77. D. hymenophyllum *Id.* 78. D. Kuhlii *Id.* 79. D. Hasseltii *Id.* 80. D. Reinwardtii *Id.*

§ 6. STACHYOBIUM.

At this point the genus assumes its greatest development, and consequently its most conspicuous brilliancy. Yellow is a prevailing colour. The species would merge in *Pedilonum,* if it were not for the large full-grown petals, and broad dilated lip, which in some cases runs inwards into a kind of sock or pouch. Two divisions are again obtainable here, by observing the differences in the form of the lip.

A. *Lip undivided.*

81. D. mutabile *Lindley.* 82. D. sclerophyllum *Id.* 83. D. triadenium *Id.*; (perhaps these three last are varieties of each other). 84. D. aduncum *Id.* 85. D. formosum *Roxburgh.* 86. D. rhombeum *Lindley.* 87. D. fimbriatum *Hooker.* 88. D. polyanthum *Wallich.* 89. D. sulcatum *Lindley.* 90. D. moschatum *Wallich*; (aliàs D. calceolus *Hooker*; aliàs D. cupreum *Herbert*; aliàs D. clavatum *Wallich*). 91. D. Dalhousianum *Paxton.* 92. D. calcaratum *A. Rich.* 93. D. flavescens *Lindley.* 94. D. nudum *Id.* 95. D. auriferum *Id.* 96. D. ramosum *Id.* 97. D. herbaceum *Id.* 98. D. japonicum *Id.* 99. D. cassythoides *Id.*

B. *Lip three-lobed.*

100. D. Heyneanum *Lindley.* 101. D. barbatulum *Id.;* (alias D. chlorops, *Lindley*). 102. D. lancifolium *A. Richard.* 103. D. bicameratum *Lindley.* 104. D. elongatum *A. Cunn.* 105. D. bicolor *Lindley.* 106. D. catenatum *Id.* 107. D. denudans *Don.* 108. D. alpestre *Royle.* 109. D. cuspidatum *Lindley.* 110. D. breviflorum *Id.*

§ 7. CERATOBIUM.

A remarkable form of the genus, with tall erect stems, flat leaves, and long racemes of flowers, conspicuous for the long narrow antennæ-like petals.

111. D. Mirbelianum *Gaudich.* 112. D. veratrifolium *Lindley.* 113. D. macranthum *A. Richard.* 114. D. antennatum *Lindley.* 115. D. taurinum *Id.* 116. D. undulatum *R. Br.*; (aliàs D. discolor *Lindley*). 117. D. affine *Lindley.*

§ 8. DENDROCORYNE.

From this point the development of the genus diminishes. The stem is contracted at the base, and club-shaped, with leaves at only the extreme end, as in the § *Spathium* among Epidendrums; the flowers are as in *Eudendrobium* and *Stachyobium.* The inflorescence may be made to constitute sectional differences.

A. *Inflorescence terminal.* (Chiefly Australian.)

118. D. speciosum *Smith.* 119. D. canaliculatum *R. Br.* 120. D. æmulum *R. Br.* 121. D. Kingianum *Bidwill.* 122. D. Veitchianum *Lindley.* 123. D. tetragonum *Cunningham.* 124. D. Macræi *Lindley.* 125. D. longicolle *Lindley.*

B. *Inflorescence lateral.*

126. D. chrysotoxum *Lindley.* 127. D. Griffithianum *Id.* 128. D. aggregatum *Roxburgh.* 129. D. compressum *Lindley.* 130. D. densiflorum *Wallich.* 131. D. Palpebræ *Lindley.*

§ 9. BOLBODIUM.

In lieu of true stems these species are furnished with pseudobulbs, sitting on a prostrate rhizome. With the exception of *D. Jenkinsii* they are all obscure plants of no horticultural value.

A. *Lip undivided.*

132. D. Jenkinsii *Wallich.* 133. D. braccatum *Lindley.* 134. D. muscicola *Id.* 135. D. pygmæum *Id.* 136. D. subacaule *Reinwardt.* 137 ? D. tricuspe *Lindley.* 138 ? D. plicatile *Id.* 139 ? D. lamellatum *Id.* 140 ? D. pusillum *Id.* 141 ? D. triflorum *Id.* 142. D. appendiculatum *Id.*

B. *Lip three-lobed.*

143. D. extinctorium *Lindley.* 144. D. microbolbon *A. Richard.* 145 ? D. angustifolium *Lindley.* 146 ? D. convexum *Id.* 147 ? D. grandiflorum *Id.* 148 ? D. cymbidioides *Id.* 149 ? D. elongatum *Id.* 150 ? D. geminatum *Id.*

§ 10. RHIZOBIUM.

Obscure species, with nothing more than a creeping rhizome, bearing solitary coriaceous leaves.

151. D. linguæforme *Swartz.* 152. D. cucumerinum *W. Macleay.* 153. D. pugioniforme *A. Cunningham.* 154. D. rigidum *R. Br.*

In addition to these, about a dozen other supposed species are to be found in books, but they are so little known as to be unworthy enumeration in this sketch. *D. amplum* of Wallich, along with some spurious Bolbophylls, forms a new genus called Sarcopodium, of which some notice will be taken hereafter.

GLEANINGS AND ORIGINAL MEMORANDA.

185. POLYGONUM CUSPIDATUM. *Siebold and Zuccarini*. A tall hardy handsome broad-leaved herbaceous plant from Japan. Flowers green, inconspicuous. Belongs to the Order of Buckwheats (*Polygonaceæ*). Introduced by the Horticultural Society about the year 1825. (Fig. 90.)

We translate the following account of this plant from Professor Morren's statement in the *Annales de Gand*, vol. v. p. 461 : "Rhizome herbaceous, stem straight, branching, flexible, smooth, round, hollow, spotted with purple. Leaves stalked, truncated or rectilinear at the base, scarcely subcordate, broadly oval, bordered with red or with a transparent edge, cuspidate, smooth on both sides, slightly rough on the under side along the nerves. Stipules obliquely truncate, smooth, naked at the edge, few-nerved, purple, finally becoming torn, deciduous. Panicles axillary, divaricatingly branched ; rachis flexible ; branches slender, scurfy haired ; bracts ochreiform, obliquely cuspidate-truncate ; flowers in twos or threes, pedicels filiform, coloured, articulated, shorter than the tube of the perianth ; stamens 8, filaments petaloid, subulate, ovary triquetrous, styles 3 divaricating, achenium elliptical, triquetrous with a 3-winged perianth, wings obcordate, opening longitudinally at the sutures. Professor De Vriese declares that this is undoubtedly one of the prettiest species of *Polygonum* known. It was introduced from Japan by M. Von Siebold. The

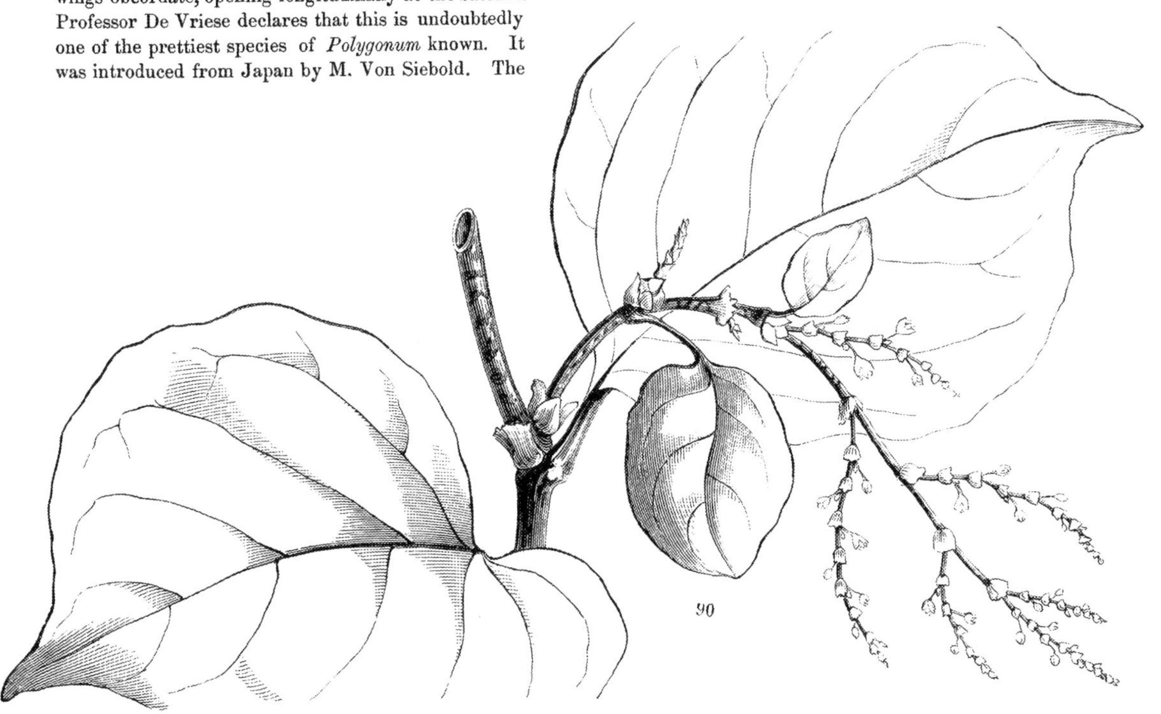

90

stem is sometimes 10 feet high and throws out numerous lateral off-shoots; the red stems and branches distinguish it immediately. The small but numerous flowers are greenish yellow and are borne on reddish pedicels. A mass of this plant produces a fine effect in gardens. It comes up in May and its stem dies in October. The root lives through the winter without either care or covering. It prefers a light soil. It can bear the hardest frosts. M. de Vriese has published an excellent drawing as well as an analysis of it; he says it is only to be found at present in M. Von Siebold's garden at Leyden. M. Von. Siebold declares that this plant is very fit for fixing loose sand, and it would be both interesting and useful to see what it is good for in this respect, especially as M. Von Siebold has seen it employed for the purpose throughout Japan."

Although unknown to botanists this plant has been cultivated in the garden of the Horticultural Society for a quarter of a century. It originally came from China as Houttuynia cordata; and for many years grew in an artificial swamp, where it formed a very handsome bush during the summer. It has since been found to thrive perfectly in dry garden ground. During the period of its cultivation it has only flowered once, and then imperfectly. From a specimen at that time preserved the annexed cut has been prepared. Where very handsome massive foliage is desired during summer only, this plant is of the greatest value, as for instance in forming rapidly a temporary screen, or in making a back ground to gaudy flowers with bad foliage. But as it dies to the ground with the first frost it makes a gap which may be unsightly. We should not have thought that it would run by the root sufficiently to hold together blowing sand, in the manner suggested by Dr. V. Siebold.

91

186. CALOCHORTUS PALLIDUS. *Schultes.* A tender bulbous plant from Mexico, belonging to the Lilyworts. Flowers dirty brown, with a deep triangular spot at the base of each petal. (Fig. 91.)

A dwarf grassy-leaved plant, with long loose few-flowered umbels of dirty pale brown flowers. Neither sepals nor petals have any gland or depression in the middle. The sepals are shorter than the petals, firmer, without any hairs. The petals are obovate, tapering to the base, rounded at the point, covered on the middle with a beard of hairs and fringed at the edge. *Annales de Gand*, t. 225.

187. CALANTHE MASUCA. *Lindley.* A beautiful terrestrial Orchid, with purple flowers. Native of various parts of India. Introduced prior to 1843.

Native of India;—according to Dr. Lindley, of "Nepal, Bengal, Ceylon, and probably Java." It blossomed in 1842 with Messrs. Rollison, at Tooting, but, though a handsome and really striking plant, it had never been figured. Our fine tuft of the plant at Kew, which blossomed in July and August, was derived from Mr. Clowes' collections.

Leaves large, herbaceous, oblong-lanceolate, tapering below, acuminated, plaited and striated. Scape erect, a foot and a half high, generally shorter than the leaves, terete, glabrous, terminated by a many-flowered raceme with handsome purple flowers. Bracts large, subulato-lanceolate, membranaceous: the upper ones coloured. Sepals and petals similar, oblong, acuminate, spreading. Lip three-parted, deep purple: lateral lobes linear oblong, subfalcate, intermediate one broadly subcuneate: the base of the lip below extends into a very long narrow spur, furrowed on one side and bifid at the point: the base of the lip above on the disc bears a five-crested tubercle, the crests transversely furrowed. This being an East Indian terrestrial Orchid, requires to be grown in a moist tropical stove. It thrives in turfy peat containing a small portion of loam. On account of its soft fleshy roots adhering to the sides of the pot, it is desirable to use a shallow wide-mouthed pot, in order to avoid tearing the roots by frequent shiftings. In summer it may be freely watered, but the pot must be well drained, so as to allow the water to pass off freely. Shading is necessary during bright sunshine. In winter it should be placed in a drier atmosphere, and especial care must be taken that no water be allowed to lodge in the folds of the young leaves.—*Bot. Mag.*, t. 4541.

Sir W. Hooker is mistaken in saying that it had not been previously figured. An excellent representation of it was given in the Botanical Register for 1844, t. 37, where will be found the following remarks :—

" From the other purple species allied to it, this is readily known by the leaves as well as by the flowers. *C. versicolor* has leaves smooth on both sides ; *C. purpurea* downy on both sides, especially beneath ; while this has down only on the under side. *C. versicolor* has white sepals and petals ; *C. purpurea*, and this, purple ones. While, however, *C. purpurea* agrees in the colour of its flowers, its lip is altogether different, being very narrow, with the lateral lobes quite round.

" *C. Masuca* should be potted in turfy heath-mould, mixed with a few pieces of potsherds. In summer it should receive an ample supply of water at its roots ; and where it can be avoided, little should be allowed to fall on its leaves, otherwise the young shoots will damp off. It enjoys a humid atmosphere and a high temperature ; but as the leaves are very delicate, they will soon become scorched if shading is not carefully attended to. In winter little water will be required ; still it is necessary to keep the soil damp enough to preserve the bulbs from shrivelling. This is one of the most difficult of Orchidaceous plants to grow well."

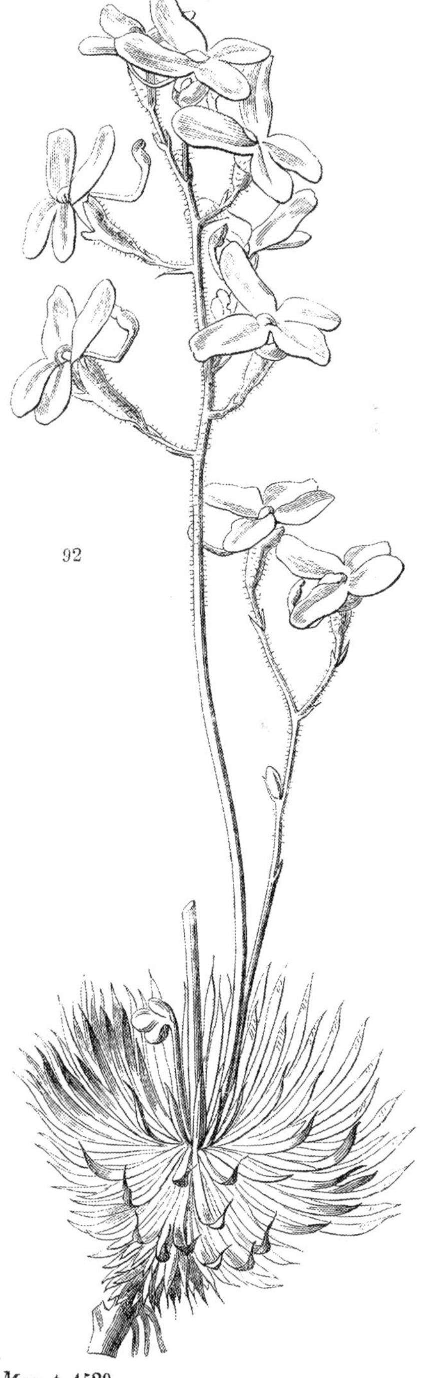

92

188. Stylidium saxifragoides. *Lindley*. A beautiful little greenhouse herbaceous plant, with lemon-coloured flowers. Belongs to the order of Styleworts. Native of Swan River. Introduced by Messrs. Veitch and Co. (Fig. 92.)

This charming greenhouse plant, raised from seeds from the Swan River Settlement, was sent by Messrs. Veitch and Sons of Exeter to the May Exhibition of the Horticultural Society for 1850, under the name of *S. ciliatum*. That plant, however, is a very different though nearly allied species, with the panicle compound, and, as well as the scaly scape, clothed with long patent hairs, tipped with dark-coloured viscid glands, and with flowers not half the size of the present one. Root perennial, dividing at the crown so as to bear copious rosettes of densely imbricated, spreading, linear leaves, slightly incurved, yellow-green tinged with purple, tapering at the base, acute at the point, and there bearing a long hair or bristle ; the margins especially roughly fringed. Scapes one or more from the centre of each rosette, a span or more high, quite smooth (except above), and there, and upon the flower-stalks and ovary, calyx and outside of the corolla, are copious, short, glandular hairs. Flower-stalks with two glandular, oblong, red bracts above the middle. Ovary oblong, green, crowned with the oblong red lobes of the calyx. Corolla large (for the size of the plant), yellow.

As regards their habit and places of growth, Styleworts may be compared to species of several British genera ; such as *Statice*, *Jasione*, *Phyteuma*, *Plantago*, *Samolus*, and even *Drosera*. This species is a native of Swan River, and must be treated as a greenhouse plant ; it requires no more artificial heat than is necessary to protect it from frost, and like many other small plants, it will thrive best when kept in a cool pit or frame ; but care must be taken that it does not suffer from damp in winter. Light peat soil is found to suit it.—*Bot. Mag.*, t. 4529.

189. Gordonia javanica. *Hooker.* A tea-like stove plant from Java. Belongs to the Natural Order of Theads. Flowers white, in the autumn. Introduced by Messrs. Rollison. (Fig. 93, A. represents the calyx, style, and stigma.)

Our Garden is indebted to Messrs. Rollison, of Tooting, for the plant of which a specimen is here figured. It was discovered by their collector in Java, probably in the mountains; and has much the general habit of *Thea* or *Camellia*, when its blossoms appear, in August and September. Our plant is about two feet high, branched, and generally glabrous. Branches terete. Leaves alternate, elliptical-lanceolate, coriaceous, evergreen, acuminated, entire, below tapering into a short petiole. Peduncles solitary, axillary, single-flowered, from the base of most of the upper leaves, and shorter than the leaves, erect, bearing two or three deciduous, spathulate, green bracteas below the calyx. Calyx of five very concave rotundato-elliptical, erect, slightly hairy sepals. Petals five, obovate, white, spreading, obliquely twisted. Stamens very numerous. Ovary globose, obscurely five-lobed, five-celled, hairy. Style columnar. Stigma peltate, of five large, rounded, somewhat leafy, rays or lobes, the centre umbilicated. Fruit the size of a large garden-pea, globose, depressed at the top, half five-valved, woody. Not being aware of its locality, we have treated it as a stove plant; but, judging from the nature of many of its allies, we may be right in presuming that it is from an elevated and temperate region, and if so, it would probably succeed in a warm greenhouse. It grows readily in loam and peat or leaf-mould, and is easily increased by cuttings.—*Bot. Mag.*, t. 4539.

190. Nymphæa micrantha. *Guillemin and Perrottet.* A Water-Lily from the Gambia, requiring a hothouse. Flowers showy, white. Introduced by the Earl of Derby.

This very pretty Water-Lily was communicated from the Tropical Aquarium of E. Silvester, Esq., the successful cultivator of *Nymphæaceæ* at North Hall, Chorley, Lancashire, in August, 1850. It was received by him from Chatsworth, but it appears to have been imported by Lord Derby, from the River Gambia. The long acuminated points of the leaves, and the viviparous axils of the lobes, are its most striking character; and in these two important particulars, as well as in some others, this species agrees with a Senegambian one to which I have referred it, viz., the *N. micrantha* of Guillemin and Perrottet. If it does not coincide in all points—such as the number of stigmatic rays—it must be remembered that aquatic plants are very variable, and we must not lay too much stress on differences of that kind. It is true the authors describe the flowers as blue, or pale blue, but native authentic specimens in my herbarium appear to be white. The leaf-stalks and flower-stalks both appear to be much lengthened (influenced, probably, by the depth of water in which they have grown), tinged with red, taper, smooth. Leaves also quite smooth, elliptic, round in outline, partly entire, partly irregularly toothed, the lower portion cut into two deep, much acuminated, moderately spreading lobes, at the re-entering angle of which, as it were from the top of the petiole, gemmæ, or little bulbs, appear and develope themselves into young plants! The underside of the leaf is pale green, tinged with pale purplish-

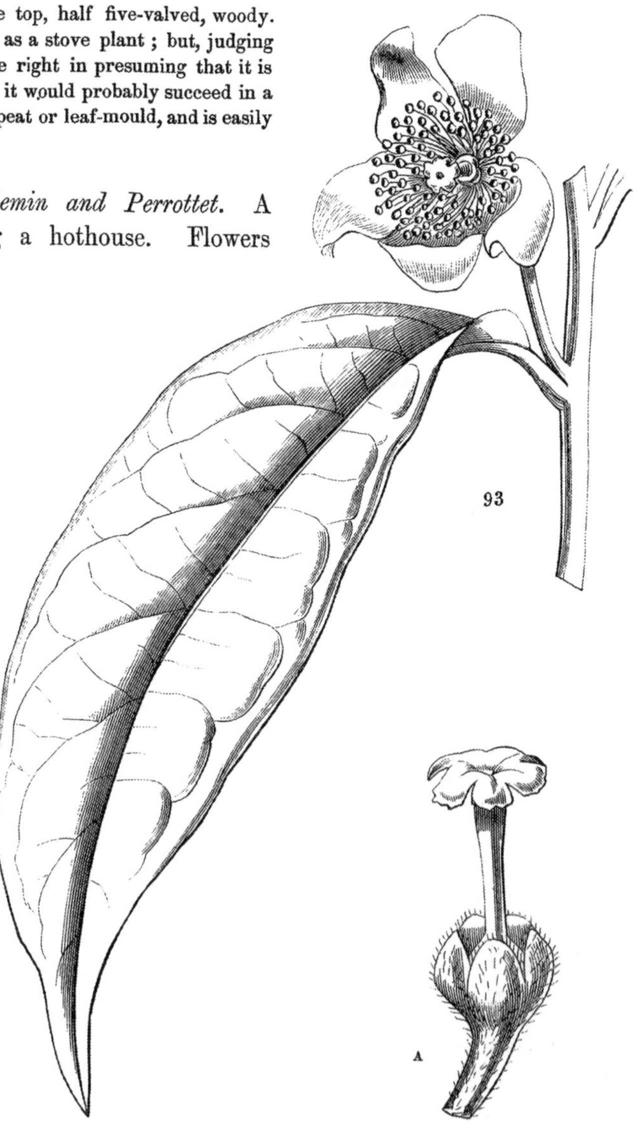

93

A

brown and minutely dotted. Flowers smaller than our common White Water-Lily, the size of *N. stellata*. Calyx of four sepals, pale yellow-green, and the numerous white or whitish petals are lanceolate and very acute, not gradually passing into stamens, though the outer stamens are more petaloid than the inner ones. Stigma in our plant with eleven incurved obtuse yellow rays. This Water-Lily, being a native of Western Africa, requires to be grown in a warm stove. It is remarkable from the circumstance of its producing a viviparous bud at the sinus of the leaf on the upper surface, which bud ultimately becomes a separate plant.—*Bot. Mag.*, t. 4535.

191. COCCOLOBA MACROPHYLLA. *Desfontaines*. A noble simple stemmed erect tree, with large leathery leaves and straight spikes of crimson flowers. Belongs to the Buck-wheat Order (*Polygonaceæ*). Native of South America? Introduced by the Royal Botanic Garden, Kew. (Fig. 94.)

One of the most striking plants which has flowered in the great stove of the Royal Gardens during the year 1850, is that here represented, of which plants were long since received from Paris, under the name of *Coccoloba macrophylla* of Desfontaines. The name is far from appropriate, for the leaves yield greatly in size to the *C. pubescens*, the latter being three or four times the size of the present. Our plant, however, equals the *pubescens* in height (our largest plant being twenty-three feet high): it tapers gracefully upwards, is leafy all the way up, and terminated at the top by a dense compact thick club-shaped raceme of flowers, of which the rachis, pedicels, and flowers are of the richest scarlet. This raceme continued in great beauty for two months, and when looked down upon from the gallery above, backed as it was by dark-green foliage, it presented a beautiful object. The drawing was made in July. A plant, with simple or scarcely divided, furrowed erect stems, twenty to thirty feet high; leafy from below to the top. Leaves alternate, distant, dark green, a foot or more long, horizontally spreading, cordate-ovate, half-stem-clasping, sessile, acute or acuminate, strongly nerved, wrinkled and reticulated, rather blistered. Raceme terminal, subsessile, erect, two or more feet long, the flowers so numerous and dense that they appear to form a compact cylindrical spike; every part of a rich scarlet colour, save the stigmas, which are yellow. Tube of the calyx funnel-shaped; limb cut into 4—6 rounded concave lobes. Stamens 8—12, monadelphous below. Ovary triquetrous, red. Styles 3. Stigmas capitate. Fruit berried, red. The genera *Coccoloba*, *Triplaris*, and *Podoptera* are the tropical representatives of the Order *Polygonaceæ*, and may be viewed as examples of the genera *Rheum*, *Rumex*, and *Polygonum*, taking the form of trees or shrubs. They are natives of the West Indies and tropical America, and often attain a considerable height. They generally have large entire coriaceous leaves, and bear spikes or racemes of flowers, succeeded by bunches of berry-like fruit, which, as many of the species inhabit the shores, have given rise to the English name, "sea-side grapes." The present species appears to be a tall-growing tree: our plant is now ten (Qu. twenty-three: see the early part of this paragraph) feet high, and with its broad stiff leaves and long erect spike of red flowers, has a very striking appearance. It requires to be kept in the stove, grows freely in light loam, and may be increased by cuttings treated in the usual way for tropical plants of like nature.—*Bot. Mag.*, t. 4536.

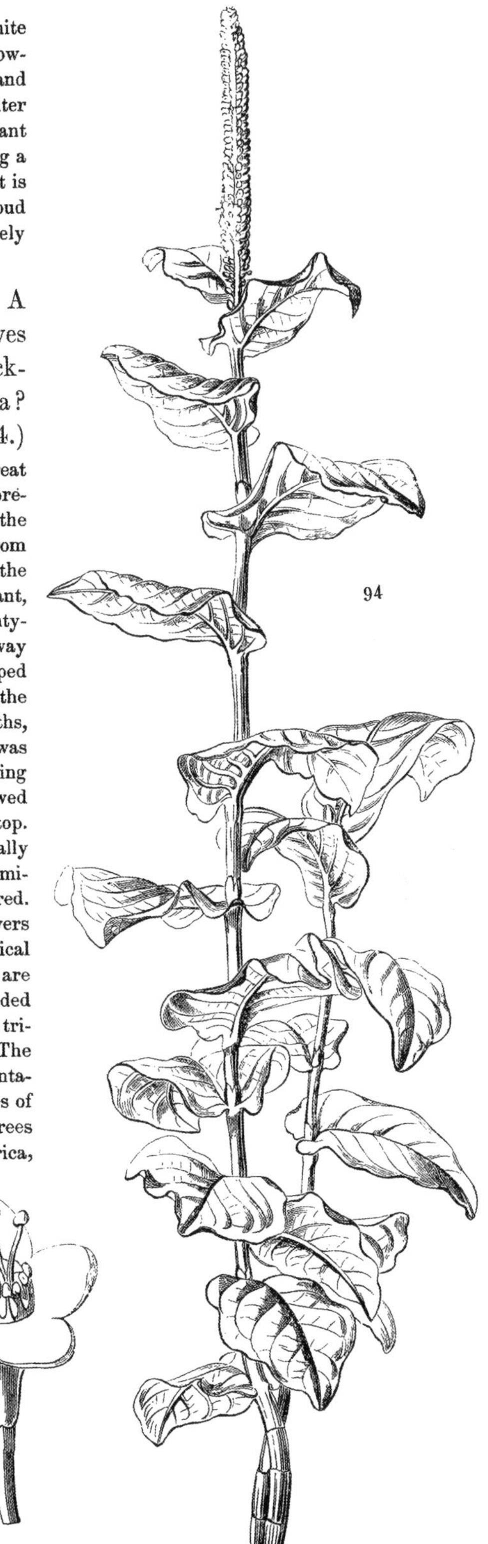

94

192. Arhynchium labrosum. An inconspicuous stove epiphyte, with small brown and yellow flowers. Introduced by Geo. Cornwall Legh, Esq., M.P. Flowered in October with Sir Philip de Malpas Grey Egerton, Bart., M.P. Native of Tropical Asia.

Arhynchium. Epiphytum; foliis distichis, coriaceis. *Sepala* et *petala* explanata, libera, basi æqualia. *Labellum* sessile calcaratum ascendens carnosum, calcare vacuo, laminâ indivisâ. *Columna* nana, teres, basi haud producta, stigmate circulari. *Anthera* subrotunda, 2-locularis, membranacea, depressa. *Pollinia* 4, geminata, æqualia; caudiculâ subulatâ, glandulâ triangulari membranaceâ semiliberâ. *Rostellum* truncatum.

A. labrosum. Labellum carnosum, basi concavum biauriculatum, calcare ascendente obtuso vacuo recurvo, ore incrassato ferè clauso; laminâ luteâ carnosâ crassissimâ rugosâ ovali, *horizontaliter* fissâ, calcaris convexitati adnatâ.

For a couple of flowers of this curious little orchid, we are indebted to Sir Philip Egerton, with whom it flowered in the middle of October. It was purchased two or three years since, by Mr. Cornwall Legh, at one of Stevens's sales of East Indian Orchids; but nothing further is known of its history. It is described as a plant with the habit of a small Vanda, or of a Sarcochile. The flowers are about an inch in diameter, placed at equal distances on a raceme. The fragment before us bore 4, about half-an-inch apart. The sepals and petals are narrow, blunt, leathery, purplish brown, spotted with dull yellow; the second smaller than the first. The lip is a hollow curved blunt horn, rising from the base of the column with its convexity upwards; on the convexity lies a flat yellow wrinkled fleshy tongue, which seems as if it consisted of two layers; at the base the lip is concave, and has on either side a short truncated ear, with which it clips the column. The column is taper, short, straight, with a nearly circular stigma.

No known genus can receive this singular plant, unless it is thrown into the crowd of Saccolabes, among which, however, it would scarcely be sought; for its thick fleshy lip is very different from the thin membrane found in that genus. Moreover it is essentially distinguished by its rostel not being extended into a long beak, as is the case in all genuine Saccolabes and Sarcanths. As for Sarcochilus, which it is said to resemble, that genus is quite different in the long narrow foot on which the lip is placed, as well as in the nature of the lip itself.

193. Pitcairnia Jacksoni. *Hooker.* A very handsome stove Bromeliad, with scarlet flowers. Native of Guatemala. Introduced by Mr. Jackson of Kingston.

This very handsome Pitcairnia was flowered by Mr. Jackson, of Kingston, who imported it in a very young state, among tufts of Orchideous plants from Guatemala. Its nearest affinity is probably with *P. bromeliæfolia.* Leaves a foot and more long, subulato-ensiform, striated, attenuated above and below, upper half only spinuloso-serrated, the rest entire, above dark green and naked, below clothed with a whitish floccose or pulverulent substance. Scape leafy below, pulverulent, bearing an erect raceme of handsome scarlet flowers. Pedicels bracteated, standing out almost horizontally, and, as well as the calyx, pulverulent. Calyx of three, imbricated, erect sepals, about three quarters of an inch long, red with a yellowish margin. Corolla scarlet, nearly three inches long, curved. Tropical America and the West Indian islands are the native places of the genus *Pitcairnia.* They generally inhabit dry places, where there is little or no soil. They increase by suckers, and ultimately become dense cæspitose tufts, sometimes found growing on trees. They appear able to bear a great degree of heat and drought, but in a state of cultivation they improve in appearance by allowing them a due share of moisture. This pretty species has flowered in the Orchid-house, under the influence of a moist and warm atmosphere, in which it appears to thrive. A soil composed of light loam and peat suits it. It is increased by taking off the young suckers, which root freely without the aid of a bell-glass.—*Bot. Mag.* t. 4540.

194. Rogiera amœna. *Planchon.* (*aliàs* Rondeletia thyrsoidea of Gardens.) A hothouse shrub, with clusters of rose-coloured flowers. Native of Guatemala. Belongs to the Cinchonads. Introduced by Mr. Skinner. (Fig. 95.)

This, and another species resembling it, appears occasionally from among the earth and rubbish hanging to the Orchids imported from Guatemala. They resemble Viburnums, and more especially Laurustines, but with red or rose-coloured flowers. That now figured is common under the name of Rondeletia thyrsoidea, and is a species of considerable beauty. All the parts are covered with soft hairs. The leaves are oblong, rather the broadest at the base, nearly sessile, with large ovate intermediate stipules. The flowers, of a bright rose a little mixed with yellow at the throat, are in very short compact roundish cymes terminating the young branches. The lobes of the calyx are five, obtuse, short; the corolla is salver-shaped with its five flat lobes oblong and emarginate, while the tube is slightly enlarged upwards. M. Planchon makes the following remarks upon the genus in the *Flore des Serres*, t. 442.

"By a great good fortune we are able to create at least two well defined genera from the chaos of different species thrown together under the name of *Rondeletia,* and to make one of these genera known by four new species, all ornamental, all recently introduced into our hothouses ; to affix, in short, to charming shrubs the name of the most active promoter of agriculture and horticulture in Belgium." (M. Charles Rogier, Minister of the Interior in the Belgian Cabinet.)

"The four species of *Rogiera,* of which we speak, inhabitants of the temperate regions of Guatemala where *Lycaste Skinneri* is found in its glory, have just produced, in M. Van Houtte's houses, their corymbs of pretty pink flowers, the limb of which, spreading like a star, encloses a tuft of golden hairs by which their throat is closed. Their want of size and brilliancy is compensated by the time which they last, their agreeable though slight odour, their profusion, and delicate colour. All four species are much alike, their differences being such as none but a botanist can appreciate. Their general appearance, their foliage, their stipules, their inflorescence, are all similar. They form a perfectly natural genus, approaching *Rondeletia,* but distinguished from it by the absence of the prominent ring in the throat of the corolla." To this M. Van Houtte adds the following remarks upon their cultivation :—

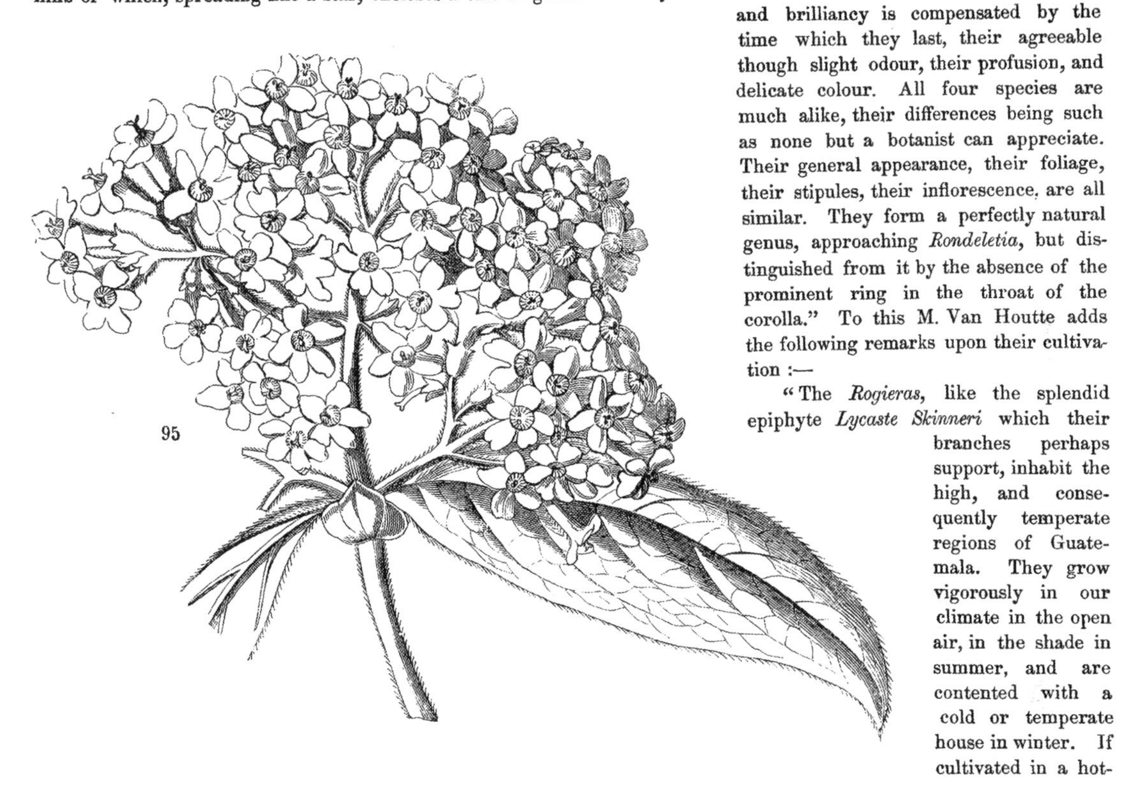
95

"The *Rogieras,* like the splendid epiphyte *Lycaste Skinneri* which their branches perhaps support, inhabit the high, and consequently temperate regions of Guatemala. They grow vigorously in our climate in the open air, in the shade in summer, and are contented with a cold or temperate house in winter. If cultivated in a hothouse their period of flowering is hastened, as it may be also by other means. The soil they prefer is a light mixture of peat or leaf-mould and a little sand. They should be frequently watered. They may be propagated by cuttings, under a bell-glass, in a moist atmosphere and on a warm bottom."

The four species which M. Planchon enumerates are *R. amœna, Menechma, Roezlii,* and *elegans ;* they seem to differ in very slight circumstances. In the same work this author proposes a genus, also cut off Rondeletia, for which he offers the name of *Arachnothryx,* and to which he refers the Rondeletias *buddleioides, laniflora,* and *reflexa* of Bentham with the *discolor* of Humboldt and some others.

195. Potentilla ochreata. *Lindley.* A hardy shrub with yellow flowers, belonging to the Roseworts. Native of the Himalayas. Flowers in September. Introduced to the Botanic Garden, Glasnevin, by Major Madden. (Fig. 96.)

This very curious and handsome plant bears a near relation to the Shrubby Potentil, so well known in Gardens. It was found in Sirmore by Capt. Gerard ; and we have a wild specimen from Dr. Royle, from some other part of the Himalayas. It forms a dwarf hairy bush, with weak spreading brown branches. The leaves are between pinnate and digitate, short-stalked, with membranous dilated brown stipules as long as the stalks ; the leaflets vary in number from five to nine, are grey, oblong, rolled back at the edge, and much wrinkled, whitish and hairy on the underside ; the uppermost pair are decurrent at the base, the others taper to the point of insertion ; some are usually two-lobed. The flowers are terminal, nearly sessile in the garden specimen, but conspicuously stalked in those found by Capt. Gerard. There are five bracts external to the calyx, linear-lanceolate, very hairy, with a distinct red scabrous keel ; the sepals are of the same length, triangular, yellow inside ; the petals nearly circular, firm and bright yellow.

When first received from the Botanic Garden, Glasnevin, it was remarked to be so much stouter in all its parts than the Ochreate Potentil, that it was mistaken for some variety of the Bush Potentil (*P. arbuscula* Don ; aliàs *P. rigida* Wallich) ; for the wild specimens of the species have very narrow leaves, white with long hairs, and a more slender manner of growth. A more careful examination, however, shows that this is really a mere garden state of the Ochreate. The Bush Potentil is a plant of more vigorous growth, with bright green, not grey foliage ; the leaflets in threes, or at most in fives, and by no means wrinkled on the under side ; its flowers are, moreover, each furnished with ten bracts, either wholly separate, or partially united in pairs, a circumstance by which it is immediately distinguishable from all the forms of the Shrubby Potentil (*P. fruticosa*). It is well figured in Wallich's Plantæ Asiaticæ ; but very ill defined by Lehmann.

The following short characters will serve to distinguish the truly fruticose Potentils, which form a very peculiar section of that great genus :—

* Flowers Yellow.

1. The Shrubby Potentil (*P. fruticosa* L. ; aliàs *P. floribunda* Pursh). Bracts five, narrow, smooth on the keel, longer than the sepals. Leaflets five, linear-lanceolate.
2. The Bush Potentil (*P. arbuscula* D. Don ; aliàs *P. nepalensis* Id.; aliàs *P. rigida* Wallich). Bracts ten, the length of the sepals.
3. The Ochreate Potentil (*P. ochreata* Lindley in Wallich's Catalogue). Bracts five, rough on the keel, the length of the sepals. Leaflets oblong, five to nine, much wrinkled beneath.

* * Flowers White.

4. The Sales of Potentil (*P. Salesovii* Steph.) An erect bush. Leaves hoary beneath, serrated at the edge.
5. The Glabrous Potentil (*P. glabra* Loddiges). A half trailing bush. Leaves smooth, entire at the edge.

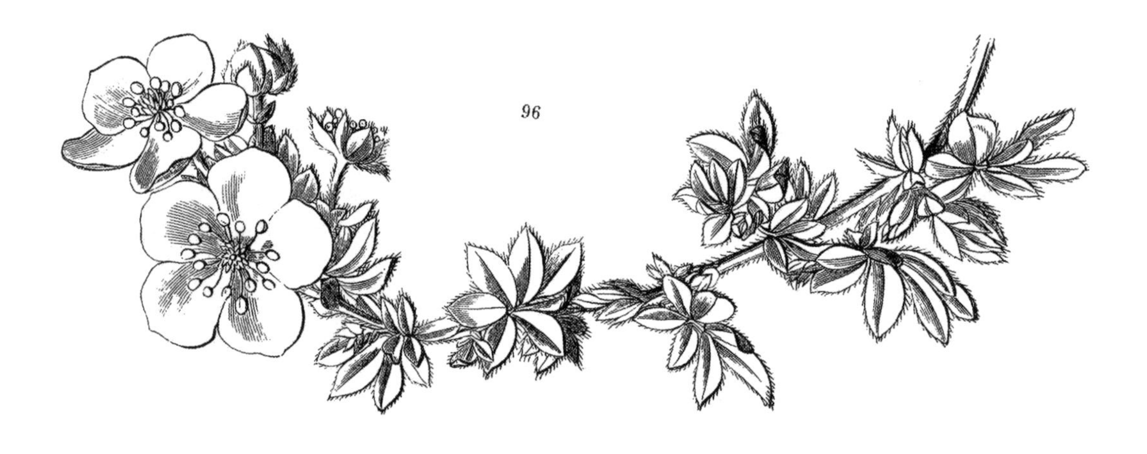

96

PLA

I. Constans, Pinx & Zinc.

Printed by C.F. Cheffins, London.

[Plate 28.]

THE GILLIES POINCIANA.

(POINCIANA GILLIESII.)

A Half-hardy Shrub, of great beauty, from Chili, *belonging to the Order of* Leguminous Plants.

Specific Character.

THE GILLIES POINCIANA.—Unarmed. Leaves bipinnate; leaflets in about twelve rows on a side, oblong. Rachis, bracts, &c., covered with a coarse brown glandular coating. Sepals fringed with hairs and glands, disarticulating at the base, closely covered when young by bracts of the same nature. Petals erect. Stamens very long, red.

POINCIANA *GILLIESII.*—Inermis, foliis bipinnatis, foliolis utrinque seriebus 12 oblongis, rachi bracteisque glandulosis, sepalis fimbriatis ciliatis et glandulosis basi articulatis, bracteis conformibus densè imbricatis, petalis erectis, staminibus longissimis declinatis sanguineis.

Poinciana Gilliesii *Hooker, Bot. Miscell.* t. 129, *Bot. Mag.* t. 4006; *aliàs* Erythrostemon Gilliesii *Link, Klotzsch, and Otto, Icones plantarum*, t. 39; *aliàs* Cæsalpinia Gilliesii *Wallich*; *aliàs* "Cæs. macrantha *Delile Ind. Sem. Monsp.* 1838, p. ."

Although this fine plant is not absolutely new, yet it is so very little known as to deserve being once more brought before the public by means of a coloured figure. According to Dr. Gillies, its discoverer in Mendoza, an arid province of the republic of Chili, it is "called by the natives *Mal de Ojos,* and is very abundant in the cultivated parts of the province, where it has the benefit of the water used in irrigation, seeming to be incapable of living on the dry arid lands, which are not under cultivation. Along the southern frontier of the province of Mendoza, between the rivers Diamante and Atuel, it is found abundantly with other shrubs in sheltered situations; also among thickets along the western side of the Rio Quarto, near the western boundary of the Pampas; those plants to be found growing in Buenos Ayres owing their origin to seeds sent from Mendoza. They do not ascend farther than to the foot of the mountains, neither are any traces of them to be seen in the province of San Juan, which follows Mendoza to the north, along the foot of the Cordillera of the Andes. The flowers have a sickly disagreeable smell, and are supposed by the common people to be injurious to the sight. Hence its vernacular name '*Mal de Ojos.*'"

It has occasionally flowered in this country, in the open air, during summer. The specimen now represented was so produced this autumn in the Nursery of Messrs. Knight and Perry, where it was trained to a wall, and blossomed in July. We cannot, however, hope to see it in beauty unless guarded from severe frosts, as when against a "conservative wall."

Sir W. Hooker, who first published it, refers it without hesitation to the genus Poinciana of Tournefort; Wallich and Delile placed it in Cæsalpinia; and Dr. Klotzsch has formed out of it a new genus called Erythrostemon, concerning which he writes: "Erythrostemon differs from Poinciana in its pod, and in its sulphur-yellow unexpanded flowers; from Cæsalpinia in the enormous length of its stamens; from Heterostemon in its long distinct stamens; from all those genera in its polygamous flowers." He also gives a description of the pod of the plant, which, although unacknowledged, is, we observe, little more than a copy of Sir William Hooker's statement concerning it.

Probably it is not a true Poinciana, that is to say a legal associate of *Poinciana elata,* from which its *deciduous* calyx, its long decurved stamens, and its erect petals, seem to separate it, independently of any peculiarity in the legume; but in the absence of a more full acquaintance with these species, we abstain from interfering with Sir William Hooker's name. As Mr. Bentham observes to us, "if *Poinciana elata* be taken as the true type of the genus, *P. Gilliesii* is scarcely a congener, and Klotzsch's name may possibly be adopted. *P. pulcherrima* cannot be generically separated from Cæsalpinia. But whether *P. Gilliesii* be really distinct or not from Cæsalpinia remains to be investigated."

I. Constans. Pinx. & Zinc.

Printed by C.F. Cheffins, London.

[PLATE 29.]

THE CRIMPED GUELDRES ROSE.

(VIBURNUM PLICATUM; *var.* DILATATA)

A Greenhouse (?) Shrub, from CHINA, *belonging to the Natural Order of* CAPRIFOILS.

Specific Character.

THE CRIMPED GUELDRES ROSE.—Leaves rounded at the base, ovate or roundish-ovate, abruptly pointed, finely serrate, closely ribbed and veined so as to appear plaited, smooth on the upper, closely downy on the under side; flowers radiating, all sterile in the cultivated plant, enlarged and collected in a globose cyme.

VIBURNUM *PLICATUM.*—Foliis e basi rotundatâ ovatis v. ovato-suborbicularibus cuspidatis argutè serratis densè venoso-costatis et plicatis supernè glabris subtus tomentosis, floribus radiantibus in plantâ cultâ omnibus sterilibus dilatatis et in cymam globosam congestis.

Viburnum plicatum *Thunberg; Siebold and Zuccarini, Fl. Japonica,* I. 81, t. 38; *Botanical Register,* 1847. t. 51.

THIS plant, procured for the Horticultural Society by Mr. Fortune, is described in their Journal as "a handsome deciduous bush, bearing some resemblance to the N. American *Viburnum dentatum.*" Mr. Fortune says that it is a native of the northern parts of the Chinese Empire, where it was found by him cultivated in the gardens of the rich, by whom it was much admired. When full grown, it makes a bush eight or ten feet high. It is a most profuse bloomer, forming numerous heads of snow-ball flowers, like the common Gueldres Rose. It is expected to prove hardy in England; but this requires to be ascertained by actual trial. At any rate, it will probably become a favourite in our gardens.

Siebold and Zuccarini speak of it thus:—"This Viburnum is one of the most beautiful plants that are cultivated in Japan. Its name, *Satsuma Temari,* indicates that it inhabits Satsuma, the most southern province of Kiusia (31° N. lat.). It was probably in the beginning imported from China. Now-a-days, it is seen in every garden. Its balls of white sterile flowers give it the appearance of the Gueldres Rose: its habit, and broad oval plaited (crimped) leaves, are more like those of the Wayfaring Tree (*Viburnum Lantana*); but it only grows from four to six feet high."

Whether or not it shall prove to be hardy, it is certainly, even as a greenhouse plant, an object of much interest, and well worth cultivating even among small selections of species.

The tendency to form distended sterile flowers, to which this owes its beauty, is one which has attracted little attention. That it does not indicate natural affinity is plain from a comparison of the very different orders in which the tendency is manifested, as in Umbellifers, Hydrangeads, and Crucifers, where it occurs in the corolla. Nor is it wholly the result of domestication; for we believe that no instance is known in which the peculiarity has been observed, unless the plant is partially thus deformed when wild. For instance, among Viburnums, the only certainly known snow-ball sorts are *V. Opulus, Oxycoccus, molle, plicatum, macrocephalum,*—all of which have sterile radiant flowers when wild. Indeed the present plant, with only a part of its flowers in this state, would, we suspect, be handsomer than the perfect monster we possess: at least the appearance of the wild specimens justifies the conjecture. This wild state was observed by Fortune, in May, 1844, both at Teintung and Ningpo, where specimens were collected.

Very few plants have yet found their way into circulation, owing to the unhealthy condition in which the originals arrived, and the length of time that elapsed before they recovered. It is probable, however, that the plant will now become common, as well as the Large-headed sort (*V. macrocephalum*).

I. Constans, Pinx & Zinc

Printed by C.F.Cheffins London.

[PLATE 30.]

THE LONG-PETALED EPIDENDRUM.

(EPIDENDRUM LONGIPETALUM.)

A Stove Epiphyte, from GUATEMALA, *belonging to the Natural Order of* ORCHIDS.

Specific Character.

THE LONG-PETALED EPIDENDRUM.—Pseudobulbs ovate. Leaves in pairs, straight, sword-shaped, blunt. Panicle loose, much longer than the leaves. Sepals and petals alike in form, spathulate, stalked, blunt. Lip posterior, free, three-lobed ; the stalk callous and concave, the segments rounded, those at the side erect, that in the middle convex, much larger, notched at the end, wavy, with numerous elevated coloured radiating veins.

EPIDENDRUM *LONGIPETALUM;* pseudobulbis ovatis, foliis binis rectis ensatis obtusis, paniculâ laxâ multò longiore, sepalis petalisque conformibus spathulatis unguiculatis obtusis, labelli postici liberi trilobi ungue concavo calloso laciniis rotundatis lateralibus erectis intermedio convexo multò majore emarginato undulato venis pluribus elevatis coloratis radiantibus.

Epidendrum aromaticum, var. *of some Gardens.*

TEN years have made a great difference in our knowledge of American Orchids. At that time they had been studied only, and very imperfectly, upon dried specimens. Lately, by the importations of the Horticultural Society and Mr. Skinner, they have become familiar to us in a living state, and opportunities have been afforded of correcting many early errors. Among those errors was the reduction of this plant to *E. aromaticum,* to which its flowers bear some resemblance when ill dried, and only then. It is, in fact, a species perfectly distinct from all others, and not very nearly related to any, except the green Epidendrum (*E. virens*), whose sepals and petals are much shorter, acute, with a white and green lip, the three lobes of which are not very different in size.

This plant is very sweet-scented, with a long straggling panicle of dull brownish-purple and green petals relieved by a white lip, beautifully marked by straight crimson veins on a yellow ground. It is a native of Guatemala, whence the Horticultural Society obtained it, and requires all the heat of a good Orchid-house, combined with a long and perfect rest for at least four months. Treated hus, it flowers abundantly, and remains in perfection for several weeks.

The species belongs to the division of Encyclian Epidendrums, having a membranous lip with three deep lobes, of which the middle one is blunt, or very slightly acute, and a smooth rachis. Of that large division the species at present known are the following:—

E. fucatum *Lindl. in Bot. Reg.*, 1828, *misc.* 17 ; (*E. polyanthum* French Gardens); pseudobulbis subrotundo-ovatis cæspitosis monophyllis, foliis ligulatis coriaceis obtusis scapo brevioribus, paniculâ nutante multiflorâ, bracteis ovatis acutis squamiformibus, sepalis petalisque lineari-oblongis tessellatis æqualibus obtusis conniventibus, labelli liberi tripartiti lobis lateralibus erectis linearibus apice rotundatis intermedio acuto ovali multò brevioribus, callo sulcato plano elevato ad basin lobi intermedii.—*Cuba.*—Flowers small, dull yellow, tessellated, with a pink spot in the centre of a white lip.

E. chloroleucum *Hooker in Bot. Mag.*, t. 3557 ; (*E. chloranthum* Lindl. in Bot. Reg., 1838, misc. 28) ; pseudobulbosum, foliis coriaceis ligulatis apice rotundatis obscurè bilobis inæqualibus, racemo erecto paniculato, sepalis petalisque subæqualibus lineari-lanceolatis obovatis, labelli trilobi liberi lobis lateralibus linearibus obtusis inflexis intermedio ovato acuminato crispulo multò brevioribus : disco venis elevatis calloso.—*Demerara.*—Flowers pale green without spots, and a white lip.

E. virgatum *Lindl. in Hooker's Journ.*, iii. 83; pseudobulbis ovatis oblongisve sub-compressis rugosis, foliis binis ternisque convexis subundulatis acutis glaucis unciam latis, paniculâ virgatâ ramis longis gracilibus, sepalis lanceolatis petalisque duplò angustioribus patentibus discoloribus, labelli hastati lobis lateralibus acutis patentibus intermedio subrotundo-obovato acuto ; callo maximo rotundato pone basin.—*Mexico.*—The habit of *E. vitellinum*, but with more glaucous leaves. Flowers small, dirty green stained with brown, arranged in a very long lax graceful panicle, the branches of which are simple, and sometimes as much as a foot long, with nearly twenty flowers on each. The lip is whitish yellow. Scape sometimes seven feet high.

E. brachiatum *A. Richard* ; "pseudobulbis ovoideis 1-phyllis ; fol. oblongo-elliptico acuto ; flor. parvulis numerosis, brunneis, paniculatis : labello albido trilobo, lobis lateralibus angustis falcatis, intermedio obovali acuto."—*Mexico.*

E. Linkianum *Klotzsch in Allg. gartenzeit, Sept.* 22, 1829 ; (*E. pastoris* Link et Otto abbild. t. 12) ; pseudobulbis fusiformibus 2-3-phyllis, foliis ensiformibus recurvis racemo paucifloro longioribus, sepalis patentissimis lineari-lanceolatis, petalis conformibus angustioribus, labelli lobis lateralibus minutis erectis intermedio ovato-oblongo crispo venis elevatis sub columnâ pubescente.—*Mexico.*—Flowers small dull yellow, streaked with purple. Lip nearly white.

E. concolor *L. no.* 12. ; foliis in pseudobulbos confertos lenticulares solitariis lato-lanceolatis-acutis, scapo filiformi 5-floro, sepalis ligulatis, petalis linearibus, labello tripartito laciniis integris intermediâ majore.—*Mexico.*—A slender plant. Flowers pale yellow, whole coloured, with a striated labellum.

E. Pastoris *L. no.* 7 ; *Klotzsch in Allg. gartenzeit, Sept.* 22, 1838 ; "caule repente radicante, pseudobulbis oblongis compressis 2-3-phyllis, foliis linearibus acutis carinatis laxiusculo-subtortuosis, floribus racemosis, perianthii foliolis patenti-subincurvis margine recurvis extus sordide flavis intus lineis longitudinalibus purpureo-fuscis striatis, sepalis lineari-subspathulatis acuminatis, petalis spathulatis acutis, labello trilobo albido dein luteo lobis lateralibus majoribus basi semilunatis integerrimis glabris lævibus basin columnæ orbiculatim amplectentibus lituris transversalibus purpureis medio cordato deflexo minore glabro acuto margine basique recurvo punctis minutis purpureis ornato, columnâ semitereti fuscâ ad apicem luteâ tridentatâ dentibus obtusis, pericarpiis elongatis acuto-triquetris."—*Mexico.*—Flowers fragrant, like Vanilla.

E. Ovulum *Lindl. in Bot. Reg.*, 1843, *misc.* 71 ; pseudobulbis oviformibus diphyllis, foliis linearibus canaliculatis acutis, scapo filiformi foliis paulò longiore 3-floro, sepalis linearibus 3-veniis, petalis angustioribus spathulatis, labelli trilobi lobis lateralibus acutis intermedio dilatato rotundato venis radiantibus glandulosis variegato, columnæ tridentatæ dentibus lateralibus rotundatis denticulatis.—*Mexico.*—A curious little plant, in the way of *E. pastoris*, or *bractescens*, or *aciculare*. The sepals and petals are olive-green ; the lip white, with crimson glandular radiating veins.

E. bractescens *Lindl. in Bot. Reg.*, 1840, *misc.* 122 ; pseudobulbis ovatis cæspitosis 3-4-phyllis, foliis linearibus, scapo debili 3-4-floro, bracteis infimis foliaceis floribus longioribus supremis obsoletis, floribus nutantibus longè pedunculatis, sepalis petalisque lineari-lanceolatis acuminatis discoloribus labello longioribus, labelli liberi lobis lateralibus apice recurvis obtusis subdentatis intermedio unguiculato subrotundo-ovato multò longiore secus unguem elevato sulcato pubescente.—*Mexico.*—This is one of the prettiest of the small species. The pseudobulbs are exactly ovate, closely clustered, and about as large as a pigeon's egg. The flowers have a beautifully but delicately painted white lip, the gay effect of which is heightened by the contrast with the dingy purple of the long narrow sepals and petals.

E. aciculare *Bateman in Bot. Reg.*, 1841, *misc.* 98 ; pseudobulbis oblongis diphyllis, foliis linearibus canaliculatis acutis racemo simplici æqualibus, sepalis petalisque lineari-lanceolatis æqualibus acutis, labelli laciniis lateralibus ascendentibus linearibus obtusis apice recurvis intermediâ ovato-oblongâ subundulatâ (pictâ) acutâ.—*Bahamas.*—A gay little species, with long narrow leaves, a slender erect raceme of six or seven flowers, whose sepals and petals are dull purple, and lip white, enlivened with rosy veins.

E. pictum *Lindl. in Bot. Reg.*, 1838, *misc.* 43; pseudobulbosum, foliis ligulatis coriaceis obtusis dorso rotundatis, racemo erecto paniculato, sepalis petalisque obovato-linearibus subæqualibus, labelli trilobi liberi lobis lateralibus linearibus acutiusculis subfalcatis columnam amplexantibus margine anteriore plicato intermedio ovali acuto crispo multò brevioribus, disco venis elevatis calloso.—*Demerara.* —Resembles *E. odoratissimum;* with dull yellow flowers, neatly striped with crimson. It is nearly related to *E. chloroleucum*, from which its leaves readily distinguish it.

E. graniticum *Lindl. in Hooker's Journ.*, iii. 83; pseudobulbis ovatis attenuatis 2-phyllis, foliis ensiformibus paniculâ multiflorâ brevioribus, sepalis petalisque patentibus lanceolatis subæqualibus acutis, labelli trilobi laciniis lateralibus lineari-oblongis obtusis intermediâ unguiculatâ obovatâ apice inflexo acuto : callo elevato acuminato secus medium canaliculato, columnâ sub apice auriculatâ.—*Guayana.* —A fine species closely allied to *E. flavum.* It has a panicle regularly branched up to the apex, nearly a foot and a half long, with each side branch having from 2-4 flowers. According to Mr. Schomburgk, the sepals and petals are green dotted with purple, the labellum white with a purple stain at its base, the flowers aromatic, the stem six feet high.

E. gracile *Lindl. in Bot. Reg.*, t. 1765; foliis in pseudobulbos ovatos corrugatos pluribus lorato-ensiformibus, racemo simplici longissimo, sepalis oblongis petalisque cuneatis patentibus, labelli ferè liberi trilobi lobis lateralibus semiovatis intermedio oblongo crispo obtusissimo duplò minoribus disco bicostato.—*Bahamas.*—Flowers green, lip yellow, lined with purple.

E. viridiflorum *Lindl. in Bot. Reg.; (Encyclia viridiflora* Hooker in Bot. Mag. xv. t. 2831; L. p. 111); pseudobulbis ovatis diphyllis, foliis ensiformibus recurvis acutis paniculâ brevioribus, sepalis lateralibus falcatis petalisque linearibus acutis erectis, labello postico apice 3-lobo laciniis lateralibus planis intermediæ ovatæ crispæ æqualibus: callo basi duplici oblongo carnoso.—*Brazil.*—Flowers dull green, marked with dull purple.

E. glutinosum *Scheidweiler in Gartenzeit*, 1843, p. 110; "foliis in pseudobulbos pyriformes tunicatos glabros, binis linearibus coriaceis oblique truncatis, racemo subsimplici pedicellisque glutinosis, sepalis oblongis acuminatis petalisque spathulatis patentibus, labelli fere liberi trilobi lobis lateralibus oblongis obtusis integris erectis, intermedio ovato crispato, disco calloso depresso, columna bidentata. Scapus terminalis pedalis, petala et sepala viridi-purpurea, extus lineis purpureis notata, labellum albo-lutescens, lobo intermedio lineis purpureis ornato."—*Rio Janeiro.*—According to Mr. Scheidweiler, very near *Epidendrum odoratissimum*, which he considers identical with the *Encyclia patens* of Hooker and *Macradenia lutescens* of Loddiges. Its scape is a foot high. The petals and sepals are greenish purple, marked outside with purple lines. The lip is whitish yellow, its middle lobe being marked with purple lines.

E. rufum *Lindl. in Bot. Reg.*, 1845, *misc.* 42; pseudobulbis pyriformibus 2-3-phyllis, foliis brevibus lanceolato-ligulatis patentibus scapo paniculato brevioribus, sepalis petalisque ovalibus acutis subcarnosis, labelli trilobi laciniis lateralibus brevibus semiovatis intermediâ obovato-oblongâ convexâ margine revolutâ apice rotundatâ basi secus axin elevatâ carnosâ, columnâ membranaceo-marginatâ.—*Brazil.*

E. flavum *Lindl. in Hooker's Journ.*, iii. 83; pseudobulbis ovatis attenuatis 3-phyllis, foliis ensiformibus paniculæ paucifloræ subæqualibus, sepalis petalisque patentibus subæqualibus lineari-oblongis obtusis, labelli trilobi laciniis lateralibus linearibus truncatis intermediâ unguiculatâ obovatâ nudâ, columnâ sub apice auriculatâ.—*Brazil.*—Leaves of this rather more than a foot long. Flowers pale yellow, about an inch and a half in diameter. The inflorescence is only panicled at the base, and is probably very often simple.

E. pachyanthum *Lindl. in Bot. Reg.*, 1838, *misc.* 42; pseudobulbosum, foliis lato-ligulatis subundulatis apice obliquè obtusis dorso rotundatis, perianthio carnoso herbaceo, sepalis lanceolatis, petalis obovato-lanceolatis apice complicatis, labelli liberi trilobi laciniis lateralibus ascendentibus truncatis intermediâ spathulatâ acutâ basi callosâ trilineatâ convexâ inappendiculatâ multò brevioribus.—*Guayana.*—A large green-flowered species. Its leaves are thinner and broader than is usual among these Epidendra, and a little wavy at the margin. The flowers are fully two inches in diameter, thick and fleshy, dull green, stained with a dirty reddish brown towards the ends of the sepals and petals. The labellum is a pale straw-colour, streaked along the middle with violet.

E. primulinum *Bateman MSS.;* pseudobulbis . . . . , foliis . . . . , scapo paniculato, sepalis petalisque patulis oblongis acutis, labelli laciniis lateralibus nanis erectis acutis intermediâ obovatâ apiculatâ; callo duplici ad basin elevato plano carnoso.—*Mexico.*—Flowers rather large, in a close erect panicle, smelling of primroses.

E. altissimum *Bateman in Bot. Reg.*, 1838, *misc.* 61; pseudobulbis elongatis teretibus 2-3-phyllis, scapis ramosis longissimis, sepalis lineari-oblongis acutis, petalis conformibus basi angustatis, labelli liberi lobis lateralibus dimidiatis erectis tortis obtusis intermedio dilatato undulato recurvo apiculato basi bicostato.—*Bahamas.*—Flowers scented with beeswax. Very like *E. oncidioides.*

E. longipetalum *of this article.*

E. Humboldtii *Reichenbach fil. in Linnæa;* "p. ph. e. oblongis acutis basi aliquid cuneatis, p. ph. i. obtusis basi valde cuneatis, sub apice dilatatis, lb. maximo trilobo, basi ima cuneato, lobis lateralibus integris obtusatis, lobo medio maximo subquadrato, antice emarginato, margine denticulato, nervis 7 medianis elevatis a basi ad centrum cristigeris, cristis crenato-serratis, gy. postice carinato, androclinii margine tridentato, interjecto dente antice rostellari."—*Puerto Caballo.*

E. virens; paniculâ laxâ erectâ angustâ, sepalis lineari-oblongis apice latioribus, petalis æquilongis spathulatis acutis, labelli laciniis subæqualibus lateralibus erectis oblongis emarginatis intermediâ convexâ plicatâ venosâ emarginatâ mucronulatâ.—*Guatemala.*—Flowers green, whole coloured, except the lip, which is white, with crimson veins in the middle lobe; the lateral lobes green, with crimson veins, but white at the point.

E. venosum *L. no.* 13; foliis ensiformibus obtusis supra et sub pseudobulbos fusiformes nascentibus, racemo striato simplici, sepalis lineari-lanceolatis petalisque angustioribus patentissimis, labello semilibero tripartito: laciniis lateralibus ovatis acutis intermediâ subrotundâ apiculatâ multo majore callo baseos et lineis tribus disci subramosis elevatis.—*Mexico.*—Scape a foot long. Lip half united to the column, white, with elevated violet veins.

E. aromaticum *Bateman, Orch. Mex.*, t. 39; (*E. incumbens* Lindl. in Bot. Reg, 1840, misc. 84); floribus densè paniculatis, sepalis linearibus patentissimis basi angustatis, petalis conformibus sed paulo latioribus, labelli postici lobis lateralibus triangularibus acuminatis intermedio subrotundo-ovato apiculato venis elevatis cristato, callis duobus oblongis secus unguem.—*Guatemala.*—Flowers very sweet; in large pale dull yellow panicles. It inhabits a climate whose temperature varies from 60° to 75°.

E. alatum *Bateman, Orch. Mex.*, t. 18.; *Bot. Reg.*, 1846, t. 53; (*Epid. calocheilum* Hooker in Bot. Mag., t. 3898); pseudobulbis ovato-oblongis diphyllis, foliis ensiformibus obtusis coriaceis obsoletè striatis paniculâ multiflorâ brevioribus, sepalis petalisque lineari-oblongis spathulatis uniformibus patentibus, labello profundè trilobo basi intùs bicarinato lobis lateralibus eroso-dentatis rotundatis intermedio oblongo undulato multò brevioribus omnium venis callosis et verrucosis, columnæ alis rotundatis.—*Guatemala.*—Its pale colour, and the peculiar markings upon its lip, at once distinguish it. These markings consist of reddish warts, plates, scales, or elevations, of various forms, arranged upon the veins, and therefore spreading from the base.

E. tripterum *Lindl. in Hooker's Journ.*, iii. 83; pseudobulbis ovalibus compressis diphyllis, foliis lineari-oblongis obtusis racemo paucifloro (4—6) subæqualibus, floribus erectis sepalis petalisque lineari-lanceolatis patulis, labelli trilobi lobis lateralibus linearibus obtusis planis intermedio subrotundo basi angustato undulato venis rugosis elevatis, capsulâ augustâ clavatâ tripterâ.—*Mexico.*—The whole plant when in bloom little more than six inches high. Flowers apparently dull purple, with a pale lip, on long peduncles, and erect not drooping.

N.B.—In the above references, *L.* signifies Lindley's Genera et Species Orchidacearum.

GLEANINGS AND ORIGINAL MEMORANDA.

196. Cestrum calycinum. *Willdenow.* (aliàs *C. viridiflorum* Hooker.) A greenhouse shrub, from Buenos Ayres, with deliciously scented green flowers. Belongs to the Nightshades. Flowers in October. (Fig. 97.)

This charming shrub would be passed by without notice, if it were not for the exquisite fragrance of its green flowers; out of flower it looks like an Oleaster (*Elæagnus*). It was originally introduced through the Glasnevin Garden. With the exception of the upper side of the leaves, the whole plant is covered with a grey starry down which gives it a dull appearance. The leaves are ovate-oblong, slightly heart-shaped at the base, on short stalks. The flowers appear in short axillary spikes, with a calyx much wider than the narrow tube of the downy corolla, which however widens upwards into a true funnel-shaped figure. The filaments are not toothed. The fragrance of the flowers is perceptible both day and night, but most so in the day.

97

Sir W. Hooker, in naming it *C. viridiflorum*, was not aware that it had been previously called *C. calycinum* by Willdenow.

197. Ungnadia speciosa. *Endlicher.* A hardy deciduous shrub, with rose-coloured flowers. Native of Texas. Belongs to the order of Soapberries. Has not yet flowered.

This plant having been lately introduced into cultivation, it is as well to quote the following memorandum concerning it from Dr. Asa Gray's valuable *Plantæ Lindheimerianæ.* It is nearly related to the genus Pavia :—" Shrub three to twenty feet high, with many long stems, one to three inches thick, branching only at the top. Fruit sweet and pleasant, but emetic (Lindheimer). Its popular name is *Spanish Buckeye.*

The fertile flowers and the fruit, although for several years known to us, have not until now been illustrated or described, except by Adolph Scheele, who has published a description from Lindheimer's specimens in the Linnæa.—The flowers which Endlicher happened to examine were pentapetalous, which is not the more usual case; and he erroneously states the plant to form a large tree, whereas it is commonly a slender shrub, of five or ten feet in height, or at most a small tree. Misled by these discrepancies, and by the differences of the two kinds of flowers, and, it would seem from his description, happening to possess tetrasepalous as well as tetrapetalous flowers, (although there are five sepals in all my Lindheimerian and other specimens,) Mr. Scheele has wrongly introduced a second species, under the name of *U. heterophylla*. The leaflets vary from five, or even three, on the earlier leaves, to seven. In seedling plants, raised in the Cambridge Botanic Garden, I have noticed a lusus of the earliest leaves, in which the leaflets are confluent."

198. Hymenocallis Borskiana. *De Vriese.* A stove bulb from La Guayra, with white flowers smelling of Vanilla. Belongs to Amaryllids. Flowered in the Botanic Garden, Leyden.

Leaves two to two and a half feet long, dull green. Scape compressed, as long as the leaves. Flowers seven, in an umbel, white, with a very thin transparent entire coronet. *De Vriese, Epimetron,* 1846.

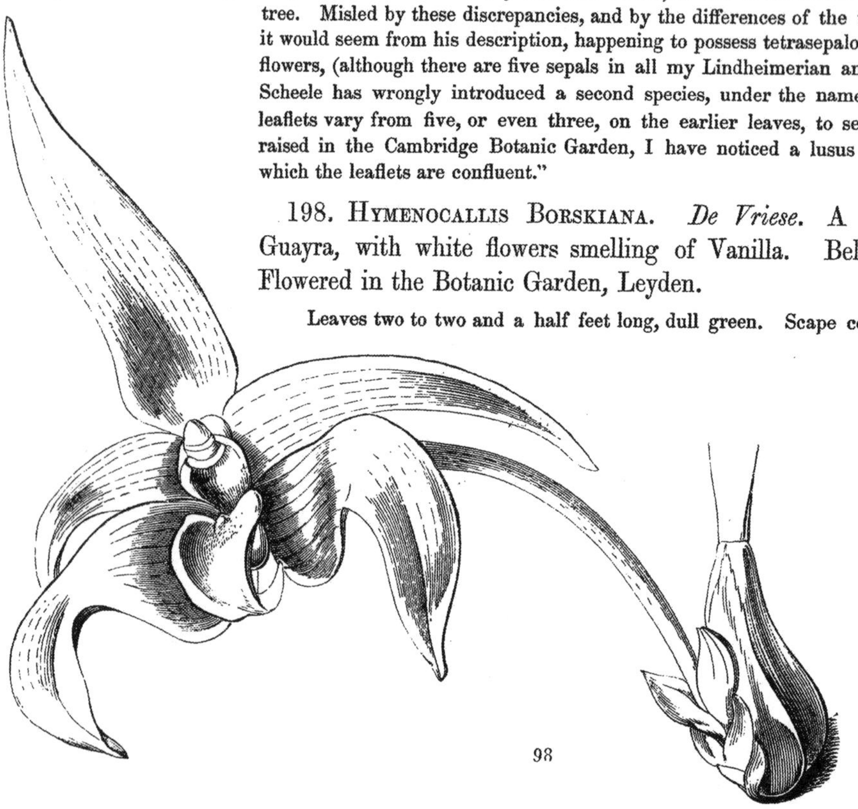

98

199. Sarcopodium Lobbii. (*aliàs* Bolbophyllum Lobbii *Lindley.*) A stove epiphyte belonging to the Natural Order of Orchids. Native of Java. Flowers nankin-yellow, large and showy. Introduced by Messrs. Veitch and Co. (Fig. 98.)

One of the many good things sent from Java to Messrs. Veitch of Exeter, by their collector, Mr. Thomas Lobb. "How fine a plant of its kind this is, may be surmised, by its having been taken for a *Cœlogyne*: the flowers are full four inches across, yellow, shaded with cinnamon, spotted with light brown, and speckled outside with brown-purple: we know of no species of the genus comparable to it for beauty." Our drawing was made from the plant of Messrs. Veitch, after it had gratified the public at the May Exhibition of the Chiswick Gardens for 1850. Pseudobulbs ovate, smooth, green, nearly as large as a pigeon's egg, springing from a scaly creeping stem terminated by a stalked, oblong, leathery, solitary leaf. Scape arising one from the side of each pseudobulb, yellowish, spotted with brown, shorter than the leaf, its base sheathed with imbricated, convex, spotted scales. Flowers large, solitary, spreading. Sepals lanceolate, acuminated, deep yellow, the upper one externally marked with purple spots running in lines; the lateral ones falcate, streaked and clouded with purple. Petals resembling the upper sepal, but smaller and streaked with purple lines, reflexo-patent. Lip cordato-ovate, acuminate, reflexed, yellow, with minute orange dots. This, like the rest of the numerous species of *Bolbophyllum*, is a tropical epiphyte, and requires to be kept in the warm division of the Orchid-house. It grows and flowers freely on a block of wood, suspended from the roof of the house, and having a piece of Sphagnum-moss attached. In winter an excess of moisture, either in the atmosphere of the house or in the moss or block of wood, is prejudicial; and in summer the plant must be shaded from the mid-day sun.—*Bot. Mag.*, t. 4532.

Between Dendrobes and Bolbophyls there exists a race having the large flowers of the former, and the peculiar habit of the latter, and hence referred to the one or the other genus according to the fancy of the observer. They agree with Dendrobes in having four pollen masses, and a hornless column; but they have coriaceous, not thin half-transparent flowers, and a tough leathery lip, enlarged not contracted at the base. If they had a caudicle and gland to their pollen masses, they would be Asiatic Maxillarias. They form neither horn nor spur, but are simply inflated and expanded at the base of the sepals. On the other hand, although they grow like Bolbophyls, yet they have no horns to their column, but two pollen masses, and their large leathery flowers afford a further difference. To these plants, consisting of the *Dendrobium amplum* of Wallich, and the *Bolbophyllum Lobbii, affine, leopardinum, Cheiri,* and *macranthum* of Lindley, the name SARCOPODIUM may be applied : with the following *distinctive* character :—

Habitus Bolbophylli. Pollinia et columna Dendrobii. Sepala coriacea, lateralia basi ventricosa. Labellum coriaceum, basi dilatatum. (Haud Bolbophyllum quod poll. 4 nec 2, et col. mutica nec cirrhata. Haud Dendrobium quod sepala et labellum coriacea basi ventricosa nec cornuta v. calcarata.)

200. RHIPSALIS PACHYPTERA. *Pfeiffer.* (*aliàs* Cereus alatus *Link and Otto ; aliàs* Cactus alatus *Bot. Mag.?*) A trailing succulent shrub, from tropical America, with leaf-like stems, small dirty white flowers, and red fruit. Belongs to the order of Indian Figs (Cactaceæ). Flowers in winter and spring. (Fig. 99 ; *a,* section of flower ; *b,* ripe fruit.)

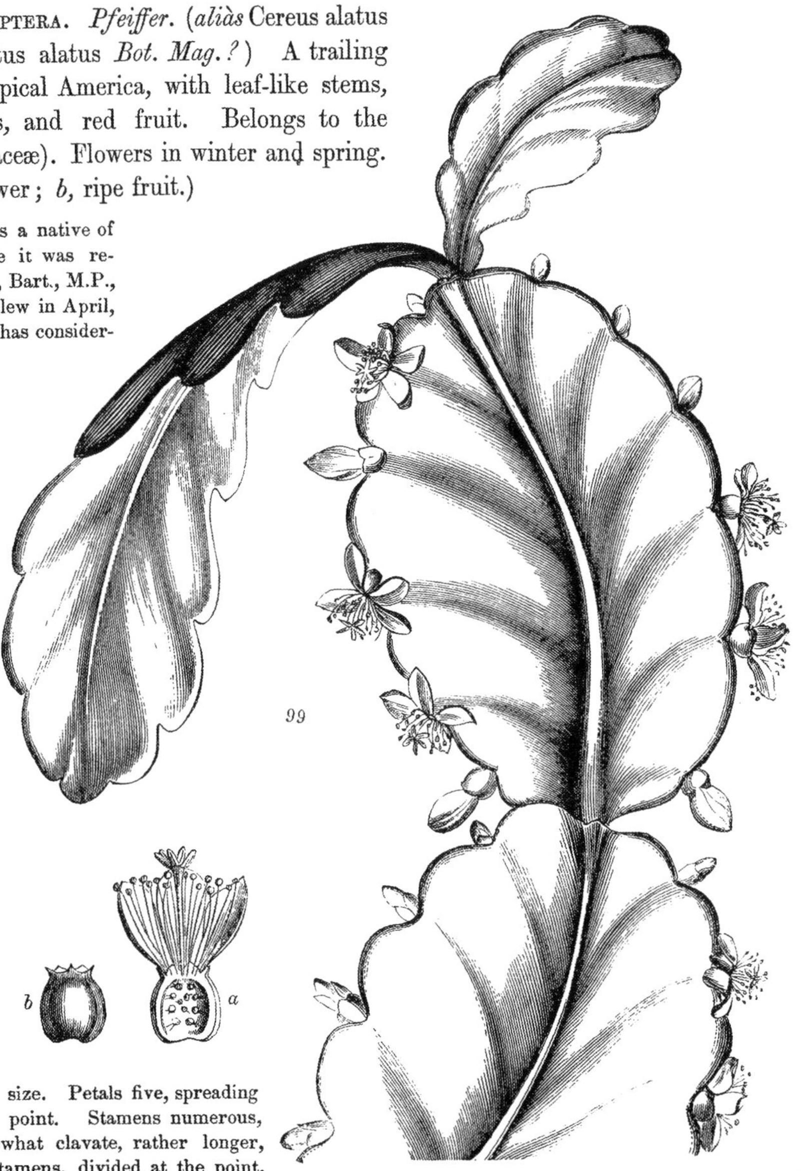

99

b a

This singular little plant is a native of Rio de Janeiro, from whence it was received by Sir Charles Lemon, Bart., M.P., in 1839, and flowered at Carclew in April, 1846. In its mode of growth it has considerable resemblance to some of the well-known showy species of Cactus with flat leaves, but on flowering it proved to be totally different. It requires a warm greenhouse or stove, and thrives very well when grown in a loamy soil with little water. Joints leafy, roundish ovate, compressed, nearly flat, hanging down, about 3 inches long and 2 inches broad, deeply crenated with a thick prominent, woody midrib, and distinct side ribs. They are of a bright green, tinged with reddish brown at the base and point, as well as along the margin, becoming, when old, of a rusty green. Flowers solitary, sessile, small, issuing from each crenature, and of a pale brownish yellow : the buds, previously to opening, being delicately tinged with pink. Sepals five, very minute and unequal in size. Petals five, spreading ovate-oblong, obtuse at the point. Stamens numerous, filiform, erect. Style somewhat clavate, rather longer, and much larger than the stamens, divided at the point, sometimes into five, but most frequently into four lobes. Fruit a small berry about the size of a red currant, and similar in colour, with numerous small jet black seeds, embedded in the pulp.

That this is the *Cereus alatus* of Link and Otto, there can be no doubt ; and consequently it is the *Rhipsalis pachyptera* of Pfeiffer ; but we are by no means satisfied that it differs specifically from the *Rh. crispata* and *rhombea* of the same author, notwithstanding the white fruit of the former. We find it, however, recognised in the Prince of Salm Dyck's latest enumeration, and we bow to so high an authority.

201. ALMEIDEA RUBRA. *Auguste de St. Hilaire.* A beautiful red-flowered hot-house shrub, from Brazil. Belongs to Rueworts (Rutaceæ). Introduced at Kew. Flowers in the autumn.

This handsome plant, with flowers of the size and colour of *Lemonia spectabilis*, but arranged in a compound raceme or thyrse, is one of six species of a shrubby genus, detected in Brazil by M. Auguste de St. Hilaire. He dedicated it to his friend and patron Don Rodriguez Pereira de Almeidea. It forms a branching shrub, three to five feet high, with leaves which are alternate, broadly lanceolate, acute at the base, acuminate at the apex, penninerved, quite entire at the margins. Panicle, or compound raceme, thyrsoid. Flowers often two or three together, moderately numerous. Calyx short, cut into five acute teeth. Petals obovato-spathulate, very obtuse, spreading, deep rose-colour (as is the calyx). Filaments linear, contracted below the anther, slightly downy, grooved towards the base, and above the groove are two hairy tubercles. Ovary of five lobes, pellucido-punctate, surrounded by an entire, cup-shaped nectary. The species of Almeidea require to be grown in a stove temperature. The one here figured flowered during the month of September in the Palm-house. It should be potted in a mixture of light loam and leaf-mould, and receive the benefit of bottom-heat, which we consider of great importance in cultivating, and maintaining in a healthy state, plants of slow growth like the present. It is increased by cuttings plunged in bottom-heat.—*Bot. Mag.*, t. 4548.

202. Acantholimon glumaceum. *Boissier*. (aliàs *Statice Ararati* of gardens.) A hardy very pretty herbaceous plant, with tufts of awl-shaped spiny leaves, and long-stalked spikes of large rose-coloured flowers. Belongs to the Order of Leadworts.

This is one of the "Hedgehog" Statices, of which an example or two are already known to gardeners. It is curious when in leaf, and very pretty while in flower. The usual treatment of "Alpine plants" suits it. Mr. Henfrey doubts whether this is or is not the species to which he refers it; we have fine specimens of it from Armenia, collected by Jas. Brant, Esq., H.M. Consul at Erzeroum, three times as large as the specimen represented in the *Gardener's Magazine of Botany*; but we do not find it among any of the authentic specimens of M. Boissier in our possession. We fear that Mr. Henfrey is right in thinking that this botanist has multiplied species too much.

203. Begonia Ingramii. *Henfrey*. A handsome garden hybrid, with loose drooping clusters of pale pink flowers. Requires a stove.

Said to have been raised by Mr. Ingram, of Frogmore, between *B. fuchsioides* and *B. nitida*. The leaves are four inches long, very oblique, half heart-shaped, dark glossy green, slightly ciliate and crenelled; the under side is green also. The male flowers have four decussating sepals, of which the inner are smaller; the females have five nearly equal sepals.—*Gard. Mag. of Bot.* ii. *p.* 153. The placentation is that of *Diploclinium*. Mr. Henfrey proposes in this article to form another subdivision of the genus Begonia, under the name of *Platyclinium*, for the well-known many-lobed placenta of *B. cinnabarina*, which however he does not connect with any other species.

204. Catasetum Lansbergii. (aliàs *Myanthus Lansbergii* Reinwardt and De Vriese.) A terrestrial stove Orchid from the Caraccas, with a long ovate raceme of thirteen to twenty green and purple flowers. Blossomed in the Garden of Leyden.

Very nearly the same as *Catasetum callosum*, from which it differs in the flowers being green, spotted with purple, and not whole coloured. It can scarcely be a distinct species.

205. Spathodea lævis. *Palisot de Beauvois*. A hothouse tree from Sierra Leone, belonging to the Order of Bignoniads. Flowers handsome, white streaked with rose. Introduced by Messrs. Lucombe and Co. Blossoms in June. (Fig. 100.)

Imperfect as are the figure and description of *Spathodea lævis* in Palisot de Beauvois, I am yet of opinion I am correct in referring it to this plant. If by the term "lævis" applied to the species it is meant that there are no glands on the calyx or corolla, I may observe, that however obscure on the *dried* specimens (from which M. de Beauvois' drawing and character were derived) they are apparent enough on the living plant. Our specimen is sixteen feet high; but it flowers when much smaller. Its stem is woody but soft. The leaves are alternate, except those below the inflorescence, which are often in whorls of three, all of them unequally pinnate, with from four to six pair of opposite, ovate, acuminate, coarsely serrated, glabrous, sessile leaflets. Panicle terminal, corymbose, with numerous large flowers. Calyx green, tipped with red, split open more than half-way down on one side, with several dark-coloured glands near the base, irregularly toothed at the apex. Corolla campanulato-infundibuliform, white, delicately spotted and streaked with rose; tube widening upwards; limb obscurely two-lipped; upper lip of two rounded lobes; lower of three similar ones, but larger and more spreading; all slightly waved. This is a tropical tree of robust growth, requiring the temperature of the stove, and growing freely in light loam. It is propagated by cuttings planted under a bell-glass in white sand, and plunged in bottom-heat.—*Bot. Mag.*, t. 4537.

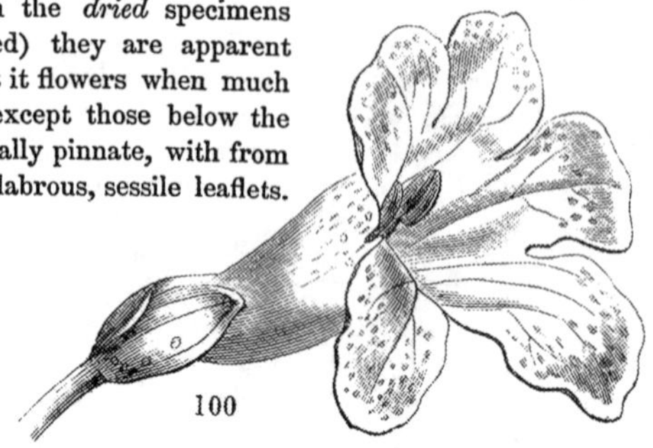

100

206. Opuntia Salmiana. *Parmentier.* A stove succulent from Brazil. Flowers, pale yellow Native of Brazil. Blossoms at Kew in September and October. (Fig. 101.)

This pretty and very distinct *Opuntia* is said to be a native of Brazil. Our collection is indebted for the possession of it to the Royal Gardens of Herrnhaussen. It blossoms freely, and the ordinary looking stems and branches are ornamented by the variegated red and yellow and rather copious flowers in September and October. Plant small, one to two feet high, erect, branched; branches erecto-patent, cylindrical, rather of an ashy-green colour, destitute of tubercles,

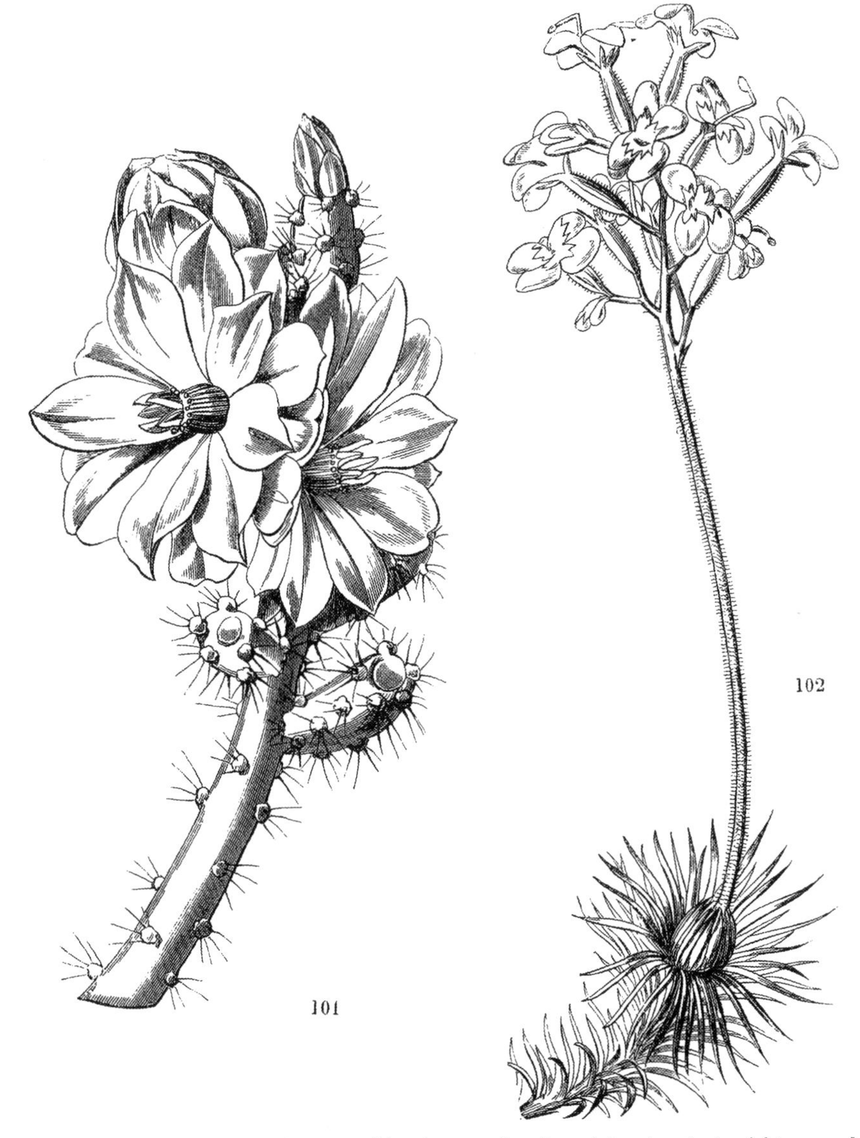

obtuse at the apex. Areoles scattered, forming white downy tufts of wool, bearing six to eight unequal, brown, small aculei, the largest less than half an inch long. Flowers moderately sized, clustered at the apex of a branch. Ovary obovate, not scaly but areolated, and bearing aculei like the branches; and, what is remarkable, after the floral coverings have fallen away, often producing young plants. Sepals and petals undistinguishable; the former gradually pass into the latter. In bud the flower is red; when fully expanded the ground-colour is sulphur-yellow, streaked with red and rose-colour in the centre. The petals are obovate, and the spread of the flower about two inches. Stamens not numerous, yellow. Rays of the stigma five or six, yellow-green. This slender straggling species grows and flowers freely if potted in light loam and leaf-mould, and placed under the full influence of the sun in summer. It should be frequently syringed in the mornings or evenings, during hot dry weather, but care must be taken that all superabundant water passes off freely, and that the soil does not remain long in a saturated state. In winter water must be given very sparingly, and the temperature of the house during the night need not at any time exceed 55°. It readily increases either by cuttings or by seeds, as also by gemmæ produced on each areole of the fruit, which ultimately form separate and distinct plants.—*Bot. Mag.*, t. 4542.

207. STYLIDIUM MUCRONIFOLIUM. *Sonder.* A greenhouse herbaceous plant, of much beauty, from the Swan River. Flowers yellow. Belongs to the Order of Styleworts. Introduced by Messrs. Lucombe and Pince. (Fig. 102.)

The plant thus called by Sonder does not wholly agree with this, for neither is the labellum in our plan "inappendiculate," nor can the leaves be said to be "radical." The first character is, indeed, easily overlooked in the dried plant, from which Sonder was likely to have drawn up his description; and with regard to the latter, tufted rosules of apparently radical leaves do, in several *Stylidia*, elongate into real leafy stems or branches. Again, the nearest natural allies of our plant are unquestionably, *S. ciliatum* Lindley, and *S. saxifragoides* Lindley; but Sonder has separated them by nearly thirty species. The present species is very pretty and produces its copious bright tufts of flowers in August. Roots wiry, brown. Stems in our plant tufted, two to three inches long, copiously leafy. Leaves glabrous, spreading, linear-subulate, broader at the base, tipped at the point with a setaceous bristle. Peduncles terminal, solitary on each branch, a span high, above, and the pedicels and calyx clothed with slender hairs tipped with glands, so delicate as to be scarcely visible to the naked eye. Panicle roundish or oval, many-flowered, rather compact. Corolla rather bright yellow, with zigzag orange lines round the mouth. Ovary or capsule much elongated, slender, cylindrical. In summer these small weak plants should be placed in a situation where they may be maintained in a moderately moist state, without having daily recourse to the water-pot; and in winter they should be placed in a dry airy place, taking care in damp weather that no water lodges amongst the fascicles of leaves, for when this happens the plant is liable to be destroyed.—*Bot. Mag.*, t. 4538.

208. BURLINGTONIA PUBESCENS. A beautiful stove Orchideous epiphyte, from Pernambuco. Flowers white. Introduced by John Knowles, Esq., of Manchester.

B. pubescens; acaulis, foliis coriaceis apice carinatis mucronatis, racemis densissimis pendulis, labello obovato bilobo breviter hastato laciniis erectis, cristæ lamellis utrinque 3 valde inæqualibus, columnæ basi pubescentis alis 2 minutis subulatis albis 2 oblongo-linearibus porrectis.

This beautiful novelty was exhibited at a meeting of the Horticultural Society in November last, when it received a silver medal. It formed a wide tuft of dark green rigid leaves, pouring forth from their bosom a profusion of bunches of snow-white blossoms. It had been sent to John Knowles, Esq., of Manchester, from some friends in Pernambuco, where it appears to be very rare. It is not now, however, introduced for the first time, for we have in our possession a dried specimen, communicated by the late Mr. George Loddiges, in November, 1846, at which time we named it *pubescens*, in allusion to the down on the column, which is not found in the other drooping white-flowered species. Of these species five are now known, of which two, *B. granadensis* and *fragrans*, have the bunches of flowers erect. The other three, *pubescens*, *candida*, and *venusta*, are thus distinguished:—

B. pubescens has a downy column, a lip with three yellow ridges on each side near the base, and a pair of erect side lobes, rendering it what is technically called hastate. Its flowers are the smallest of the three.

B. venusta has a smooth column, a lip in no degree hastate, with many shallow ridges on each side near the base. Its flowers are larger than in the last, and the flowers more loosely arranged.

B. candida has a smooth column, a lip very slightly hastate, with a stalk two-thirds as long as the column, and only one ridge on each side, forming a broken row of callosities. The flowers are much fewer in each bunch, but twice as large as in the last.

209. FRANCISCEA EXIMIA. *Scheidweiler.* A handsome stove shrub from Brazil, with large deep violet flowers. Belongs to the Linariads. Introduced by M. de Jonghe, of Brussels.

Habit of *Fr. latifolia.* Branches downy. Leaves oblong-lanceolate, not shining. Flowers terminal, about two together, very deep purple, two and a half inches across the limb.

In Belgium this Franciscea eximia is spoken of as the finest species of the genus yet in cultivation; and we learn also that it proves to be a free flowerer,—plants of the height of two feet and a half producing successively through the blooming season upwards of two hundred blossoms, of the size and colour represented in our plate. The first blossoms borne in Europe were produced in March, 1849; and the original plant again commenced flowering in January, 1850, and continued to produce blossoms till the end of June. Young plants are also reported to flower freely.—*Gardener's Magazine of Botany*, ii. p. 177.

210. Tillandsia inanis. A stove epiphyte belonging to Bromeliads, with scurfy, dry, twisted leaves, and violet flowers issuing from crimson bracts. Native of the province of Buenos Ayres. (Fig. 103, *a piece of the inflorescence; 104, a diminished figure of the plant.*)

Commodore Sulivan, C.B., who brought it to this country in 1841, on his return from the command of the South American station, presented it to Sir Charles Lemon, Bart., M.P., with whom it flowered in March, 1846. It is a native of the interior provinces of Buenos Ayres, high up the Parana, and is stated to be greatly prized there for its delicious perfume, although at no period could Mr. Booth discover that it possessed any fragrance; and it is probable that the statement referred to T. xiphiifolia,—a very different species. Like the rest of its tribe, it requires the constant heat of a warm damp stove, and similar treatment to that which is usually given to epiphytal Orchids. It thrives very well when attached to a branch of any soft-wooded tree, and suspended from the roof of the stove. In winter it must be kept dry, but during the rest of the year it can scarcely have too much water. Mr. Booth describes the recent plant thus:—

"Roots numerous, round and slender, deep brown, partly adhering to the branches of trees, or spreading horizontally, as if to draw nourishment from the air. Leaves broad at the base, closely imbricated, so as to have a sort of bulbous appearance; but otherwise flexuose and recurved, narrow, much longer than the scape, spreading and twisted, with the edges so much incurved as to leave only a deep groove from one end to the other. They vary from 9 inches to a foot in length, and are of a deep green, closely covered with brownish red blotches, and speckled with minute white scurfs. The scape rises from the centre of the leaves, and is about 6 inches high, round at the base, and covered with several sheathing leaves, which closely embrace it. Near the top, it enlarges, and becomes two-sided, with moderately large oblong acuminate sheathing, imbricated bracts, of a brilliant red, tinged with brownish green at the base. The flowers, which appear to be only two in number, issue from underneath the third and fourth bract from the top. They are erect, of a purplish lilac colour, and rather more than an inch long. Sepals? Petals three, united at the base, but so arranged, from being convolute as to form a kind of tube, very slightly recurved at the point. Filaments of the same purplish colour as the petals, comparatively broad and thin, and projecting about a quarter of an inch beyond the tube. Style the same length as the filaments, but round, and of a pale colour, excepting at the extremity, which is a greenish yellow, and 3-lobed."

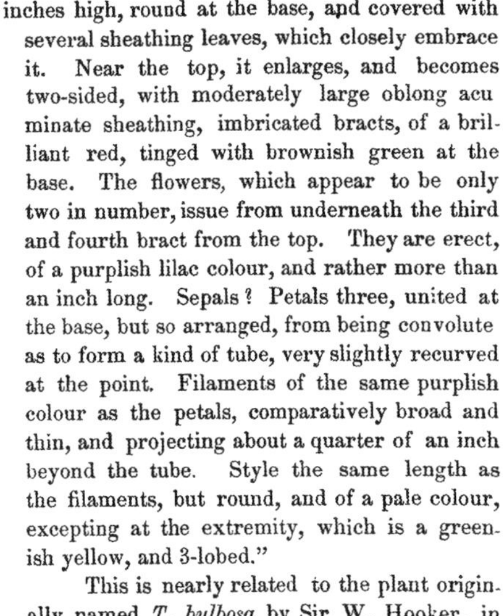

103

This is nearly related to the plant originally named *T. bulbosa* by Sir W. Hooker, in his "Exotic Flora," t. 173, from a poor specimen obtained from Trinidad. But we can scarcely regard it as the same species, any more than a very handsome plant, with long spreading crimson bracts, obtained from Jamaica by Sir W. Hooker, and figured in the "Botanical Magazine," t. 4288, under the name of *T. bulbosa*, variety *picta*. There appears to be several species of Tillandsia possessing the peculiarity of having the bases of the enlarged leaves collected into a kind of bulb, but otherwise differing as much among each other as species of the same genus generally do. Since some are beautiful things, and very likely to reach our gardens, we take the present opportunity of pointing out in what we conceive their peculiarities to reside. In the first place, there is the original *T. bulbosa*, whose spike has all the bracts green and fertile, with some tendency to branch. Next it stands our *T. inanis*, with a perfectly simple spike, whose bracts are coloured red, and all flowerless, except the two uppermost. Another is the supposed variety of *T. bulbosa*, already mentioned, with the upper leaves and bracts very long, deep crimson, apparently not scurfy, and a spike distinctly branched; the corolla being longer and white-edged: this we would call *T. erythræa*; we have the same species from Para. A fourth, *T. eminens*, is a St. Domingo plant, with the leaves much shorter than the spike, which is leafless, branched, and composed of numerous two-ranked crimson-keeled naked bracts; it may be compared to *T. polystachya*, although very different. A fifth is from Para, and is readily distinguished by a peculiar lumpish habit, an abundance of very coarse loose scurfs, spreading up to the very points of the outer bracts, which are not coloured, and a nearly simple spike sessile among the leaves, which, nevertheless, scarcely overtop it; this may be named *T. pumila*. For the convenience of our scientific readers, we put these distinctions into technical language :—

Folia radicalia basi dilatata bulbum simulantia.

T. inanis; scapo foliis breviore, spicâ simplici basi foliosâ, bracteis viridi-purpureis lepidotis inferioribus omnibus inanibus.—*Buenos Ayres.*

211. *T. bulbosa* (Hook. Exot. Fl., t. 173); scapo foliis breviore, spicâ aphyllâ basi ramosâ, bracteis herbaceis arctè lepidotis.—*Trinidad.*

212. *T. erythræa* (aliàs *T. bulbosa picta* Hooker, Bot. Mag., t. 4288); scapo foliis breviore, spicâ ramosâ, bracteis foliaceis coccineis nudis (?) infimis spicâ longioribus.—*Jamaica; Para.*

213. *T. eminens;* scapo foliis altiore, spicâ aphyllâ ramosâ, bracteis nudis coccineis distichis carinatis apice uncinatis.—*St. Domingo.* The inflorescence is almost that of a branched Vriesia.

214. *T. pumila;* scapo inter folia sessili, spicâ subsimplici aphyllâ, bracteis herbaceis coriaceis ventricosis laxissimè lepidotis.—*Para.* Valves of the fruit straight, and chesnut brown; not pitch black, as in *T. erythræa.*

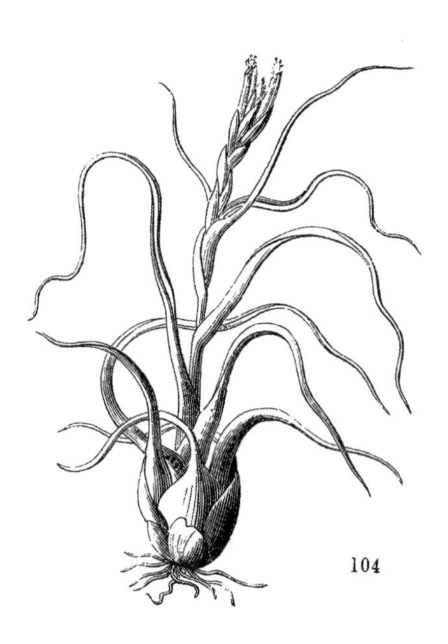

104

PLATE 31.

L. Constans, Pinx & Zinc.

Printed by C. F. Cheffins, London.

[PLATE 31.]

THE DEEP BLOOD-COLOURED MOUTAN.

(MOUTAN OFFICINALIS; ATROSANGUINEA.)

A Hardy Undershrub, from CHINA, *belonging to the Natural Order of* CROWFOOTS.

Pæonia Moutan, atrosanguinea : *Journal of the Horticultural Society*, vol. iv., p. 225.

IT will probably be admitted, without any difference of opinion, that this is the finest of the Moutans introduced by the Horticultural Society. It is a plant with a vigorous growth, a deep green foliage tinged with red, and very large, very double flowers, with dark blood-coloured petals, which are nearly as broad in the centre as at the edge. In foliage it is much like the common *Moutan papyracea*.

And now a word respecting the genus Moutan, which we propose to separate from Pæonia. We need not say that all the Moutans are furnished with a tough leathery coat which is drawn tightly round the carpels, of which it allows nothing but the stigmas to project. This organ has no existence in PÆONIA, or in that part of it which one of us formerly proposed to call ONÆPIA, containing P. Brownii and another. It is of somewhat uncertain nature; wherefore it has received from different persons the names of Disk, Nectary, Perigynium, Paracorolla, &c. Upon this organ the genus MOUTAN is founded; and thus it differs from Pæonia as much as Ranunculus from Adonis, Actæa from Thalictrum, Trollius from Helleborus, all genera of the same order, that is to say, because of the presence of a part which does not appear in others.

Of the nature of this part there is little room for doubt. It is in all probability an innermost row of abortive stamens, the filaments of which are united into a cup, while the anthers refuse to appear; and therefore it is referable to that part of the flower which botanists now call disk. D. Don said he found anthers upon its edge, and if he was not mistaken that would be conclusive as to its nature; but we have never been able to find anthers upon it, nor does it appear that any-one except Mr. Don ever did.

In one of his interesting letters, Mr. Fortune gives the following account of the manner in which the Chinese propagate Moutans:—

"The propagation and management of the Moutan seem to be perfectly understood by the Chinese at Shanghae, much better than they are in England.

"In the beginning of October, large quantities of the roots of a herbaceous Pæony * are seen heaped up in sheds and other outhouses, and are intended to be used as stocks for the Moutan. The bundle of tubers which forms the root of a herbaceous Pæony is pulled to pieces, and each of the finger-like rootlets forms a stock upon which the Moutan is destined to be grafted. Having thrown a large number of these rootlets upon the potting bench, the scions are then brought from the plants which it is desirable to increase. Each scion used is not more than an inch and a half or two inches in length, and is the point of a shoot formed during the bygone summer. Its base is cut in the form of a wedge, and inserted in the crown of the finger-like tuber just noticed. This is tied up or clayed round in the usual way, and the operation is completed. When a large number of plants has been prepared in this manner they are taken to the nursery, where they are planted in rows about a foot and a half apart, and the same distance between the rows. In planting, the bud or point of the scion is the only part which is left above ground; the point between the stock and the scion, where the union is destined to take place, is always buried beneath the surface. Kæmpfer states that the Chinese propagate the Moutan by budding; but this must have been a mistake, as budding is never practised in the country, and is not understood. He was probably deceived by the small portion of scion which is employed, and which generally has only a single bud at its apex.

"Many thousands of plants are grafted in this manner every autumn, and the few vacant spaces which one sees in the rows, attest the success which attends the system; indeed it is rare that a graft fails to grow. In about a fortnight the union between the root and the scion is complete, and in the following spring the plants are well-established and strong. They frequently bloom the first spring, and are rarely later than the second, when they are dug up and taken to the markets for sale in the manner I have described. When each has only one stem and one flower-bud, it is of more value in the eyes of the Shanghae nurserymen than when it becomes larger. In this state it is more saleable; it produces a very large flower, and it is easily dug up and carried to the market. I could always buy large plants at a cheaper rate than small ones, owing to these circumstances.

"In the gardens of the Mandarins it is not unusual to meet with the tree Pæony of great size. There was one plant near Shanghae which produced between three and four hundred blooms every year. The proprietor of it was as careful of it as the Tulip fancier is of his bed of Tulips. When in bloom it was carefully shaded from the bright rays of the sun by a canvas awning, and a seat was placed in front, on which the visitor could sit down and enjoy the sight of its gorgeous flowers."

* A variety with small single flowers.

PLATE 32.

L. Constans Pinx & Zinc

Printed by C. F. Cheffins, London.

[PLATE 32.]

THE ASOCA.

(JONESIA ASOCA.)

A Stove Tree, Native of the EAST INDIES, *belonging to* LEGUMINOUS PLANTS.

Specific Character.

THE ASOCA.—A tree. Leaves in 3—5 pairs, with smooth lanceolate wavy acuminate leaflets rather acute at the base. Flowers in terminal fasciculate corymbs, hexandrous.

JONESIA *ASOCA*:—Arborea; foliis 3—5-jugis foliolis lævibus lanceolatis undulatis acuminatis basi acutis, corymbis terminalibus fasciculatis, floribus hexandris.

Jonesia Asoca, *Roxburgh in Asiatic Researches, vol.* 4, *p.* 355.

THIS beautiful tree, with glowing fragrant flowers, blossomed, in June last, at Chatsworth, in the aquatic house, whence our specimen was obtained. It is a native of various parts of the East Indies, where it is also much cultivated in gardens. Roxburgh says it is—

"Found in gardens about Calcutta, where it grows to be a very handsome, middling-sized, ramous tree; flowering time, the beginning of the hot season; seeds ripen during the rains. The plants and seeds were, I am informed, originally brought from the interior parts of the country, where it is indigenous."

Sir W. Jones himself, after whom the genus was named, states that—

"The number of stamens varies considerably in the same plant: they are from six to seven, to eight or nine, but the regular number seems eight,—one in the interstices of the corol (calyx), and one before the centre of each division. Most of the flowers, indeed, have one abortive stamen, and some only mark its place, but many are perfect, and Van Rheede speaks of eight as the constant number; in fact, no part of the plant is constant. Flowers fascicled, fragrant just after sunset and before sunrise, when they are fresh with evening and morning dew; beautifully diversified with tints of orange-scarlet, of pale yellow, and of bright orange, which grows deeper every day, and forms a variety of shades, according to the age of each blossom that opens in the fascicle. The vegetable world scarce exhibits a richer sight than an Asoca tree in full bloom; it is about as high as an ordinary Cherry-tree. A Brahmin informs me, that one species of the Asoca is a creeper, and Jayadéva gives it the epithet "voluble;" the Sanscrit name will, I hope, be retained by botanists, as it perpetually occurs in the old Indian poems, and in treatises on religious rites."

Mr. Harrington writes of it thus:—

"ASOCA: This is the true name of a charming tree, inaccurately named Asjogam in the Hort. Malab., vol. 5, tab. 59. It is a plant of the eighth class and first order, bearing flowers of exquisite beauty; and its fruit, which Van Rheede had not seen, is a legume, compressed, incurved, long, pointed, with six, seven, or eight seeds; it will be described very fully in a paper intended for the Society. The Brahmins, who adore beautiful objects, have consecrated the lovely Asoca: they plant it near the temples of Siva, and frequently mention a grove of it, in which Rávan confined the unfortunate Síta. The eighth day from the new moon of Chaitra, inclusive, is called Asocashtami."

We suspect that more species than one are mixed under the common name of Asoca. The late Mr. Griffith found in Burmah, cultivated, a tree with very dense corymbs of flowers, and leaves in 3-pairs, the lowest of which is distinctly heart-shaped. This is scarcely the Asoca of Bengal, but is much nearer the Java plant, called by Zollinger, *Jonesia minor*, without being the same. Then again the plant now figured is surely not what Sir W. Hooker has given in the Botanical Magazine, t. 3018, with small whole-coloured flowers, having a reflexed limb, and leaves in 5-pairs; nor do either sufficiently correspond with Roxburgh's figure in the Asiatic Researches. In short, the question requires that elucidation at the hands of an Indian botanist, which a European cannot undertake.

Those who assert that the wholesome law of priority in deciding the validity of botanical names is immutable, will do well to consult the history of this plant, first called by Linnæus *Saraca indica*, then by Burmann *Saraca arborescens*, and twenty-seven years later, *Jonesia Asoca*, by Roxburgh, whose name is, nevertheless, universally adopted.

PLATE 33.

L. Constans, Hax & Zinc.

Printed by C.F. Cheffins. London.

[PLATE 33.]

THE VARIEGATED ONCID.

(ONCIDIUM VARIEGATUM.)

A Stove Epiphyte, from the WEST INDIES, *belonging to the Natural Order of* ORCHIDS.

Specific Character.

THE VARIEGATED ONCID.—Leaves acuminate, fleshy, equitant, serrulate. Flowers panicled ; lower sepals united into one spoon-shaped body. Petals obovate, emarginate, unguiculate, cuspidate. Lip with small acute lateral lobes, a broad 2-lobed middle lobe with a denticulate unguis, and a double fleshy crest, the upper half consisting of two lobes, the lower of three. Wings of column hatchet-shaped, acuminate, entire.

ONCIDIUM *VARIEGATUM*—(Equitantia); foliis carnosis acuminatis serrulatis ; floribus paniculatis, sepalis inferioribus in usum cochleatum connatis, petalis obovatis unguiculatis emarginatis cuspidatis, labelli laciniis lateralibus nanis acutis intermediâ latâ bilobâ ungue denticulato, cristâ duplici supernè 2-lobâ infernè 3-lobâ, alis columnæ acinaciformibus acuminatis integerrimis.

Oncidium variegatum : *Swartz act. holm.* 1800; p. 240. *Lindl. gen. et. sp. Orch.* p. 198.

THIS charming little plant was first introduced from the Havannah, by Sir Charles Lemon, Bart.; more recently it has been put into circulation by Linden, who gave a plant to the Horticultural Society, in whose garden the materials for the accompanying figure, aided by native specimens, were obtained. It is a small species, growing ill on wood, and hitherto, in cultivation, not more than a quarter of the natural size.

When in health the leaves are fleshy, 3 or 4 inches long, equitant, sharp pointed, and very much broken at the edge. The panicle is a foot and a half high, erect, and decorated with flat, pink flowers, richly stained with cinnamon-red on the sepals, and at the base of the sepals and lip. The lower sepals form a blunt spoon-shaped body ; the petals are large, obovate, almost retuse, with an intermediate point ; the lip has the middle lobe distinctly placed upon a somewhat serrated unguis ; the crest consists of two sets of tubercles, one lying on the other, the upper set made up of two large lateral ones, and a minute one in the middle, the lower set, of three equal blunt ones, the intermediate of which is curved upwards.

This Variegated Oncid is very like the Tetrapetalous Oncid, from which it differs in having the leaves broken up at the edge, petals coloured, broad and cuspidate, not herbaceous, blunt and serrulate, in the double sepal being blunt and spoon-shaped, not divided into two taper-pointed divisions, and in its richer colours.

But this does not apply to the Cuba specimens referred to the Variegated Oncid in the *Orchidaceæ Lindenianæ*, which certainly belong, at least in part, to a distinct species. It is the more necessary to mention this, because it is possible that Mr. Linden may have circulated plants of them under the name erroneously applied to it in the work above quoted, by the writer of the present article, who looked upon them as mere varieties of the Variegated Oncid. In general appearance, they wholly correspond with it, and also in the ragged edge of the foliage; but they differ in the flowers being downy, the wings of the column blunt, the middle lobe of the lip perfectly sessile, and the lateral lobes joining it by a broad base. The crest, too, consists of five tubercles, of which the uppermost are much the longest. The plant is stated by Mr. Linden to vary with white or rose-coloured flowers, as well as in stature—a large form growing in the Pine forests of Yatara, in Cuba; the smaller on Coffee trees in the Sierra Maestre, and on the Liban mountain. But it is probable that this applies to both the species in question.

In order to enable those who may possess the second species to identify it, if indeed it does occur in living collections, we subjoin the following:—

Specific Character.

THE VELVETY ONCID.—Leaves acute, fleshy, equitant, serrulate. Flowers velvety, panicled. Back sepal obcordate, lower united into one spoon-shaped body. Petals nearly orbicular, a little narrowed to the base. Lip with rounded lateral divisions much smaller than the petals, abruptly passing into the broad 2-lobed middle division, without the intervention of any unguis; crest consisting of two long posterior cylindrical lobes, and three smaller short ones in front. Wings of column hatchet-shaped, blunt, entire.

ONCIDIUM *VELUTINUM*—(Equitantia); foliis acutis carnosis equitantibus serrulatis, floribus velutinis paniculatis, sepalo dorsali obcordato lateralibus in unum obtusum cochleatum connatis, petalis suborbicularibus basi paululum angustatis, labelli laciniis lateralibus rotundatis quam petala multò minoribus in intermediam decurrentibus latam sessilem bilobam; cristæ tuberculis 2-posticis elongatis tribusque minoribus anticis, alis columnæ acinaciformibus obtusis integerrimis.

In some respects this approaches *O. pulchellum*, which, however, is readily distinguished by the petals being much smaller than the lateral lobes of the lip.

GLEANINGS AND ORIGINAL MEMORANDA.

215. Cupressus torulosa. *D. Don.* A large evergreen tree, with glaucous leaves. Belongs to Conifers. Native of the Himalayas. (Fig. 105.)

It would seem that there is but one species of Cypress inhabiting the North of India, and that the *Cupressus torulosa*—why so called we cannot discover. For the native country of this plant Bhotan was first given by the late Prof. Don, upon the authority of Mr. Webb. Afterwards Dr. Royle stated that it appeared to be the plant called *theelo* by the natives, seen between Simla and Phagoo, and near Jangkee Ke Ghat, a high hill to the southward of Rol. "It is also found in Kemaon, near Neetee, Simla, and in Kunawur." Endlicher says that it occurs in Butan and Nepal, as high as 8500 feet of elevation. Dr. Wallich adds the southern mountains of Oude. Is it really true that there is but one Indian Cypress, and that the Torulosa? And is the Torulosa what is spoken of by all these writers? We doubt it much. In the first place *Cupr. horizontalis* occurs in Persia; why not then in India? In the next place, there are such differences among the specimens of Indian Cypresses raised in England, and between them and the wild specimens, as to suggest reasonable doubts concerning their identity. As far as we can investigate the matter, Indian evidence seems to fail us, and home evidence is inconclusive. All that can be affirmed with confidence is, that in this country, raised from Himalayan seeds, exists a glaucous, upright, graceful Cypress, which is distinct from all European kinds, and to which the name

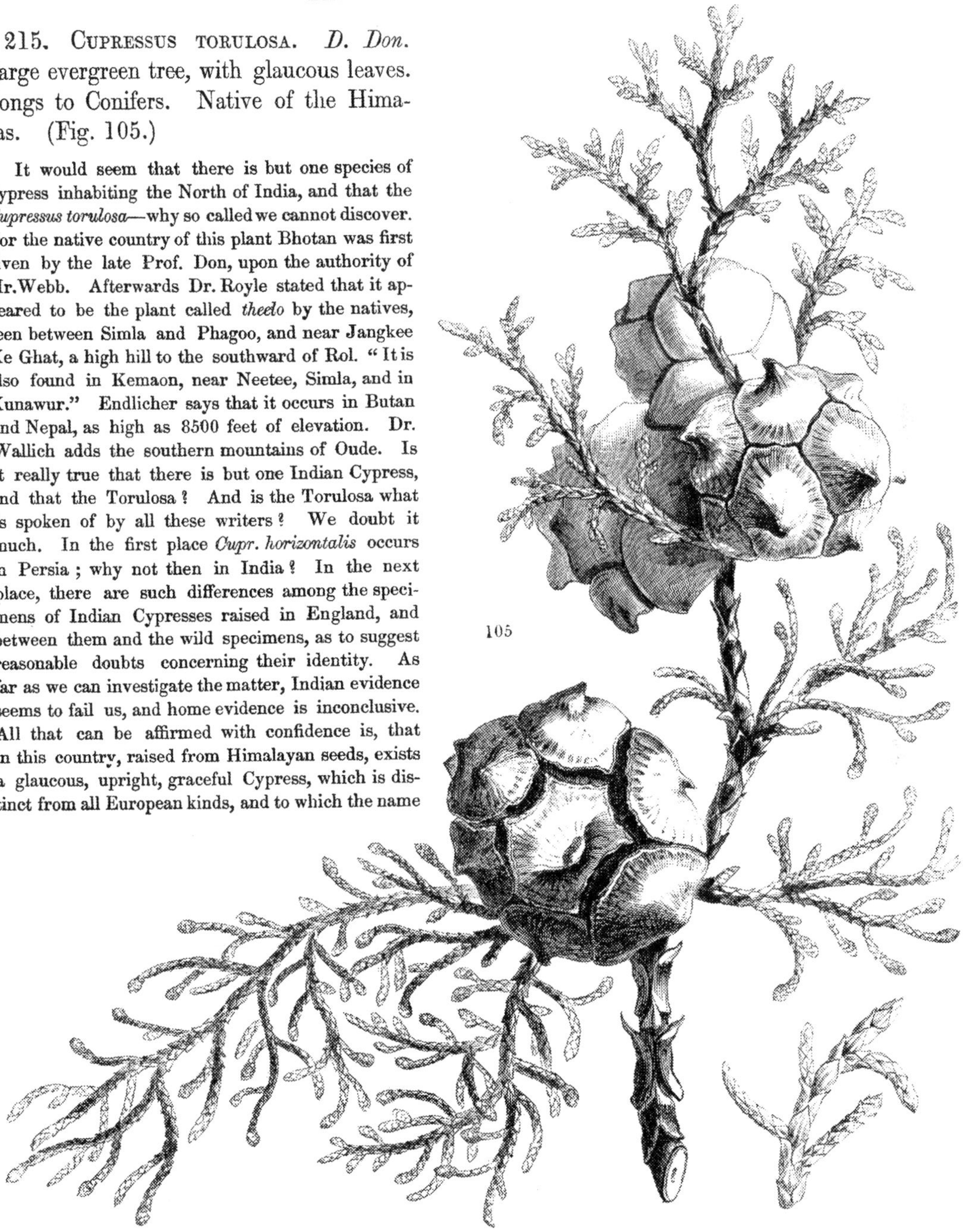

105

of *torulosa* is applied. It has a perfectly straight stem, and, when young, a compact conical growth, by which it is known at first sight. Its cones are, as usual, globular, and are made up of four pairs of hard woody scales, with a hexagonal mucronate extremity of about two more pairs. The leaves when the plant is old are blunt, in four rows, and so uniformly imbricated, that they give the young branches a regular four-sided appearance. The old wood is deep purplish brown, and perfectly smooth; whereas the branches of the Evergreen Cypress and its varieties have more or less of a cinnamon brown appearance.

Is this the one and sole Indian Cypress ? Among the specimens distributed by the East India Company, we have one (named *Thuja orientalis ?*)) which to the foliage of this adds cones not more than one-fourth the size, the scales being scarcely mucronate ; and a second found by Blinkworth in the Himalayas, without cones, the foliage of which also corresponds with this. Are these really the same ? That is what we cannot answer.

Such difficulties render it impossible to tell with certainty what the stature and habit of our garden Torulosa may become. Endlicher says the tree is sometimes forty feet high ; Don, that it is handsome and pyramidal ; Griffith, who calls the Bhotan plant *C. pendula*, that it is eighty feet high, and extremely handsome (*elegantissima*) ; the last traveller also represents the Bhotan Cypress as a tall tree running to a sharp point, like a Spruce fir, with gracefully drooping branches. (See his Private Journals, p. 272, where is a figure of it as it was seen in the village of Chindupjie, a place more than 7800 feet above the sea.) Let us hope that Major Maddox will bring his local knowledge and acute criticism to the explanation of these difficulties, in a future number of the Transactions of the Agri-horticultural Society of India.

The accompanying figure was taken from specimens produced in the garden of the Hon. W. F. Strangways, at Abbotsbury.

216. Bertolonia maculata. (*Martius.*) (See p. 27, fig. 14.)

Upon the *Eriocnema marmoratum*, given above upon the authority of M. Naudin, who has specially studied the *Melastomads*, Sir W. Hooker makes the following observations, "Botanical Magazine," t. 4551 :—

"But the plant is no *Eriocnema*. It belongs to the curious and beautiful genus *Bertolonia*,—'dont le caractère essentiel consiste,' as M. Naudin has himself well expressed, 'dans la forme tout-à-fait insolite du calyce et de la capsule;' and it is equally certain that it is the *B. maculata* of De Candolle and of Martius above quoted, t. 257. This fruit or capsule is an elegant object, especially when the eye is aided by a small power of the microscope ; for it is singularly inflated, with three very prominent angles and several ribs, and every rib, as well as the margin of the lobes of the calyx, is beset with bristles, terminated by a gland."

217. Consolida Aconiti. *Lindley.* (*aliàs* Aconitum monogynum *Forskahl; aliàs* Delphinium Aconiti *Linnæus.*) A hardy annual, with finely divided leaves, and purple flowers of little beauty. Native of Erzeroum. Belongs to the Crowfoots. Introduced by H. H. Calvert, Esq. (Fig. 106 *a*, a single flower magnified; *b*, the two united petals.)

A weak erect annual, about one and a half foot high, with a very slight covering of silky hairs upon all the green parts. The leaves are divided into from three to five pedate linear taper pointed lobes. The flowers form a loose straggling somewhat zigzag raceme, the peduncles of which are from one and a half to two inches long, with about one awl-shaped bract above the middle. The flowers, which grow singly, are of a deep bluish lavender colour, with the following structure. The calyx consists of five coloured oblong sepals, of which four hang downwards, the side ones being the broadest ; and the fifth, which is turned in an exactly opposite direction, is extended into a horizontal blunt hairy spur with a short narrow ovate acute limb. The corolla consists of two petals united by their back edge into one simple somewhat fleshy spur, enclosed within that of the fifth sepal, and with a hooded limb, having four small round lobes at its point, and two larger oblong lateral ones. The solitary carpel slightly projects beyond the declinate stamens. De Candolle and others speak of the petaline spur being slit on the upper side, a structure of which I find no trace.

Forskahl regarded this curious plant, it is said, as an Aconitum ; Linnæus considered it a Delphinium. In reality it is neither the one nor the other. Its united petals, and long sepaline spur, are at variance with the distinct hammer-headed petals and convex back sepal of Aconite. Its petals being reduced to two, and these completely combined into one, equally remove it from Delphinium. That the petaline body is really composed of two parts only seems to be proved by its origin, which looks as if opposite the back sepal, in consequence of the union of the two contiguous edges of the lateral petals. But it is completely separated from the front sepals, with which it does not in any degree alternate. These considerations lead to the conclusion that the old genus Consolida should be re-established, and by no means confounded with Delphinium proper.

In a scientific point of view this is a highly interesting species ; but its growth is too feeble, and its flowers and leaves too diminutive and straggling to give it any horticultural value.—*Journal of Hort. Soc.*, vol. vi.

The following short generic character will serve to render the above statement more precise in the eyes of systematical botanists :—

CONSOLIDA. *Bauhin.* Sepala 5, colorata, supremo refracto unguiculato calcarato. Petala 2, in unum coalita calcaratum lobatum intra sepalum superius intrusum. Stamina declinata. Carpellum solitarium.

A crowd of Oriental Annuals, including our "Branching Larkspurs," will be found to belong to this genus.

218. VERBENA TRIFIDA. *Kunth.* A sweet-scented perennial, with white flowers, from the temperate parts of America. Blossoms in the autumn. Introduced from Santa Martha by His Grace Hugh Duke of Northumberland. (Fig. 107.)

A dwarf herbaceous plant, growing about a foot high, with the habit of *V. tuberosa;* covered all over with short hairs, which give a grey tint to the deep green surface. The stems are four-cornered. The leaves are stalkless, opposite, rather curved downwards, nearly 3-lobed or 5-lobed, in consequence of the middle lobes having two lateral divisions. From the axils of the principal leaves several smaller regularly 3-lobed ones also arise, producing the condition which botanists call fasciculated. The flowers are pure white, extremely sweet, in oblong hairy simple or compound heads. The lobes of the calyx are awl-shaped, those of the corolla are oblong, nearly equal, and blunt or retuse. The species is found wild both in Mexico and New Grenada, but can hardly be called a shrub, as it is stated to be by M. Schauer. It possesses little beauty, but its fragrance is delicious, and it seems destined to aid in founding a family of sweet-scented brilliant bedding plants; for there is no reason to suppose that it will refuse to cross with the gay varieties now such universal favourites.

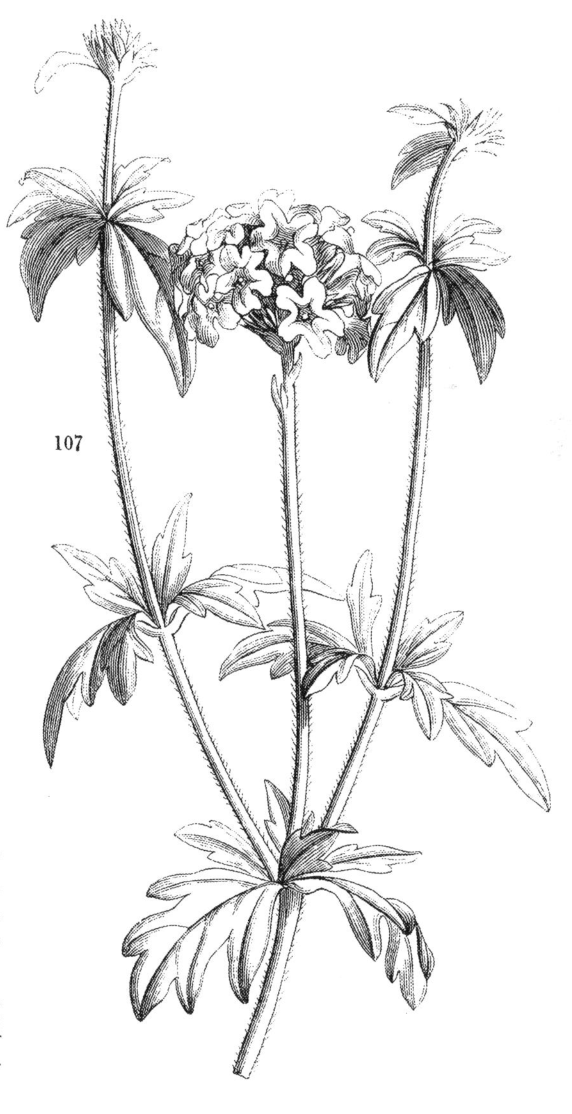

219. OXYSPORA VAGANS. *Wallich.* (*aliàs* Melastoma rugosa, *Roxburgh.*) A very handsome stove shrub, with panicles of crimson and purple flowers. Native of the Himalayas. Flowers in autumn.

Raised from seeds sent by Dr. Hooker from hilly country bordering on the plains in the approach to Darjeeling. If less showy, it is a more graceful plant than the *O. paniculata,* being truly subscandent and the panicles all very drooping. Three to five feet high, loosely branched; the branches long and weak, drooping, obscurely four-angular, the younger ones downy. Leaves ovate or cordate-ovate, acuminate, five to seven-nerved, smooth above, obsoletely downy with short hairs, or quite smooth below, where also the nerves are very prominent and red. Panicles terminal, drooping, often a foot long. Petals four, of a bright rose-colour, obovate, acute. Stamens eight, four long and four short; the four smaller anthers are pale-coloured, and have a distinct spur pointing downwards at the back of the connectivum; the four longer ones are deep purple, much curved, and have a small spur. Grows freely in light loam and leaf-mould, in a moderately warm stove. —*Botanical Magazine,* t. 4553.

220. ONCIDIUM PLANILABRE. *Lindley.* A hothouse orchid from Brazil, with yellow and brown flowers. Introduced by the Horticultural Society. Flowers in August.

O. planilabre (Plurituberculata); pseudobulbis ancipitibus tenuibus costatis, foliis ensatis recurvantibus racemo brevioribus, racemo simplici, sepalis petalisq. lanceolatis unguiculatis undulatis subæqualibus, labelli laciniis lateralibus oblongis parvis intermediâ semicirculari planâ emarginata, cristâ rhomboideâ cuspidatâ margine erosâ verrucis 2 inæqualibus utrinque versus cuspidem, dente forti obtuso faciei columnæ adnato, columnæ brevibus carnosis inflexis.

This plant has the foliage of *O. flexuosum*, and flowers much like those of *O. Suttoni*. The pseudo-bulbs are thin, sharp edged, and ribbed at the side. The leaves are sword-shaped, lorate, recurved, and shorter than the raceme. The raceme is long and narrow like that of the Sutton Oncid (*O. Suttoni*), and the flowers are as nearly as possible of the same colour; that is to say, the sepals and petals are dull brown tipped with yellow, and the lip is clear yellow stained with cinnamon brown at the base. The sepals and petals are nearly of the same size and form, rhomboid-lanceolate, acuminate, wavy, very distinctly stalked. The lip is three-lobed, with the side lobes nearly as wide as that in the centre, which is slightly stalked, nearly hemispherical, emarginate, and perfectly flat. The crest consists of a broad lozenge-shaped rugged-edged cuspidate process, beneath which, near the point, on either side, are two small unequal tubercles; in addition to which there is a stout blunt tooth which rises in front of the column, forming part of it. The wings of the column are roundish, dwarf, and incurved. There is no published Brazilian species with which this can be usefully compared. From the Sutton Oncid and similar Mexican forms it differs in the form of the crest, and especially in the strong tooth already mentioned as standing in front of the column. It is rather a pretty species, of the third class in point of personal appearance.—*Journal of Hort. Soc.*, vol. vi.

221. Daphne Houtteana. (*aliàs* Daphne Mezereum, foliis atropurpureis *of Gardens.*) A hardy evergreen bush, with vernal purple flowers. Belongs to Daphnads. Origin unknown.

That this plant is not a Mezereum is evident; in Mezereum the flowers precede the leaves; but here they appear simultaneously. In Mezereum the leaves are obovate-lanceolate, gradually extended into a wedge-shaped base, thin, glaucous beneath, downy in the bud, fringed at the edges when full grown; in this plant the leaves are lanceolate, taper-pointed, half leathery, with no trace of glaucousness or down. The flowers of Mezereum are bright carmine, and seem to come out of the very wood of the stem; those of the present plant are violet-lilac, and grow in little stalked cymes, the ramifications of which remain behind after the fruit has fallen. Is this, then, a new species? It is scarcely probable. M. Planchon suggests that it may be the *D. papyracea* of Wallich, a Himalayan species, introduced many years since into England, according to Sweet's "Hortus Britannicus;" and of which the short diagnosis in Walpers agrees pretty well with our plant. This can be ascertained by those who have access to the figure of that species, published by M. Decaisne, in the botanical part of "Jacquemont's Voyage." Be that as it may, this plant is well worth growing, for it is perfectly hardy, and flowers in March, rather later than D. Mezereum.—*Flore des Serres*, t. 592.

This is a handsome evergreen, with deep purple leaves, occasionally met with in English gardens. Can it be a mule, between the Mezereum and the Spurge Laurel (*D. Laureola* ?)

222. Eria acervata. *Lindley*. A white-flowered hothouse orchid from India, of no beauty. Introduced by the Horticultural Society.

E. acervata; pseudobulbis compressis uno super alterum cumulatis collo brevi diphyllis, foliis rectis ensatis, racemis axillaribus 2-3-floris, bracteis pluribus super pedunculum ovatis acuminatis revolutis, sepalis petalisq. ovatis acutis, labelli trilobi 3-lamellati lobis acutis intermedio oblongo multò longiore.

This little Eria is scarcely known in gardens. The peculiarity of it consists in the stem when fully formed being nothing more than a collection of pseudobulbs or compressed bodies, in form not unlike a flat flask, and piled one over the other in a very singular manner. The flowers are white, smooth, with a slight tinge of green, but otherwise colourless. The lip is 3-lobed, with 3 elevated parallel lines, the middle lobe the longest, oblong and acute. The foot of the column is neither chambered nor toothed. In all respects this plant is so entirely an Eria that it is referred to that genus, although, in the flowers examined, the number of its pollen masses was only 4, instead of 8. But this may have been accidental. In its 3-ridged lip, and reflexed bracts, it so strongly calls to mind that genus, as to raise a reasonable presumption that the number of pollen masses would, in more perfect flowers, be as usual.—*Journal of Hort. Soc.*, vol. vi.

223. Lonicera tatarica, *var.* punicea. A hardy shrub from Siberia, with crimson flowers. Belongs to Caprifoils. Introduced by the Horticultural Society.

This plant does not seem to differ in any essential particular from the old Tartarian Honeysuckle, except that its flowers are larger, later, and of a deep rose colour. In these respects it has much more value for gardens; for it is not so apt to be cut off by spring frosts. If uninjured, the rich tints of its flowers give the bush quite a handsome appearance among early flowering plants. It is worthy of note, that although this seems to differ from the common Tartarian Honeysuckle in no essential circumstance beyond what has been just mentioned, yet *it comes true* from imported seeds. It is reported that the berries are yellow, but of this we have no evidence.—*Journal of Hort. Soc.*, vol. vi.

224. SOLANDRA LÆVIS. *Hooker*. A fine stove shrub, with very large pale green flowers. Belongs to Nightshades. Native of Guatemala. (Fig. 108.)

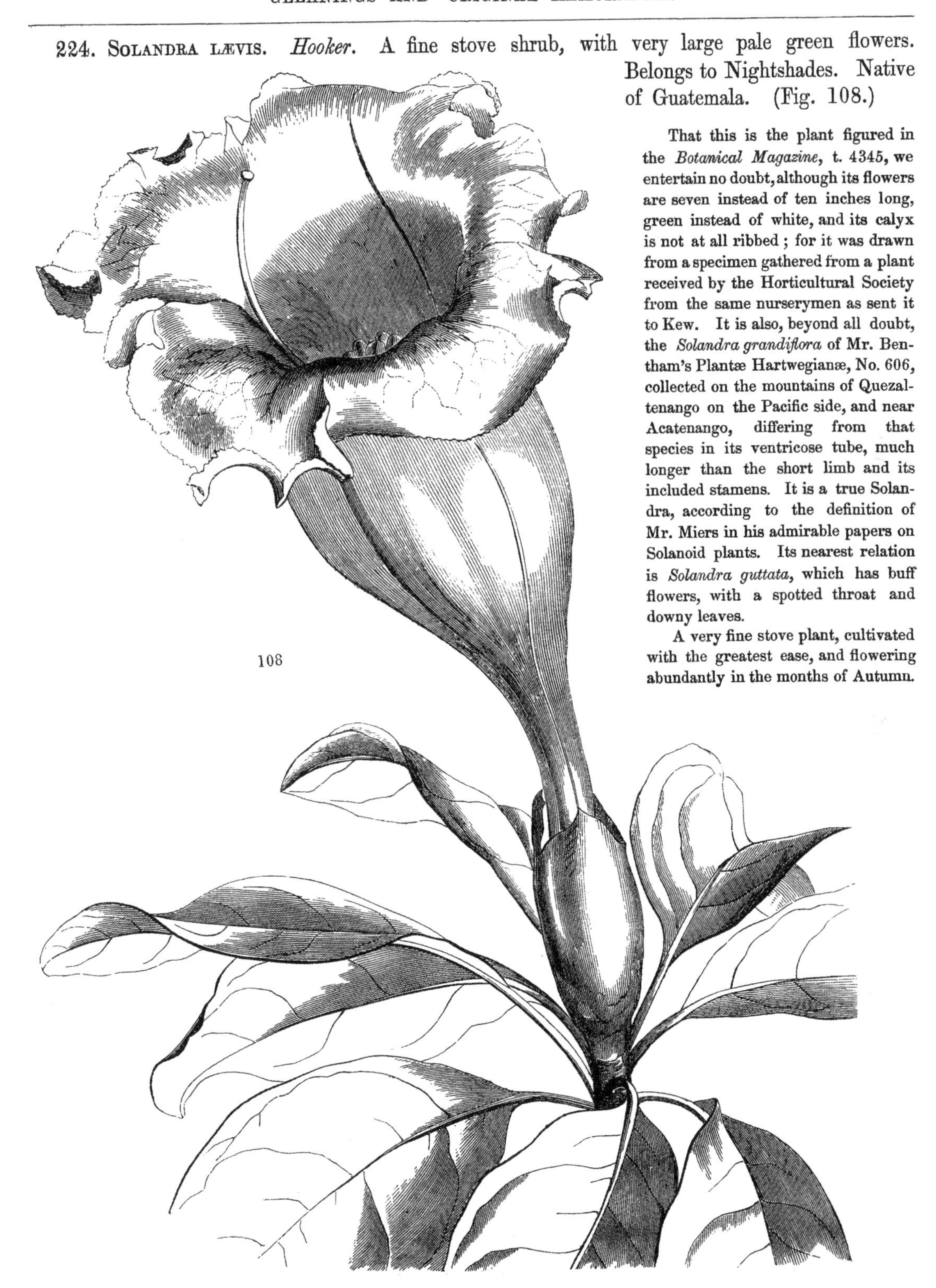

108

That this is the plant figured in the *Botanical Magazine*, t. 4345, we entertain no doubt, although its flowers are seven instead of ten inches long, green instead of white, and its calyx is not at all ribbed; for it was drawn from a specimen gathered from a plant received by the Horticultural Society from the same nurserymen as sent it to Kew. It is also, beyond all doubt, the *Solandra grandiflora* of Mr. Bentham's Plantæ Hartwegianæ, No. 606, collected on the mountains of Quezaltenango on the Pacific side, and near Acatenango, differing from that species in its ventricose tube, much longer than the short limb and its included stamens. It is a true Solandra, according to the definition of Mr. Miers in his admirable papers on Solanoid plants. Its nearest relation is *Solandra guttata*, which has buff flowers, with a spotted throat and downy leaves.

A very fine stove plant, cultivated with the greatest ease, and flowering abundantly in the months of Autumn.

225. Primula capitata. *Hooker.* A hardy herbaceous plant, with close round heads of deep purple blossoms. Native of the Himalayas. Introduced to Kew. Flowers in October.

Raised at the Royal Gardens of Kew, from seeds sent by Dr. Hooker, which were gathered in June, 1849, from plants growing on gravelly banks at Lachen, Sikkim-Himalaya, one of the Passes into Thibet; elevation 10,000 feet above the level of the sea. It is, although of the same group of *Primulæ* with the *P. denticulata* of the Nepal mountains and our own *P. farinosa* of the north of England and Scotland,—a remarkable and well-defined species, the flowers being actually sessile, and so crowded as to form a compact globose head, like that of many species of *Allium* or *Armeria*. Dr. Hooker observed that it yields a faint fragrance, which it does in cultivation; but this, in part at least, is derived from the farinaceous substance of the leaves and flowers. It flowers with us in a pot in the rock-border. Scape often a foot long, moderately stout and thickened upwards, mealy, terminated by a dense globose head of flowers, bracteated at the base, the outer bracteas lanceolate, and forming a small reflexed involucre. Calyx sessile, mealy, large, campanulate, deeply five-fid, the segments ovate, acuminate, subpatent. Corolla with the tube nearly twice as long as the calyx, almost white, mealy, a little inflated upwards, and transversely wrinkled; limb of five, obcordate, spreading lobes, deep purple above, pale beneath. In habit this approaches our native species, *P. farinosa* and *P. Scotica;* and although it is a native of a high region, and consequently subjected to a great degree of cold, yet, like other Alpine species of the genus, it will probably require some slight protection in this climate, especially under our artificial mode of cultivation. During the past summer we had a number of plants growing very luxuriantly,—apparently too much so, for not one of them has yet shown any appearance of flowering. The figure (in the Bot. Mag.) was drawn from a plant that had not been so well taken care of, and was stunted in its growth. Several of the vigorous plants suddenly died: it is therefore safest, till we become better acquainted with this species, to grow it in a frame during winter; and in summer to set it in a shady place, that it may escape the heat of the sun in the middle of the day. It appears to suffer from frequent watering overhead; the pot should, therefore, be placed in a pan, so as to receive water from the bottom.—*Botanical Magazine*, t. 4550.

This is illustrated by one of the happiest of Mr. Fitch's always beautiful figures.

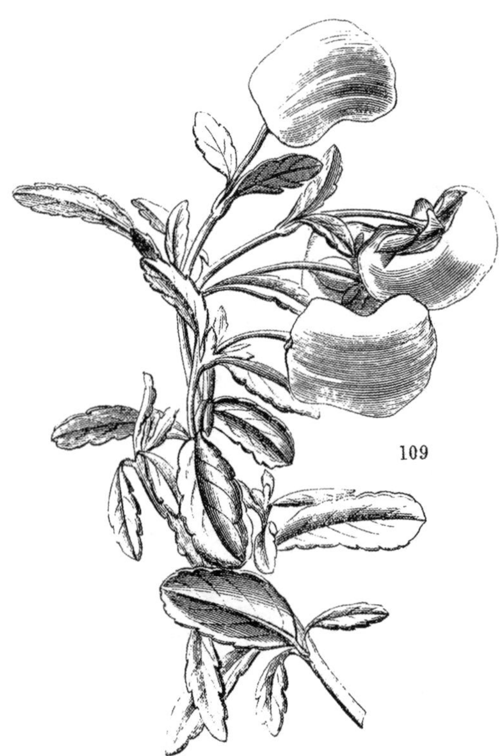
109

226. Calceolaria cuneiformis. *Ruiz and Pavon.* A greenhouse shrub, with pale lemon-coloured flowers, from Bolivia. Blossoms during all the autumn and winter. Introduced by the Horticultural Society. (Fig. 109.)

Raised from seeds purchased from Mr. Thomas Bridges, in 1846. This, in its wild state, is a stiff, short-branched bush, with small wedge-shaped leaves, covered with white hairs on the under side. It bears two or three flowers at the end of each branch, which is closely covered with short, rough hairs. In its cultivated state it has much larger and softer leaves, and weaker branches. The flowers are about as large as those of *C. integrifolia*, and of a pale lemon-colour. It is a very pretty greenhouse plant, with a better habit than the old shrubby Calceolarias. *Journal of Hort. Soc.*, iii. p. 242.

227. Cordyline Sieboldii. *Planchon.* (*aliàs* Dracæna javanica *Kunth; aliàs* Sanseviera javanica *Blume.*) A stove shrub, with small panicles of pale green flowers, and rich spotted leaves. Belongs to Lilyworts. Native of Java. Flowered by Mr. Van Houtte.

This plant has been recently introduced from Java, by Dr. von Siebold. The leaves are of a very dark green colour, firm, convex, recurved, and beautifully variegated with pale green roundish blotches. The flowers are something like those of a Hyacinth in form, but are much smaller, and in terminal bunches. It gained a prize at the Exhibition of Flowers by the Horticultural Society of Ghent. The species is very handsome, and would look well among a collection of Orchids, the climate of which is precisely what it wants.—*Flore des Serres*, t. 569.

M. Planchon, in the article from which this extract is taken, and some others, treats at length of the plants usually combined under the name of Dracæna. He forms the new genus Dracænopsis upon *Dracæna australis* of Hooker; points out *D. ferrea* of Linnæus, or *D. terminalis* of Jacquin, as the type of another which he afterwards names Calodracon; and he adopts the genus Charlwoodia.

228. Portlandia platantha. *Hooker*. A handsome white-flowered hothouse shrub of unknown origin. Belongs to the Cinchonads. Blossoms in July. (Fig. 110.)

Messrs. Lucombe and Co. received, and have cultivated this in the stove, under the name of "*Portlandia grandiflora*, fine variety;" but they remark, that both in its foliage and in the flowers it differs considerably from that species. "It flowers," say these nurserymen, "in a very dwarf state, and is almost always in blossom," an observation confirmed by the continual flowering, during the summer of 1849, of a small plant not more than a foot and a half high, which they sent to the Royal Gardens, and from which a figure was taken in July, 1850. A shrub, a foot and a half high, erect, branched, smooth. Leaves opposite, nearly sessile, elliptical-obovate, acute, evergreen, leathery, full glossy green, entire. Stipules broadly triangular, obtuse. Pedicels very short, axillary, solitary, often opposite. Ovary long 4-angled, 2-celled; cells with many ovules. Limb of the calyx of four spreading, leafy, lanceolate lobes. Corolla white, not more than half the length of that of *P. grandiflora*, broadly funnel-shaped, approaching to bell-shaped, 5-ribbed. Limb of five spreading ovate lobes, their margins revolute. Filaments downy in their lower half. A tropical shrub with fine glossy leaves and showy white flowers, worthy of a place in every collection of woody stove-plants. It grows freely in a mixture of loam and leaf-mould or peat soil. It must be kept in a moist tropical stove, the necessary precautions of watering and shading during clear summer sunshine being carefully attended to. It is propagated by cuttings placed under a bell-glass, and plunged in moist bottom-heat.—*Bot. Mag.*, t. 4534.

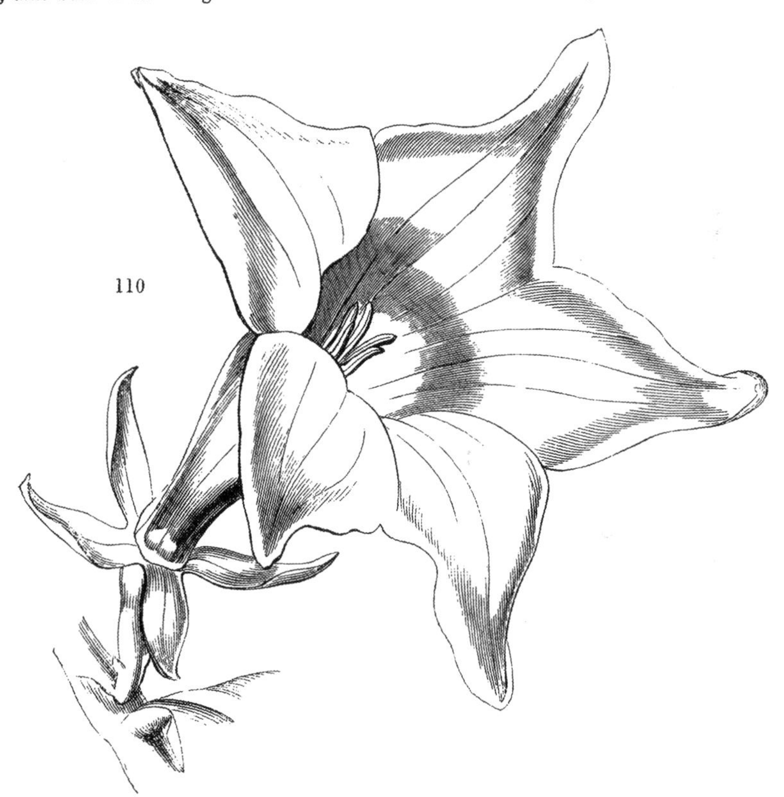
110

229. Fortune's Double Yellow Rose. A deciduous half-hardy scrambling plant, with buff semi-double flowers. Found cultivated in China. Introduced by the Horticultural Society.

This is a straggling plant, with the habit of *R. arvensis*, but with handsomer though deciduous leaves. The branches are dull green, strongly defended by numerous short hooked prickles, without setæ. The leaves are smooth, in about three pairs, bright shining green above, rather glaucous beneath. The flowers are as large as those of the Common China Rose, semi-double, solitary, dull buff, tinged with purple. The petals are loose, and the whole aspect of the flower that of a slightly domesticated wilding. The bush looks like a cross between the China Rose and some scrambling species, such as our European *R. arvensis*. That species being however unknown in Asia, the plant before us must have had some other origin, concerning which it is fruitless to inquire. In its present state this variety has little claim to English notice; but it may be a good breeder, and would certainly be much handsomer in a warmer climate than ours.

Mr. Fortune continues to speak highly of its beauty in China, where it is said to be loaded with buff blossoms; in England, however, its wood is easily killed by frost, and it cannot be regarded as being hardier than a Tea Rose.—*Journal of Hort. Soc.*, vol. vi.

230. VICTORIA REGIA.

For many years this plant has been allowed to bear the name which was first given to it by an authority which we at least shall not presume to question. But some attempts have been lately made at effecting an alteration, which he, to whom the high honour was assigned of rendering the plant known under the name of *Victoria regia*, is bound to resist.

Sir William Hooker, in announcing his intention of publishing certain plates by Mr. Fitch, in illustration of the plant, speaks of it under the name of VICTORIA REGINÆ. We presume he has been led to do so by trusting to the accuracy of a statement made in *The Annals of Natural History* for August 1850, p. 146; to which statement attention is now requested. The author, Mr. John Edward Gray, a zoological officer in the British Museum, writes thus :—

" This plant has three names very nearly alike, and two of them appear to have originated from errors of the press.

" Mr. Schomburgk, on the 11th of May, 1827, sent, through the Geographical Society, a letter to the Botanical Society of London, containing the description of this beautiful Water Lily, accompanied by two drawings and a leaf of the plant. He proposed to call it *Nymphœa Victoria*, but before the paper was read it was observed that the plant appeared to form a genus intermediate between *Nymphœa* and *Euryale*. The paper was slightly altered to make this change, and in a Report of the Proceedings of the Botanical Society, which appeared in the Athenæum Journal of the *9th of September*, 1837 (p. 661), Mr. Schomburgk's description is printed entire, as that of a 'new genus of Water Lily named *Victoria Regina*, by permission of Her Majesty.' Mr. Schomburgk's paper was again read, and his drawings exhibited at the Meeting of the British Association on the 11th of September, 1837, by me, and I am reported to have 'remarked, that this splendid plant would form a new genus with characters intermediate between *Nymphœa* and *Euryale*, and proposed to name it *Victoria Regina*:' see Report in Mag. Zool. and Bot. for October 1837, vol. ii. p. 373. Schomburgk's description, and an engraving of the plant, copied from his drawing, appeared in the next number of that Journal, which came out on the 1st of November, 1837 (vol. ii. p. 441, tab. 12). The description was reprinted again, with copies of Mr. Schomburgk's drawing of the plant and his details of the flower, in the Proceedings of the Botanical Society, p. 44. t. 1 & 2. So much for the name *Victoria Regina*, Schomburgk.

" In the Magazine of Zoology and Botany, by a mistake of the engraver, the plate is lettered '*Victoria Regalis Schomburgh*,' though the proper name is used in the text. This second name has not been anywhere adopted. In the Index to the Athenæum Journal for 1837, p. vii., under the head of Botanical Society, occurs, 'Schomburgk on the *Victoria regia*, p. 661,' which is evidently an error of the press, as the name in the page referred to is *V. Regina*.

" Shortly *after the appearance of the description and figure* in the Annals of Zoology and Botany, and after Sir William Jardine had returned them, Captain Washington, R.N., then Secretary of the Geographical Society, *borrowed from the Botanical Society* the original description and drawing of the plant made by Mr. Schomburgk, with the intention of their appearing in the Journal of the Geographical Society with Mr. Schomburgk's Journal of his Travels. Instead of this being done, the papers *found their way into the hands of Dr. Lindley*, who printed, for private distribution, twenty-five copies of an essay on this plant, entirely derived from Mr. Schomburgk's paper, and illustrated with highly embellished copies of Schomburgk's drawing. In the essay he *adopted* the view which had been stated before the Botanical Society and British Association, *that it formed a genus intermediate between Euryale and Nymphœa* (see Bot. Reg. 1838, p. 11), but he called the plant *Victoria regia*, thus *continuing the error* of the printer of the Athenæum.

" In Miscellaneous Notices attached to the Botanical Register for 1838, p. 9—18, Dr. Lindley having been enabled to examine a specimen of the flower in a bad state, which Mr. Schomburgk had sent home in salt, gave some further details, and for the first time published an account of the plant under the above name, and this name has been adopted by several succeeding botanists, who have quoted it as *V. regia* of Lindley. I think, however, that this account proves that the name of *Victoria Regina, which received the sanction of Her Majesty*, was the one first used and published, *and has the undoubted right of priority*."

The italics are our own; and we beg the reader's particular attention to them while comparing with Mr. Gray's statement the following *précis* of the letters, &c., relative to this transaction, as they appear in the records of the Letter-book of the Geographical Society .—.

1837, *July* 18.—Letter received from Mr. Schomburgk, dated Berbice, 11th May, 1837, announcing the discovery of a Water Lily on that river, on the 1st of January, 1837, stating that he has sent two sets of drawings home, with a request that, if a new genus, he might be permitted to append to it the name of Victoria.

July.—Three days later, a packet, containing two sets of drawings and descriptions, arrives.

The President of the Royal Geographical Society communicates on the subject with Sir Henry Wheatley.

July 26.—Sir H. Wheatley signifies the Queen's commands that the drawings be sent to the palace for inspection.

July 27.—The President, Sir H. Wheatley, sending drawings, and adding request that the flower may bear the name Victoria.

July 29.—Sir H. Wheatley to the President, signifying Her Majesty's pleasure, that the name of *Victoria Regia* should be affixed to the flower. Drawing returned for the purpose of enabling this to be done.

July 30.—The Secretary of the Royal Geographical Society to the Secretary of Botanical Society, forwarding, at the request of Mr. Schomburgk, one copy of the drawings and descriptions, and adding, that as Mr. Schomburgk was travelling entirely under the control, and at the cost, of the Geographical Society, the Council were of opinion, that whatever drawing he may wish to present to Her Majesty should pass directly to the Queen through the hands of the Royal Geographical Society, and they will therefore relieve the Botanical Society from any further trouble on that account.

Aug. 1.—Secretary of Royal Geographical Society to Mr. Schomburgk, stating that his drawing had been presented to the Queen, that Her Majesty had accepted the dedication under the name of *Victoria Regia*, as it would prove to be a new genus; and that it would be placed in proper train for being suitably published.

Aug. 3.—Secretary of Royal Geographical Society to Dr. Lindley, transmitting the Queen's copy of the drawings, and requesting him to superintend the publication of the flower, and a correct description of it. Also stating, that the Queen had been pleased to accept the dedication of it, and to signify her pleasure that it should bear the name of *Victoria Regia*, if, as believed, the flower should prove to be an undescribed genus.

Thus it is manifest that Mr. Gray's statement is a tissue of mistakes; as he has, indeed, been subsequently obliged to admit in the Annals of Natural History for December last. 1. The plant received the name it bears, by Her Majesty's permission, before Mr. Schomburgk's drawings were even in the hands of the Botanical Society. We may add, that it was generally known to the Council of the Royal Geographical Society, and to the numerous visitors that called to see the drawings within the first fortnight, by the name of *Victoria Regia*, and by no other; and that, consequently, Mr. Gray might have informed himself of that circumstance had he made any inquiry, as we think he was called upon to do, before he ventured to make public a document which the Botanical Society had been officially informed was forwarded by a traveller "entirely under the control and at the cost of the Geographical Society,"—a tolerably intelligible, although courteous hint, which most men would have known how to receive. 2. That the Editor of the Athenæum, in changing the words *Victoria Regina* to *Victoria Regia*, in the Index of the year 1837, did not commit "an error of the press," but silently corrected one, by employing the name which he, as a well-informed man, knew was that by which the plant would be in future called. Possibly, too, as a scholar, he saw the absurdity of the name *Victoria Regin*A. 3. That Mr. Schomburgk's papers did not "find their way into the hands of Dr. Lindley," as Mr. Gray pretends, but were officially communicated to him for the express purpose of publication, and by the only Society which had any property in them. 4. That the Geographical Society could scarcely have afterwards borrowed drawings which they already possessed, and most certainly did not do so, if they borrowed them at all, for any such purpose as Mr. Gray asserts.

But Mr. Gray's inaccuracy does not terminate here. He says, that Dr. Lindley *adopted* his view, that the plant forms a genus intermediate between Euryale and Nymphæa; and in support of this assertion he quotes the Botanical Register for 1838, p. 11. But if the reader will consult that work, he will find nothing of the sort. Dr. Lindley's statement, before examining the plant personally, and judging merely from Mr. Schomburgk's drawings, was this:—"This noble plant corresponds with the genus Euryale in the spiny character of the leaves and stalks, and to a certain extent in the great development of the former organs; but it is, in fact, most nearly related to Nymphæa itself." At p. 12, where the result is given of an examination of some decayed flowers, it is stated that "Victoria is quite distinct from Euryale;" and the whole of the succeeding observations are made for the purpose of showing that Victoria is very different from Euryale; the last words of the little dissertation referred to being these—"notwithstanding a *primâ facie* resemblance to Euryale, Victoria is, in fact, more nearly allied to Nymphæa."

So much for Mr. John Edward Gray. Another proposal, made by Mr. Sowerby, to change the name of *Victoria regia* to that of *V. amazonica*, because it now appears that the plant was originally called *Euryale amazonica*, we do not think worth serious consideration.

231. Gynerium argenteum. *Nees.* (*aliàs* Arundo dioica *Sprengel; alias* Arundo Selloana *Schultes.*) A tall reedy perennial, with harsh serrated leaves, and large erect silky plumes of flowers. Belongs to Grasses. Native of Brazil and Montevideo. (Fig. 111.)

This noble plant, now called the Pampas Grass, in consequence of its inhabiting the vast plains of S. America so named, has been introduced within a few years through Mr. Moore, of the Glasnevin Botanic Garden. Although but a Grass it will probably form one of the most useful objects of garden decoration obtained for many years. In stature it rivals the Bamboo, being described as growing in its native plains several times as high as a man. The leaves are hard, wiry, very rough at the edge, not half an inch broad at the widest part, of a dull grey green colour, much paler below. They are edged by sharp points or teeth, little less hard than the teeth of a file. The flowers appear in panicles from $1\frac{1}{2}$ to $2\frac{1}{2}$ feet long, resembling those of the common reed, but of a silvery whiteness, owing to their being covered with very long colourless hairs, and themselves consisting of colourless membranous glumes and pales.

According to Prof. Kunth this species is an Arundo. But to us it appears quite as different from that genus as from Gynerium. And although it is by no means one of the same genus as *G. saccharoides*, yet it may as well preserve its common name, faulty though it be, as be transferred to Arundo, from which it must be expelled. The inflexed hook of its pales is extremely remarkable, and, together with its diœcious character, leads to the inference that it may be a genus distinct from either.

The plant appears to be hardy. The annexed sketch was made in the garden of Robert Hutton, Esq., of Putney Park; the species exists also in that of the Horticultural Society, to which it was presented by the Botanic Garden, Glasnevin.

111

PLATE 34

L. Constans, Pinx & Zinc.

Printed by C. F. Cheffins, London.

[PLATE 34.]

THE ANGLEBEARING LEAF-CACTUS.

(PHYLLOCACTUS ANGULIGER.)

A Fine Greenhouse Shrub, with White Flowers, from the WEST OF MEXICO, *belonging to the Order of* INDIAN FIGS.

Specific Character.

THE ANGLEBEARING CACTUS.—Branches leafy, stiff, flat, thick, pinnatifid, the lobes being nearly right-angled triangles. Flowers brown without, white within. Sepals longer than the petals. Stigmas 9-10.

PHYLLOCACTUS *ANGULIGER;* ramis foliaceis rigidis planis crassis pinnatifidis, lobis ferè rectangulari-triangularibus, floribus extus fuscis intus candidis, sepalis quam petala longioribus, stigmatibus 9-10.

Phyllocactus anguliger, "*Lemaire, Jardin fleuriste,* 1, 6;" *according to the Gardeners' Magazine of Botany.*

THIS noble plant is nearly related to the *Cereus crenatus* of the Botanical Register, which itself stands in close affinity to the *Cereus Phyllanthus* of the Botanical Magazine, which is very different from the *Cactus Phyllanthus* of Linnæus. Of the three, the last is the least showy, but all must rank among the most striking of the white-flowered species of this great order. The present opens its flowers by day, retains them in beauty and fragrance for several hours, and yields a succession for days together; they are less white than in the other two species, on account of the dark brown tinge of the sepals; but, on that very account, the petals, which are much sharper pointed than in *C. crenatus,* are, perhaps, more conspicuously fair.

In Hartweg's meagre account of his Journey to California, this plant is first mentioned as occurring near Matanejo, a village in the west of Mexico, at no great distance from Tepic.

"The vegetation," says this collector, "as far as the small village of Matanejo, where we arrived in the evening, affords little interest at this season. The copsewood covering the sides of the ravines is composed of deciduous leafless shrubs, only relieved by a giant Cereus, forming a singular tree; this generally has a single stem, two or four feet high, by eighteen inches in diameter, when it divides into numerous triangular branches, rising perpendicularly to the height of twenty to thirty feet. In May it yields a delicious fruit, called Pitaya, when it is much sought after by the natives. Leaving Matanejo early the following morning (Jan. 22nd), we soon entered a forest of oaks; here I found two species of Epidendrum, an Oncidium, Odontoglossum, and an EPIPHYLLUM, the latter, like E. Ackermanni, inhabiting trees. Although I have not seen it in flower, yet, judging from its broad, deeply-cut leaves, or rather stems, it will prove a valuable acquisition to that interesting tribe of plants."—*Journal of Horticultural Society*, vol. i., p. 184.

The plant called an Epiphyllum in this extract is what we now represent. It would seem, from its being associated with oaks, that it will require no greater protection than a good greenhouse; and, in fact, it proves to be one of the hardier species of its order. Nevertheless, like others of the leafy kind, the atmosphere of a stove is best suited to it while making its growth.

In deference to the opinion of Prince Joseph of Salm-Dyck, we call this a Phyllocactus rather than a Cereus; for it must be owned that, if such genera as Echinocactus, Mammillaria, and Opuntia, deserve to be adopted, because of the peculiar form of their stems, so also must Phyllocactus, whose jointed stems are very different from the uninterrupted stems of the true Cerei. Under the former genus are now collected the following additional species, viz., *Cereus phyllanthoides* of the Botanical Magazine; *Epiphyllum Ackermanni* of the Botanical Register; *Cereus latifrons* of Pfeiffer; and *Cactus Phyllanthus* of Linnæus; to which are to be added two new species of Phyllocactus, viz., *stenopetalus* of Salm-Dyck, and *grandis* of Lemaire.

In strict justice, the generic name of Phyllocactus, now employed, and first applied by Link in 1833, ought to be surrendered for that of Phyllarthrus, proposed by Necker in 1791; but custom and convenience disregard the laws of dogmatists, and refuse to be fettered by maxims which, however just and useful in the main, are never to be allowed to bend to expediency.

The accompanying drawing was made in the Garden of the Horticultural Society last October.

PLATE 35.

L. Constans. Pinx & Zinc.

Printed by Cheffins London.

[PLATE 35.]

THE OCCIDENTAL BANKSIA.

(BANKSIA OCCIDENTALIS.)

A Greenhouse Shrub, from KING GEORGE'S SOUND, NEW HOLLAND, *belonging to the Natural Order of* PROTEADS.

Specific Character.

THE OCCIDENTAL BANKSIA.—A shrub. Branches smooth. Leaves long-linear, with spiny teeth beyond the middle, veinless and white with down beneath. Spike long, cylindrical. Bracts broadly triangular, acute, smooth at the point, the lowermost long and awl-shaped. Calyxes shrivelling, silky, with the claws downy at the base on the inside. Style very long, with a small withered stigma. Follicles ventricose, downy, somewhat compressed and naked at the point.

BANKSIA *OCCIDENTALIS;* fruticosa, ramulis glabris; foliis elongato-linearibus, extra medium spinuloso-serratis, subtus aveniis niveo-tomentosis; amento elongato, cylindrico, bracteis late triangularibus, acutis, apice glabris, infimis elongatis subulatis; calycibus marcescentibus, sericeo-puberulis, unguibus basi intus pubescentibus; stylo prælongo, stigmate minuto sphacelato; folliculis ventricosis, tomentosis, apice compressiusculo nudis.—*Meisner.*

Banksia occidentalis: *R. Brown, Prodromus Floræ N. Hollandiæ, p.* 392.

THIS shrub, from the west of New Holland, is described by Preiss as growing from 6 to 8 feet high, erect, on the sandy peaty grounds, which are overflowed in winter, near Seven Miles Bridge, in the Swan River Colony. It has been long in gardens, but we had never seen the flowers till they were produced in the Glasnevin collection, under the care of Mr. Moore.

This gentleman describes it as "an elegant species; the bush from which the specimen was cut, is not above three feet high, with half-a-dozen of such pretty flowers on it as are here represented. The seeds from which the plants were produced were presented to the Garden by his Grace the Archbishop of Dublin, who received them from the district of King George's Sound."

There is some difficulty in distinguishing this from the Littoral Banksia, in which also the leaves are occasionally verticillate. Professor Meisner has probably pointed out the essential peculiarities, which consist in the branches of the Occidental Banksia being smooth and brown, not downy, in the bracts being smooth at the point, and in the calyxes hanging on after flowering instead of dropping off. The leaves of the Littoral Banksia are longer, too, and somewhat broader.

As to the Cunningham Banksia, figured in the Botanical Register under the false name of B. littoralis, whose leaves are also somewhat verticillate, the branches of that species are hairy, and the leaves shorter, with scarcely any marginal serratures, unless quite at the point.

Constans, Prix & Zinc.

Printed by C.F. Cheffins, London.

[PLATE 36.]

THE BLUE VANDA.

(VANDA CŒRULEA.)

A Stove Epiphyte, from Woods on the KHASYA HILLS OF INDIA, *belonging to* ORCHIDS.

Specific Character.

THE BLUE VANDA.—Leaves distichous, leathery, equal-ended, truncate, with a concave notch and acute lateral lobes. Spikes close, erect, many-flowered. Bracts oblong, concave, very blunt, membranous. Sepals and petals light blue, membranous, oblong, very blunt, flat, with a short claw. Lip leathery, deep blue, linear-oblong, obtuse at the point with two diverging lobes, three plates along the middle, and a pair of triangular acuminate lobes at the base. Spur short, blunt.

VANDA *CŒRULEA ;* foliis distichis coriaceis apice æqualibus truncatis sinu concavo lobis lateralibus acutis, spicis densis erectis multifloris, bracteis oblongis concavis obtusissimis membranaceis, sepalis petalisque azureis membranaceis oblongis obtusissimis planis subunguiculatis, labello coriaceo lineari-oblongo apice divergenti-bilobo obtuso per axin trilamellato laciniis basilaribus triangularibus acuminatis, calcare brevi obtuso.

Vanda cœrulea : *Griffith MSS. : Lindl. in Bot. Reg.*, 1847, *sub t.* 30. : No. 1284 *Griffith, Itinerary Notes, p.* 88.

"THIS glorious plant, perhaps the noblest of the Indian race, was called *Vanda cœrulea* by Mr. Griffith, who found it among the Khasya or Cossya Hills, and sent us dried specimens. Its flowers are as large as those of *Vanda teres*, and the foliage is as good as that of *Aerides odoratum.* It is to be regretted that we should have no more exact information as to where it may be found, but we can hardly suppose that it could be missed by any plant-collector who might be sent after it into Sylhet.

"The leaves of this wonderful plant are five inches long by nearly one inch wide ; at their end they are two-lobed equally, and each lobe is sharp-pointed, so that the end looks as if a piece had been struck off by a circular punch. The flowers grow in upright spikes. A piece of a stem but four inches long bears four such spikes, which are from six to nine inches long, and carry from nine to twelve flowers. Each dried flower is between three and four inches in diameter, and if allowance is made for their having shrunk in drying, they may be estimated as at least a foot in circumference. The lip is, as is usual among Vandas, small; it is barely three-quarters of an inch long, narrow, with a short spur and a two-lobed point. Its surface is broken by three deep parallel perpendicular plates, and the lateral lobes of the base are triangular and acuminated."

It was thus that one of us spoke of the present plant three years ago. The accompanying plate is witness of its arrival, and of the extraordinary beauty that belongs to it. The colour of the flowers is of a rich tender lilac, their texture is as delicate as that of Phalænopsis, and their dimensions are at least equal to what was stated in the above paragraph. In short the species is a dangerous rival of Phalænopsis itself.

Its exact residence is now known. Mr. Griffith tells us that it occurs near the River Borpanee on trees of Gordonia, in the Pine and Oak forests of that region.* It is, however, not a little remarkable that his Journal contains no allusion to it; but we find that the district produces Bauhinias, Randia, Phyllanthus Emblica, and Sugar canes, all indications of a tropical region. The woods are described as delightful, reminding one of England. The elevation of the Borpanee above the sea is 2508 feet; the temperature 74°; the neighbouring vegetation Castanea (tropical species of course), Kydia, Camellia oleifera, Rhododendron punctatum (whatever that may be) and Cuscuta.

The honour of having introduced this glorious plant belongs to Messrs. Veitch who received it from their invaluable traveller Mr. Thomas Lobb. The accompanying figure does scanty justice to it: for although it represents faithfully the beautiful tender blue of the flowers, it by no means equals the magnitude of the wild plant. We have a dried specimen now before us with nine flowers open at the same time.

* We transcribe his note upon the plant, as published in his Itinerary, the blunders of the editor and transcriber being corrected. "Caule altiusculo interdum 2-pedal.; foliis distichis loriformibus, canaliculatis, apice profundè et inæqualiter emarginatis, quam maximè coriaceis. Racemis axillaribus folia longè excedent., flexuosis, supra bracteis adpressis livido maculat.; bract. florum membranaceis reflexis fuscescent.; floribus resupinatis maximis, diametro 2½ uncial., pulcherrimis, cœrulescentibus saturatiore colore tessellatis; labelli lobis lateral., albis, columnaque alba. Perianth. patentiss. lacin. obovat.; sepalis undulatis uti petala; petalis sepaloque postico paulo minoribus; labelli trilobi lobis lateral. dentiformibus, medio emarginato, apice bicalloso tricarinato, calcare brevi recto. Color cœruleo-purp. Columna albida, nana, basi ad junctionem labell. macula lutea. Anth. simplex. Pollinia 2 complanat. posticè fissa; caudiculâ latâ; glandulâ maximâ trigonâ."

GLEANINGS AND ORIGINAL MEMORANDA.

232. CYPRIPEDIUM GUTTATUM. *Swartz.* A hardy terrestrial Orchid, with white flowers spotted with purple. Native of Northern Russia, Siberia, and North America. (Fig. 112.)

This charming plant has lately flowered with Mr. Van Houtte, of Ghent. M. Planchon may well call it "une vraie perle pour les jardins." It is one of the most exquisitely beautiful little things imaginable. A diminutive stem, a few inches high, with a pair of broad plaited leaves, bears one solitary flower as large as a pigeon's egg, most curiously painted with rich deep purple upon a pure white ground. This plant has been occasionally received from Russia in a living state, but no English gardener has managed even to keep it alive. Mr. Van Houtte does not say how he proceeded; but since it is clearly cultivable, it will be easy for those who have friends in Canada or at Moscow, to procure supplies with which further experiments may be tried. We can only say that it grows in morasses and bogs.—See *Flore des Serres*, t. 573.

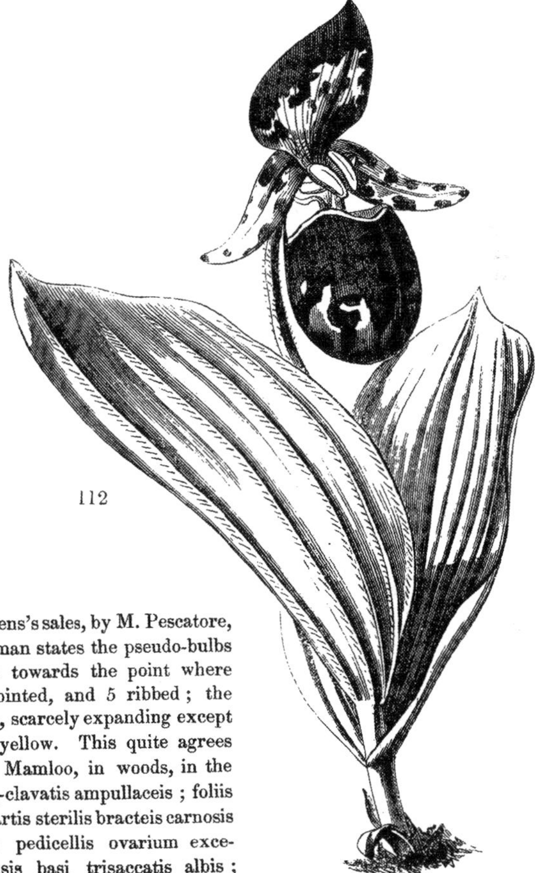
112

233. CŒLOGYNE TRISACCATA. *Griffith.* An Orchidaceous epiphyte from tropical India, with large white flowers. Blossomed with Mons. Pescatore.

C. trisaccata, Griffith (*Itinerary notes*, p. 72); pseudobulbis elongatis, foliis membranaceis obovato-lanceolatis 5-nerviis, racemis recurvantibus, bracteis latis ovatis obtusis cucullatis sterilibus carnosis floriferis minoribus membranaceis, floribus conniventi-clausis basi trisaccatis, petalis linearibus, labelli apice 3-lobi lamellis 2 carnosis flexuosis perax in laciniis subserrulatis rotundatis minutè ciliatis lateralibus rotundatis intermediâ nanâ bilobâ basi dilatatâ.

This plant was purchased last February, at one of Stevens's sales, by M. Pescatore, in whose fine collection it has lately flowered. M. Luddeman states the pseudo-bulbs to be dilated at the base, much lengthened and narrowed towards the point where they are quadrangular; the leaves to be lanceolate, pointed, and 5 ribbed; the flowers to be arranged 6-8 in nodding racemes, distichous, scarcely expanding except at the point, pure white with the end of the lip sulphur-yellow. This quite agrees with Griffith's statement; that botanist who found it at Mamloo, in woods, in the Khasijah hills, describes it thus:—"Pseudobulbis obovato-clavatis ampullaceis; foliis plicatis repandis; racemis basilaribus a medio pendulis, partis sterilis bracteis carnosis adpressis florescentiæ concavissimis submembranaceis; pedicellis ovarium excedentibus; floribus amplâ longitudine conniventi-clausis basi trisaccatis albis; labelli lobo medio cristisq. lutescentibus."

234. EPIDENDRUM ANTENNIFERUM. A singular Orchid with inconspicuous long-tailed flowers. Native of Xalapa. Introduced by M. Quesnel. Flowered by M. Pescatore. (Fig. 113. *a* diminished, *b* magnified.)

E. antenniferum (Amphiglottium) foliis coriaceis oblongis acutis, pedunculo gracillimo apice subpaniculato, petalis longissimis filiformibus, labello ovato leviter dentato basi trituberculato.

This plant was originally found near Xalapa, by Henchman, who brought home a small dried specimen without leaves, which was given us by the late Mr. George Loddiges. Among the plants purchased of M. Quesnel by M. Pescatore, it was found alive marked as a native of Gabon, a place in the province of Rio Janeiro ; but this locality is doubtful. A short time since it flowered with the latter gentleman at Celle St. Cloud, when we were favoured with a specimen. In many other species of the genus, especially among the Amphiglots, there is that tendency to lengthen the petals, of which so striking an example was given in the long-tailed Lady's Slipper (our tab. 9.) ; but in no other known species does it occur in anything like the same degree as here ; and it is to be observed that in this Epidendrum the lengthening is an after-growth, the petals being straight and short before the flowers expand. In the annexed cut the flower at *b* is magnified : its real size is that of Epidendrum elongatum.

235. PACHIRA MACROCARPA. *Hooker*. (*aliàs* Carolinea macrocarpa *Chamisso* and *Schlechtendahl.*) A noble stove tree, with huge white and yellow flowers. Belongs to Sterculiads. Native of Mexico.

The flowers are truly magnificent, and yet produced from a young and small plant. As a species, it comes very near the *P. aquatica* of Aublet, and may probably prove identical with it. Of the ordinary size of the native plant we are ignorant. Our flowering specimen had not attained a greater height than four feet. Leaves large, smooth, digitate, with from seven to eleven leaflets, which are oblong-ovate, entire, acuminate, cuneate, and tapering at the base into a short footstalk. Calyx short-cylindrical, truncated, thick and leathery, clothed with minute velvety down, bearing a circle of conspicuous glands at the base. Petals full six inches long, linear-strap-shaped, the upper half reflexed, white and smooth within, pale greyish or greenish-brown, and slightly velvety externally. Staminal tube rather short, divided into innumerable parcels, each again divided into eight or ten filaments, which are yellow below, the rest deep red. This is a tall tree of rapid growth ; and, as it requires the temperature of a stove, it is adapted only for growing in lofty hothouses, such as the Palm-house in the Royal Gardens, in which a plant has quickly attained the height of twenty-five feet, and, according to the present rate of growth, will soon double that height. In our cultivation it appears to have no season of rest. It will grow freely in any kind of light loam, kept in a proper state of moisture.—*Botanical Magazine*, t. 4549.

We are at a loss to know what it is intended that this plant should be called. At the head of the article in the "Botanical Magazine" it is named *Pachira longifolia*,—but this is translated long-*flowered ;* and then it is immediately afterwards styled *P. macrocarpa*. We presume, however, that the latter is what is meant.

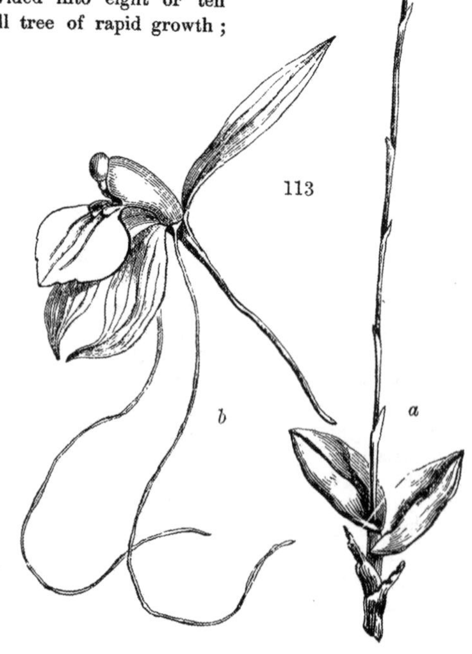

236. HYDROMESTUS MACULATUS. *Scheidweiler*. A yellow flowered stove shrub, belonging to the order of Acanthads. Native of Mexico. Introduced by Messrs. Lowe & Co.

Is really a handsome plant, with very glossy leaves, bright yellow flowers, and a singularly shining imbricated spike of large bracts, from which the flowers spring. An under-shrub, with terete purplish branches, and opposite, large, very glossy, ovate or ovate-lanceolate, entire leaves. Bracts broad-ovate, keeled, bright green, imbricated like the scales of a cone, but in four rows. Corolla protruded much beyond the bracts. Tube narrow, funnel-shaped, a little inflated, yet laterally compressed at the mouth ; limb large, two-lipped ; lips spreading : upper one two-lobed, the lower three-lobed, all the lobes emarginate. Stamens four, included.—*Bot. Mag.* t. 4556.

237. POSOQUERIA FORMOSA. *Planchon.* (*aliàs* Stannia formosa *Karsten*). A very fine stove plant, from the Caraccas, with long white flowers. Belongs to Cinchonads. Introduced by M. Karsten. Flowered by M. Van Houtte. (Fig. 114.)

A fine tree, from the virgin forests of the mountains of Tovar, at the elevation of 5000 to 6000 feet above the sea. It grows from 12 to 20 feet high. Its leaves are broad, oblong-lanceolate, wavy, leathery like a laurel. The flowers are 3 to 4 inches long, pure white, slender-tubed, and highly fragrant. When in fruit it is said to resemble an apple tree. Nearly related to the Gardenias, as which it requires the same cultivation.—*Flore des Serres*, t. 587.

238. ONCIDIUM LURIDUM ATRATUM. *Lindley.* A handsome orchidaceous Epiphyte, from Mexico, with rich crimson flowers. Introduced by the Horticultural Society.

Whether or not *O. luridum* is really a mere variety of the Carthagena Oncid becomes more and more doubtful as our knowledge of such plants extends. In the present instance it is unnecessary to open that question, the plant now mentioned being undoubtedly a very fine form of the lurid Oncid, whatever the relation of the latter to the Carthagena Oncid may finally prove to be. With the habit of the common form of the species this combines flowers smaller than usual, very flat, with olive and rose-coloured sepals and petals, and a rich crimson lip furnished at the base with 5 purple-black tubercles, four of which surround the fifth ; of these tubercles the central and two anterior are oblong and simple, the two posterior are concave, or almost kidney-shaped, with the concavity backwards. The wings of the column are oblong truncated fleshy bodies attached by the narrowest end. It is a fine variety, in some respects like the purple-lipped Oncid (*O. hæmatochilum*), and requiring the same treatment as *O. luridum* itself.—*Journal of Hort. Soc.*, vol. vi.

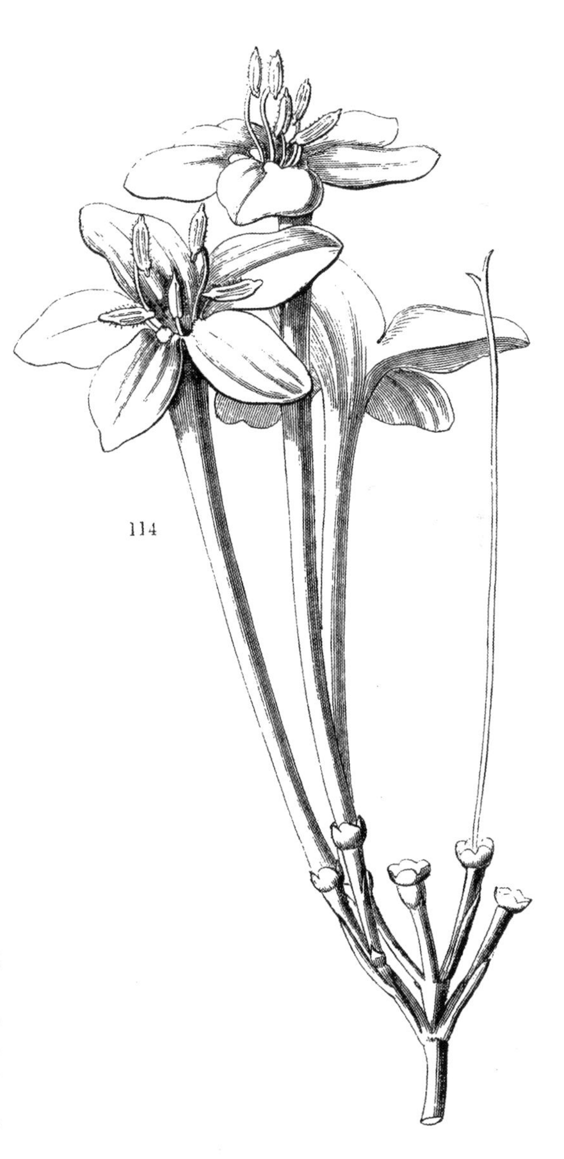
114

239. ADENOSTOMA FASCICULATA. *Hooker and Arnott.* A hardy, heath-like evergreen bush, with small white flowers. Belongs to Roseworts. Native of California. Introduced by the Horticultural Society.

A small heath-like bush, with erect weak branches. Leaves linear, sharp-pointed, concavo-convex, arising in fascicles from the axil of primordial leaves of the same form, but dying early and leaving behind a pair of spine-pointed stipules ; in this arrangement they may be compared to Berberries and similar plants. Flowers white, small, in terminal panicles, with much the appearance of the Alpine Spiræa. The leaves of the cultivated plant continually evince a tendency to become 2- or 3-lobed near the point. It is said to grow 2 feet high, in open exposed places near Monterey. In point of beauty it is inferior to the worst of the Spiræas, and is a mere botanical curiosity. Whether it is hardy or not has not been at present ascertained.—*Journal of Hort. Soc.*, vol. vi.

240. CENTROSOLENIA GLABRA. *Bentham.* A hothouse plant from La Guayra, with pale yellow fringed flowers. Belongs to Gesnerads. Introduced at Kew. Flowers in autumn.

A plant imported through Mr. Wagener, a German collector. It forms a stove plant, and keeps up a succession of flowers with us through the autumnal and early winter months. We submitted the figure to Mr. Bentham for his opinion, as he had paid much attention to the family to which it belongs, and has published the result of his observations in the 5th volume of the 'London Journal of Botany,' p. 357, &c. That gentleman considers the plant as clearly constituting a second species of his new genus *Centrosolenia* (l. c., p. 362). Decaisne's *Trichanthe*, since published, probably in the 'Revue Horticole,' for 1848, he believes to be identical with *Centrosolenia*. If so, it must give place to the latter name, which appeared in 1846, and consequently has the right of priority. An erect plant, with a succulent reddish-brown, terete stem, a foot or more high. Leaves succulent, smooth, the lower ones six to eight inches long, opposite ; each pair singularly unequal in size, one being small, lanceolate, and acuminate ; the other large, ovate, tapering at the base into a stout petiole, and acuminate at the apex ; the margin serrated. Corolla tubular, enlarged upwards, projected below into a short obtuse spur, the whole tube about an inch and a half long, clothed outside with a short thin down, the limb divided into five broad short lobes, of which the three lower are fringed with long thread-like laciniæ ; inside of the corolla smooth. Annular disc nearly obsolete, with a large posterior gland. (Mr. Fitch represents two glands,—one anterior, the other posterior, and of nearly equal size.) Ovary wholly superior, with two lamelliform, bipartite, parietal placentæ. Style smooth, thick, somewhat clavate, with the stigmatic extremity rarely emarginate.—*Botanical Magazine*, t. 4552.

241. GERANIUM THUNBERGII. *Siebold.* A prostrate annual, with small purple flowers. Native of Japan. (Fig. 115.)

An annual, with hairy prostrate stems ; leaves long-stalked, with long spreading hairs, rather fleshy, 5-lobed, flat, the lower lobes much the smallest, the others 3-lobed, and slightly serrate. Peduncles 2-flowered, longer than the leaves. Petals deep purple, undivided, obovate, larger than the mucronate sepals. Probably the *G. palustre* of Thunberg. A mere weed.

242. ECHINOCACTUS VISNAGA. *Hooker.* (*aliàs*? E. ingens *Zuccarini.*) A noble plant from Mexico, belonging to the Natural Order of Indian Figs (Cactaceæ). Flowers bright yellow, produced at Kew.

115

Of this singular species, Sir William Hooker gives the following account :—" One of the most remarkable plants in the Cactus-house of the Royal Gardens of Kew, and that which chiefly attracts the attention of strangers, is the subject of the present plate. It bears the name of *Visnaga* with us (*Visnaga* means a tooth-pick among the Mexican settlers, and the plant is so called because that little instrument is commonly made of its spines), and under that name, believing it to be a new species, we had described it, and it was figured in the *Illustrated News* for 1846. I had, at one time, been disposed to refer the species to the *Echinocactus ingens*, of which a brief and most unsatisfactory character is drawn up by Pfeiffer (for Zuccarini does not appear to have noticed it) from some 'dried flowers,' and a living specimen 'six inches high ;' but it can scarcely be that, for the angles of the plant are said to be eight, the aculei nine in a cluster, and the petals obtuse. Our plate represents a very diminished figure of a specimen, unfortunately no longer existing, but which, in 1846, was an inmate of our Cactus-house, and apparently in high health and vigour. Its height was nine feet, and it measured nine feet and a half in circumference, its weight a ton. After a year of apparent health and vigour, it exhibited symptoms of internal injury. The inside became a putrid mass, and the crust, or shell, fell in with its own weight. Other lesser ones were already, and are still, in the collection ; and the one, from which one small flowering portion is represented of the natural size, weighs seven hundred and thirteen pounds, its height is four feet six inches, its longitudinal circumference ten feet nine inches, and its transverse ditto eight feet seven inches, its ribs amount to forty-four. All our plants were procured with great labour, and sent many hundred miles, over the roughest country in the world, from San Luis Potosi, Mexico, to the coast, for shipping, and presented to the Royal Gardens by Fred. Staines, Esq. It flowers through a good part of the year, but, in comparison with the bulky trunk, the blossoms are quite inconsiderable and void of beauty." The summit of the trunk is crowned with a dense mass of tawny wool, concerning which it is remarked, that "this wool covers the whole *crown* of the plant, and is a few inches deep, and we are much mistaken if it is not a tuft of this substance, taken from an *Echinocactus Visnaga*, which constitutes that botanical curiosity from Mexico, long in the possession of the late Mr. Lambert (now at the British Museum), known under the name of the '*Muff Cactus*.' A small quantity taken off the plant may, by handling and admitting air within the *staple*, be distended to a considerable size. An entire mass from a good-sized plant, thus treated, might be made to assume the cylindrical form of the specimen alluded to."—*Bot. Mag.*, t. 4559.

243. ACONITUM SINENSE. *Siebold.* A hardy plant of the order of Crowfoots. Flowers deep violet, appearing in the autumn. Native of Japan. (Fig. 116 ; *a* represents a flower of *A. autumnale* by way of contrast.)

We have now two perfectly distinct autumnal Asiatic Monkshoods in cultivation ; one the *A. autumnale*, the other Siebold's *A. sinense.* The latter forms a stem from one and a half to two feet high, slightly downy, round, with regularly 5-parted leaves, the segments of which are incised, marked with a deep middle vein, and recurved a little ; the flowers few, deep violet, on woolly and glandular peduncles ; the helmet hemispherical, with no visible peak. The former is similar in foliage, except that the lobes of the leaves are much longer, and quite falcate, the flowers larger, in a close erect raceme, pale violet, with a pubescent stalk, and a more compressed helmet, with a long curved peak. (This is not shown at *a*, in consequence of the foreshortening.) Either of them may be the *A. Napellus* of Thunberg. Both are distinguished from the *A. japonicum* by the deep falcate divisions of the leaves, and long racemes of flowers

They are very useful autumn plants, are quite hardy, but worth a greenhouse, in which, in England, they are seen to most advantage. The specimen figured is a very small one. We have one before us from Prof. De Vriese, with a branched inflorescence, and eight flowers open at once.

244. ORNITHARIUM STRIATULUM. (*aliàs* Ornithochilus striatulus *Hort. Calcutt.*) An Indian epiphytal Orchid of little beauty. Flowers yellow with a white lip. (Fig. 117.)

ORNITHARIUM. Caulescens, foliis distichis. Flores spicati, resupinati, clausi, carnosi. Sepala lateralia basi imâ connata, cum labello parallela, dorsale paulò sejunctum. Petala conformia. Labellum liberum, unguiculatum, carnosum, a basi sagittatâ cuniculatum. Columna semiteres, brevis, stigmate verticali. Pollinia 2, solida, caudiculâ obovatâ, glandulâ triangulari, rostello reflexo.

O. *striatulum*. Sepala et Petala obtusa, carnosa, lutea, maculis quibusdam interiùs. Labellum spongiosum, candidum, oblongum, rugosum, minutissimè scabrum, apice appendice sphærico cavo atropurpureo auctum, intra cuniculum læve.

Of this curious little plant, which flowered last October with W. F. G. Farmer, Esq., of Nonsuch Park, we have only seen a few flowers. They were about as large as those of the Egerton Odontoglot, arranged along a slender narrow rachis. The petals and sepals were waxy-yellow, with a few bars of red inside. The lip was white with a few violet stains and a deep purple round knob at the end, giving the flower the appearance of concealing within it a tiny bird with a white body and purple head. Mr. Carson, the gardener at Nonsuch Park, gives us the following account of it :—

"The *Ornithochilus striatulus* came from India in the autumn of 1847, sent by Dr. M'Clelland of the Botanical Garden, Calcutta, and was so named and labelled in the invoice. In habit it has a resemblance to Camarotis atropurpurea in its slender stem, with an abundance of aerial roots, yet the leaves are much larger ; they are flat, fleshy, disposed in two opposite uniform rows, of a pale green colour, notched at the end, about five inches long by one and a half broad, and not unlike small leaves of Aërides odoratum. The plant is epiphytal, sending out at every joint its slightly tortuous tail-like flower spikes, some of which are above a foot in length. Although the plant is small, not more than six inches in height, it is remarkable that after one flower-spike has grown eight or ten inches, another pushes from the under side of it, so that it produces two spikes from the same point. I think it must prove an interesting plant in the Orchid-house from its very singular appearance."

We have never seen this in any of the numerous Indian collections which have come into our possession, nor can we trace the name by which it was received. It is certainly no Ornithochilus, whether the plant so named by Dr. Wallich be retained as a distinct genus or merges in Aërides ; nor does it seem referable to any other published genus. From Arhynchium, Camarotis and the like, its simple pollen masses and unguiculate lip clearly separate it. It can be no Micropera, because of its unguiculate lip, short rostel, &c. ; nor do we find among the species referred to Saccolabium anything that approaches it at all nearly. In the following cut, *a* represents a flower seen in front ; and *b* the same from the side, both magnified ; *c* is the lip and column deprived of the sepals and petals ; *d* is the lip only seen from above ; *e* the column ; and *f* the pollen-apparatus.

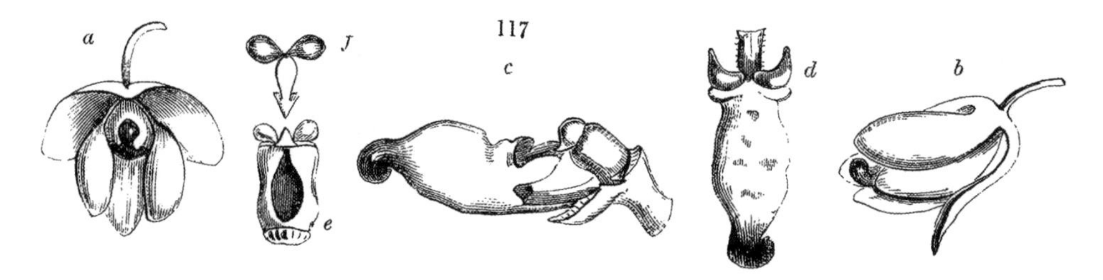

245. ASTRAPÆA VISCOSA. *Sweet.* (*aliàs* Dombeya Ameliæ *Guillemin.*) A soft sticky-leaved stove plant, with clusters of white and pink flowers. Belongs to Byttneriads. Native of Madagascar. Introduced in 1823. (Fig. 118.)

A noble plant or tree, thirty feet in height, as now seen in the great stove of the Royal Gardens of Kew, with a large rounded head of copious branches, and dense foliage, studded, in the spring months, with numerous snowball-like heads of flowers, each flower stained with a deep blood-coloured eye. The flowers have a honey-like smell. The young herbaceous branches and nascent leaves, accompanied by large, cordate, afterwards deciduous stipules, are exceedingly viscid. Leaves on long stalks, the largest a span and more long, heart-shaped, roundish, five-angled (the smaller ones three-angled), the angles or lobes acuminate, the margins serrated. The young flower-head is clothed by large deciduous bracteas, and at the base of the head three or four such bracteas form an imperfect involucre. These bracteas disappear on the full expansion of the many flowers into a globose head, four inches and more in diameter. Sepals ovate, acuminate, hairy externally. Petals five, twisted broad-wedge-shaped, pure white, the base deeply dyed with crimson. Staminal tube urceolate, bearing five perfect short stamens, and five long sterile filaments. Style divided at the top into five reflexed branches. This is a tree, of quick and robust growth, soon arriving at a height that renders it unsuitable for hothouses of the ordinary dimensions. In the Royal Gardens it has rapidly attained the height of upwards of twenty feet; but, as it branches freely, it may, with management, be kept within bounds by frequently cutting back the leading shoots. It grows readily in light loam, and should be rather freely supplied with water, as its numerous fibrous roots take it up very quickly, and the size and texture of its leaves present a large and free evaporating surface. It is easily increased by cuttings, planted under a bell-glass, the pot being plunged in bottom-heat.—*Bot. Mag.* t. 4544.

118

246. FREZIERA THEOIDES. *Swartz.* (*aliàs* Eroteum theoides *Swartz.*) A green-house shrub from Jamaica, with the aspect of a tea-plant. Flowers white. Belongs to Theads. Blossomed at Kew in September. (Fig. 119.)

A Jamaica shrub or small tree, inhabiting the higher mountains of that island, and remarkable for its very near resemblance, both in the leaves and flowers, to the black tea of China. Dr. M'Fadyen informs us, in his useful 'Flora of Jamaica,' that the leaves are astringent, and in taste resemble those of the green tea. A smooth shrub four or five feet high in our stove; in Jamaica, it attains a height of twenty feet. Leaves alternate, on short stalks, leathery, very dark green, elliptical-lanceolate, acute, serrated. Peduncles all solitary, axillary, curved down, single-flowered. Flower an inch and a half across. Calyx bibracteolate at the base, five-sepaled; sepals broad ovate, acute, green, margined with red. Petals cream-white, obcordate. Stamens numerous, attached to the base of the petals. Anthers oblong, opening by two pores, furnished with a tuft or pencil of hairs at the back. Fruit "a berry, the size of a small cherry, globose, purple, juicy, three- or four-celled. Although not a showy plant, its neat evergreen habit renders it worthy of a place in general collections. It resembles the well-known *Ardisia crenulata*, but grows more luxuriantly; as, however, it bears cutting back, it may be kept to a proper size, and will form a neat bush. It should be grown in a moderate stove temperature, and will thrive in any kind of light loam, water being freely given it during dry weather in summer. It is readily propagated by cuttings, planted in sand, under a bell-glass, and plunged in a moderate bottom-heat. — *Bot. Mag.* t. 4546.

119

247. Didymocarpus crinita. *Jack.* (*aliàs* Henckelia crinita *Sprengel.*) A yellow-flowered herbaceous plant from Malacca, with dark green leaves purple beneath. Belongs to Gesnerads. Flowers at Kew in August.

A lovely plant, its beauty rather depending on the leaves (which have a rich velvety hue, as well as a richness of *colour*, especially beneath) than upon anything striking in the flowers. The latter are pale yellow white with us (Jack says, in their native country suffused with blush), and they contrast well with the dark foliage. We possess, in our herbarium, fine native specimens, gathered by Mr. Thomas Lobb at Singapore, given to us by Mr. Veitch (No. 311 of Lobb's collection), and we find, too, that this distinguished cultivator exhibited flowering plants at the Horticultural Society's Rooms in June, 1847. Stem erect, scarcely a span high, densely shaggy with purplish hairs. Leaves opposite, broad-lanceolate, acute, finely dentato-serrate, all over hairy, above dark coppery green with a velvety lustre, beneath rich purple-red, penninerved, nerves prominent beneath. Corolla funnel-shaped, ventricose below the broad spreading five-lobed white lip, yellow, with the tube two inches long. Should be cultivated in a warm stove, in a temperature such as is suited to tropical *Orchidaceæ*, *Gesneriaceæ*, and other sub-epiphytal plants, that require a warm and moist atmosphere during their season of growth. It appears to be of dwarf growth, and produces short lateral shoots from amongst the leaves, which strike root readily when treated as cuttings. —*Bot. Mag.*, t. 4554.

We fear that cultivators will be disappointed who expect to find *much beauty* in this plant beyond what belongs to the foliage, which is very handsome.

120

248. Calceolaria alba. *Ruiz and Pavon.* A shrubby white-flowered slender plant, from Chili. Introduced by Messrs. Veitch & Co. Flowers in July. (Fig. 120.)

A slender, smooth, viscid shrub, with linear leaves arched downwards, and furnished with distant simple teeth at the edge. The flowers are pure *white*, and form loose thyrso-like panicles. The lower lip of the corolla is nearly spherical. In a genus the species of which are so generally either yellow or purple, a white-flowered species is a horticultural acquisition. The present, if well cultivated, is one of the prettiest greenhouse shrubs of modern introduction, and may be expected to find much favour among gardeners. Except in colour, it is very like the yellow-flowered *C. thyrsiflora*, from the same country.

INDEX OF VOLUME I.

[*Plate* signifies the coloured representations; *No.* the number of the Gleanings and Memoranda; *fig.* the woodcuts.]

END OF VOLUME I.

LONDON:
BRADBURY AND EVANS, PRINTERS, WHITEFRIARS.

Printed in the United States
By Bookmasters